FOR ALL PRACTICAL PURPOSES

Project Director

Solomon Garfunkel, Consortium for Mathematics and Its Applications

Coordinating Editor

Lynn A. Steen, St. Olaf College

Contributing Authors

PART 1 MANAGEMENT SCIENCE

Joseph Malkevitch, York College, CUNY
Walter Meyer, Adelphi University

PART 2 STATISTICS

David S. Moore, Purdue University

PART 3 SOCIAL CHOICE

William F. Lucas, Claremont Graduate School

PART 4 ON SIZE AND SHAPE

Donald Albers, Menlo College
Paul J. Campbell, Beloit College
Donald Crowe, University of Wisconsin
Seymour Schuster, Carleton College
Maynard Thompson, Indiana University

PART 5 COMPUTER SCIENCE

Wayne Carlson, Cranston/Csuri Productions, Inc.
Zaven Karian, Denison University
Sartaj Sahni, University of Minnesota
Paul Wang, Kent State University

Joseph Blatt, Chedd-Angier Production Company

FOR ALL PRACTICAL PURPOSES

INTRODUCTION TO

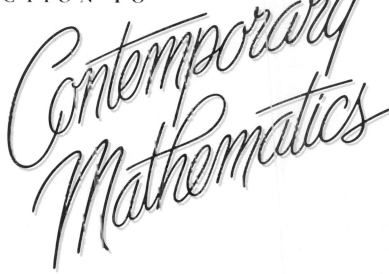

W. H. FREEMAN AND COMPANY
NEW YORK

Back Cover Left: Mandelbrot set, "Piece of Tail of the Seahorse." (From H.-O. Peitgen and P. H. Richter, *The Beauty of Fractals*, Springer-Verlag, Heidelberg, 1986.) *Back Cover Middle, Right*: Computer-generated images. (Computer graphics by Philip Zucco on equipment by Symbolics, Inc.)

Library of Congress Cataloging-in-Publication Data

For all practical purposes: introduction to contemporary mathematics
    [project director, Solomon Garfunkel; coordinating editor, Lynn
    A. Steen; contributing authors, Joseph Malkevitch . . . et al.].
    p. cm.
A project of Consortium for Mathematics and Its Applications,
    Arlington, Mass.
    Bibliography: p.
    Includes index.
    ISBN 0-7167-1830-8
    1. Mathematics—1961– I. Garfunkel, Solomon A., 1943–
II. Steen, Lynn Arthur, 1941– . III. Malkevitch, Joseph, 1942–
. IV. Consortium for Mathematics and Its Applications (U.S.)
QA7.F68 1987   87-21453
510—dc19   CIP

Printed in the United States of America

2 3 4 5 6 7 8 9 0   RRD   6 5 4 3 2 1 0 8 9 8

Contents

Preface

Every mathematician at some time has been called upon to answer the innocent question, "Just what is mathematics used for?" With understandable frequency, usually at social gatherings, the question is raised in similar ways: "What do mathematicians do, practice, or believe in?" At a time when success in our society depends heavily on satisfying the need for developing quantitative skills and reasoning ability, the mystique surrounding mathematics persists. *For All Practical Purposes: Introduction to Contemporary Mathematics* is our response to these questions and our attempt to fill this need.

For All Practical Purposes represents our effort to bring the excitement of contemporary mathematical thinking to the nonspecialist, as well as help him or her develop the capacity to engage in logical thinking and to read critically the technical information with which our contemporary society abounds. We attempt to implement for the study of mathematics Thomas Jefferson's notion of an "enlightened citizenry," in which individuals having acquired a broad knowledge of topics exercise sound judgment in making personal and political decisions. Environmental and economic issues dominate modern life, and behind these issues are complex matters of science, technology, and mathematics that call for an awareness of fundamental principles.

To encourage achievement of these goals, *For All Practical Purposes* stresses the connections between contemporary mathematics and modern society. Since the technological explosion that followed World War II, mathematics has become a cluster of mathematical sciences encompassing statistics, computer science, operations research, and decision science, as well as the more traditional areas. In science and industry mathematical models are the tools par excellence for solving complex problems. In this book our goal is to convey the power of mathematics as illustrated by the great variety of problems that can be modeled and solved by quantitative means.

This book is designed for use in a one-term course in liberal arts mathematics or in courses that survey mathematical ideas. The assumed background of our audience is varied. We expect some ability in arithmetic, geometry, and elementary algebra. Even though a few of the topics included here are traditionally thought of as part of advanced mathematics, we aim to develop in the reader strong conceptual understanding and appreciation, not computational expertise. Our stated bias throughout the book is on presenting the subject through its contemporary applications.

In determining the selection of topics for this book, the authors posed a question to leading mathematicians and educators nationwide: "What would you teach students if they took only one semester of math during their entire college career?" Their answers are best demonstrated by the main topic selections for the text: management science, statistics, social choice, the geometry of size and shape, and mathematics for computer science. These topics were chosen both for their basic mathematical importance and for the critical role their applications play in a person's economic, political, and personal life.

Because readers will approach this book with a great range of expertise in using mathematical symbols, the material has been designed in layers to accommodate different objectives and diverse backgrounds. *For All Practical Purposes* concentrates on discussions about mathematics—about its nature, its content, its applications. At various places, in separate boxes or optional sections, certain mathematical issues are pursued with explicit use of equations and appropriate mathematical symbols. Problems are also divided by type, with some emphasizing descriptive mat-

ters and others providing practice in calculation and symbolic manipulation.

Additionally, *For All Practical Purposes* includes interviews with practitioners—the people who put mathematics to work. Each major topic section is complete and self-contained so that instructors can adjust the topic ordering to suit their particular needs. A large number of photographs and 69 full-color illustrations (in 24 pages of inserts) have been incorporated to emphasize the fact that contemporary mathematics is visually alive. Many of the color images are computer-generated and represent the accomplishments at the forefront of some mathematical research.

Perhaps the most distinctive new feature of this text is its relation to the 26 half-hour television programs that constitute the series "For All Practical Purposes." In 1983, the Consortium for Mathematics and Its Applications (COMAP) received a grant from The Annenberg/CPB Project, with co-funding from the Carnegie Corporation of New York, to produce a telecourse in mathematics for public broadcast on PBS. Development of the textbook was funded by the Alfred P. Sloan Foundation. An impressive array of people committed to educational excellence in mathematics as well as experienced television producers and computer graphics specialists were gathered to work on this project from the original outlines to final shows and text. Though this book stands alone from the video series, the authors worked simultaneously on developing television scripts and book chapters.

Some instructors will use the text and television shows in concert. Some will use the text alone in a traditional lecture course. Still others will make the videos available to students on tape for enrichment or special credit. The choice of how best to use this mix of resources will depend on each individual's particular preference and course requirements. An Instructor's Manual that relates the video programs with the text is available. Information about television-course licensing, off-air taping rights, and prerecorded video cassettes can be obtained by calling (202) 955-5251 (collect), or by writing The Annenberg/CPB Project at 1111 Sixteenth Street NW, Washington, DC 20036.

Students enrolled in the telecourse will view the videos and read the text on their own. To aid them a Course Guide has been developed by the text authors that provides an overview of each program, skill objectives, and a sample short-answer examination. We hope that these materials will provide a rich and exciting environment in which to learn more about the power and centrality of mathematics in our world.

Acknowledgments

First, it is a pleasure to acknowledge the support of the Alfred P. Sloan Foundation, which provided a generous grant to help fund the production of this book. Over the almost four years of this undertaking, a remarkable number of people have made significant contributions. It is difficult to find words expressing our gratitude and appreciation. For the authors, this was no ordinary writing task. The revision process was enriched but certainly made a great deal more complicated by the relationship with the television series. With so many contributors, coordination was of necessity a key factor. The cluster leaders—Donald Albers, Zaven Karian, William F. Lucas, David S. Moore, and Joseph Malkevitch—did yeoman duty.

Professor Malkevitch truly deserves a very special recognition. To a great extent the underlying philosophy of this book is a reflection of his ideas, beliefs, and dedication.

We also owe a huge debt to Stephanie Stewart, John Rubin, and David Gifford, the writers, producers, and directors of the Chedd-Angier Production Company, who created "For All Practical Purposes" as a television series. The selection of locations, the interviews, and the scripts were all their domain. In particular, we acknowledge the tireless efforts of Joseph Blatt, the series' producer. His ingenuity and creativity, along with those of producer Olga Rakich, permeate this book. Their words and their spirit are found throughout the text.

We are indebted to the many instructors at colleges and universities around the country who offered us their critical comments of the manuscript during the development and production of For All Practical Purposes. Their efforts helped improve the conceptual and pedagogical quality of the text.

Larry A. Curnutt, Bellevue Community College, Washington

Jane Edgar, Brevard County Community College, Florida

Marjorie A. Fitting, San Jose State University, California

Jerome A. Goldstein, Tulane University, Louisiana

Elizabeth Hodes, Santa Barbara Community College District, California

Dale Hoffman, Bellevue Community College, Washington

Frederick Hoffman, Florida Atlantic University, Florida

Barbara Juister, Elgin Community College, Illinois

Peter Lindstrom, North Lake College, Texas

Francis Masat, Glassboro State College, New Jersey

Charles Nelson, University of Florida, Florida

Virginia Taylor, University of Lowell, Massachusetts

Special thanks is extended to Larry A. Curnutt and Dale Hoffman of Bellevue Community College and Frederick Hoffman of Florida Atlantic University who class tested prepublication portions of this book.

Many of the outstanding full color images were obtained from research scientists. We thank them for allowing us to reproduce some of their important work: H.E. Benzinger, S.A. Burns, and J. Palmore, University of Illinois (chaotic basins of attraction); Douglas Dunham, University of Minnesota (hyperbolic patterns); David Hoffman, University of Massachusetts (minimal surfaces); Benoit Mandelbrot, IBM (fractal images); Heinz-Otto Peitgen, University of Santa Cruz (Mandelbrot sets); Roger Penrose, Oxford University (Penrose tilings).

Other images appearing in the color plates are courtesy of Philip Zucco (computer graphics) and Cordon Art (Escher patterns). Color renderings were done by Tom Moore.

The writing of this book by our author team took place simultaneously with the production of the television series, and the complications and deadlines were numerous. The manuscript and complex illustrations would not have come together without the expert guidance of the W. H. Freeman and Company staff. In particular we wish to thank Susan Moran, project editor, Susan Stetzer, production coordinator, Mike Suh, art director, and Bill Page, art coordinator. Lloyd Black, senior development editor, arrived on the scene just in time to pull the words and figures together and guide the book through the production process. We send a special note of thanks to Jeremiah Lyons, senior mathematics editor. His faith in the project and efforts over two plus years have brought order out of chaos.

The final acknowledgments must go to the COMAP staff. To the production and administrative staff—Philip McGaw, Nancy Hawley, and Annemarie Morgan—go all our thanks. And finally, we recognize the contribution of Laurie Aragon, the COMAP business, development, personnel, etc. manager, who kept this project, as she does all of COMAP, running smoothly and efficiently. To everyone who helped make our purposes practical, we offer our appreciation for an exciting, exhausting, and exhilarating time.

Solomon Garfunkel
COMAP

Lynn A. Steen
St. Olaf College

FOR ALL PRACTICAL PURPOSES

PART I

MANAGEMENT SCIENCE

The first steps on the moon represented a great leap for management science. Here, Buzz Aldrin poses for Neil Armstrong. [NASA.]

Those who were watching live television one night in July 1969 will never forget the spectacle of seeing the first man walk on the moon. The element of danger and uncertainty, heightened by a history of trial and error, added to the suspense of the lunar landing. Before the 1960s, no one really knew if rockets would ever be able to carry humans into space.

Neil Armstrong's first step onto the moon's surface was a triumph for American science and technology and the culmination of a national quest that had begun in the office of President John F. Kennedy. It was Kennedy's goal to put a man on the moon before the decade was out, a goal realized in the Nixon administration.

In the eight years from 1961 to 1969 we moved from a president's vision to the reality of a lunar landing. Most of us think of this achievement in terms of the tremendous scientific advances it represented — in physics, engineering, chemistry, and associated technologies. But there was another side to this far-reaching project. Someone had to set the objectives, commission the work, suffer the setbacks, overcome unforeseen obstacles, and tie together a project with thousands of disparate components. The kind of science responsible for such details is a branch of mathematics called **management science.**

NASA administrators faced many new problems in putting a man on the moon: they had to choose the best design for the spacecraft, design realistic ground simulations, and weigh the priorities of conducting experiments with immediate returns against carrying out tests that would serve long-term goals. When NASA commissioned the Apollo module, it was asking several hundred companies to design, build, test, and deliver components and systems that had never been built before.

Supporting these space age goals, however, were the nuts and bolts issues that make up the major concerns of management science, namely, finding ways to make the operations as productive and economical as possible (see the accompanying box on the Apollo 11

launch). Issues of efficiency are important in all organizations. In business, industry, or government, operating efficiently is hardly a novel idea. Directors of large corporations and those in the high ranks of the military have always pursued efficiency. What is new about management science is that it distinguishes between trial-and-error — "seat of the pants" — approaches and new ideas guided by systematic mathematical analysis.

The need to establish scientific principles for operations management arose in World War II. The progenitors of management science were mathematicians and industrial technicians associated with the armed services who worked together to improve military operations. In applying quantitative techniques to project planning, these pioneers founded a new science.

Management science, or **operations research,** as it sometimes is called, turned out to be a powerful notion with wide-ranging applications. It enabled mathematicians to bring a long history of pure research to bear on practical problems. We will explore some of these problems and solutions in the sections ahead.

Apollo 11 Launch Owes Success to Management Science

Today, Captain Robert F. Freitag is director of Policy and Plans for NASA's space station. Back in 1969, however, Freitag headed the team responsible for landing the Apollo 11 safely on the moon. The success of the lunar mission can be traced to management-science techniques that ensured that thousands of small tasks would come together to meet a single giant objective. Freitag shares his observations about that historic event:

> I think the feeling most of us in NASA shared was, "My gosh, now we really have to do it." When you think that the enterprise we were about to undertake was ten times larger than any that had ever been undertaken, including the Manhattan Project, it was a pretty awesome event. But we knew it was the kind of thing that could be broken down into manageable pieces and that if we could get the right people and the right arrangement of these people, it would be possible.

Captain Robert F. Freitag, NASA.

In the case of the Apollo program, it was very important that we take a comprehensive system engineering approach. We had to analyze in a very strict sense exactly what the mission was going to be, what each piece of equipment needed was and how it would perform, and all the elements of the system from the concept on through to the execution of the mission, to its recovery back on earth.

We started out, in a very logical way, by having a space station in earth orbit. We would then take the lunar space craft and build it in orbit, and then send it off to the moon and bring it back. It turned out that this approach was probably a little more risky and took a lot longer, so with the analyses we made, we shifted our whole operation to building a rocket that would go all the way to the moon after it took off from Cape Canaveral. It would then go into orbit around the moon rather than landing on the moon, and from orbit around the moon would descend to the surface of the moon and perform its exploration. Then it would return to its orbit around the moon and come back home.

Well, that was a very comprehensive analysis job. It was probably more deepseated than the kind of job one would do for building an airplane or a dam because there were so many variables involved. What you do is break it down into pieces: the launch site, the launch vehicles, the space craft, the lunar module, and worldwide tracking networks, for example. Then, once these pieces are broken down, you assign them to one organization or another. They, in turn, take those small pieces, like the rocket, and break it down into engines or structures or guidance equipment. And this breakdown, or "tree," is the really tough part about managing.

In the Apollo program, it was decided that three NASA centers would do the work. One was Huntsville, where Dr. Von Braun and his team built the rocket. The other was Houston, where Dr. Gerous and his team built the space craft and controlled the flight operations. The third was Cape Canaveral, where Dr. Debries and his team did the launching and the preparation of the rocket.

Those three centers were pieces, and they could break their pieces down into about 10 or 20 major industrial contractors who would build pieces of the rocket. And then each of those industrial contractors would break them down into maybe 20 to 30 or 50 subcontractors — and they, in turn, would break them down into perhaps 300,000 or 400,000 pieces, each of which would end up being the job of one person. But you need to be sure that the pieces come together at the right time, and that they work when put together. Management science helps with that. The total number of people who worked on the Apollo was about 400,000 to 500,000, all working toward a single objective. But that objective was clear when President Kennedy said, "I want to land a man on the moon and have him safely returned to the earth, and to do so within the decade." Of course, Congress set aside $20 billion. So you had cost, performance, and schedule, and you knew what the job was in one simple sentence. It took a lot of effort to make that happen.

Street Networks

The underlying theme of management science is finding the best method for solving some problem — what mathematicians call the **optimal solution.** In some cases, it may be to finish a job as quickly as possible. In other situations, the goal might be to maximize profit or minimize cost. What we define as "optimal" depends on the nature of the goal.

Whatever the aim, the optimal solution is directly linked to the goal of efficiency in managing a complex activity. Complex activities arise in many places, and management-science techniques are not limited to multimillion-dollar corporations. They also turn out to be useful in planning the public services we depend on. In a city, for example, such techniques might benefit the public works department, the sanitation authority, or the parking department.

Let's concentrate on the parking department. Most cities and many small towns have

parking meters that must be checked regularly for damage or parking violations. We will use an imaginary town to show how management-science techniques can help to make parking control more efficient.

Euler Circuits

The street map in Figure 1.1 is typical of many towns across the United States, with streets, residential blocks, and a village green. There are almost unlimited possibilities for parking-control routes. Our job, or that of the commissioner of parking, is to find the optimal solution, the most efficient route for the parking-control officers, who travel on foot, checking the meters in an area.

The commissioner has two goals in mind: (1) the parking-control officer must cover all the sidewalks that have parking meters with-

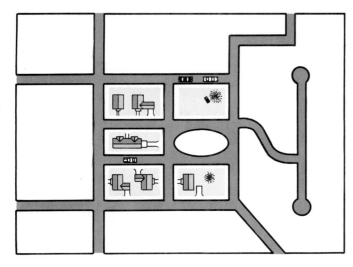

Figure 1.1   A street map for part of a town.

out retracing steps any more than is necessary; (2) the route should end at the same point it began, perhaps where the officer's patrol car is parked.

We can think of this problem in terms of a structure called a **graph,** one of the many mathematical models that can help to simplify complex problems. A graph is a finite set of

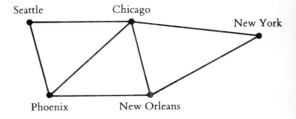

Figure 1.2   The edges of this graph show nonstop routes that an airline might offer.

dots and connecting links. The dots are called **vertices** (a singled dot is called a **vertex**), and the links are called **edges.** Each edge must connect two different vertices. A graph can represent our city map, a communications network, or even a system of air routes (see Figure 1.2).

In the case of parking control, we can represent the whole territory to be patrolled by a graph: we can think of each street intersection as a vertex and each sidewalk that contains a meter as an edge (see Figure 1.3). Notice in Figure 1.3b that the street separating the blocks is not explicitly represented; it has been shrunk to nothing. In effect, we are simplifying our problem by ignoring any distance traveled in crossing streets at the street corners.

The numbered sequence of edges in Figure 1.4a shows one route that covers all the meters and returns to its starting point. But Figure 1.4b shows another solution that is better because its route covers every edge (sidewalk) exactly once. In Figure 1.4b there is no reuse of

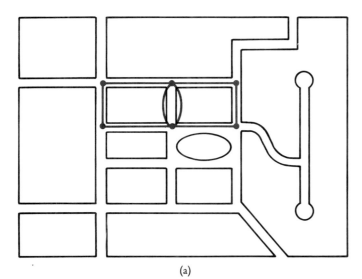

(a)

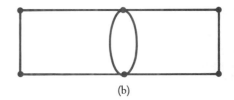

(b)

Figure 1.3   (a) A graph superimposed upon a street map. The edges show which sidewalks have parking meters. (b) The same graph.

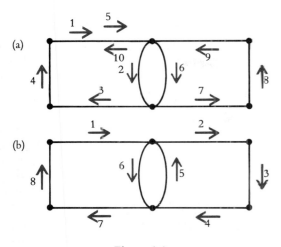

(a)

(b)

Figure 1.4

an edge already covered, or no *deadheading* (a term borrowed from shipping, which means making a return trip without a load). A route that covers every edge only once is called an **Euler circuit**, after the great eighteenth-century mathematician Leonhard Euler (pronounced oy'lur) who first studied them (see Box 1.1). In any graph, a route is called an Euler circuit if two conditions are satisfied:

1. Each edge of the graph is traversed once and only once.

2. The route starts and ends at the same vertex.

BOX 1.1 Leonhard Euler, 1707–1783

Portrait of Leonhard Euler.

Although born in Switzerland, Leonhard Euler spent a large part of his life in St. Petersburg, Russia. He was extremely prolific, publishing over 500 works in his lifetime. His collected works, which are now being published, are expected to exceed 70 volumes. Euler made major contributions to many areas of mathematics, including algebra and the theory of functions. A contemporary claimed that Euler could calculate effortlessly, "just as men breathe, as eagles sustain themselves in the air."

Human interest stories about Euler have been handed down through two centuries. He was extremely fond of children and had thirteen of his own, of whom only five survived childhood. It is said that he often wrote difficult mathematical works with a child or two in his lap. He was a prodigy at doing complex mathematical calculations under less than ideal conditions and continued to do them even after he became totally blind later in life. His blindness diminished neither the quantity nor the quality of his output. Throughout his life, he was able to mentally calculate in a short time what would have taken ordinary mathematicians hours of pencil and paper work.

Euler divided his working life between the St. Petersburg Academy in Russia and the Berlin Academy. He moved to Berlin largely because of difficult political conditions in Russia. When asked in Berlin why he spoke so little, Euler replied, "Madam, I come from a country where if you speak, you are hanged."

Figure 1.4b shows an Euler circuit. You can see that the circuit in Figure 1.4a is not an Euler circuit. It does not satisfy the first condition because one of its edges is reused.

One of the first discoveries made in the theory of graphs was that some graphs have no Euler circuits at all. For example, in the graph in Figure 1.5, it would be impossible to start at one point and cover all the edges without retracing some steps: since once we reach the center (or even if we begin at the center), no matter which edge we decide to follow, we will always have to return to the center to reach the remaining edge. But this is not allowed in an Euler circuit.

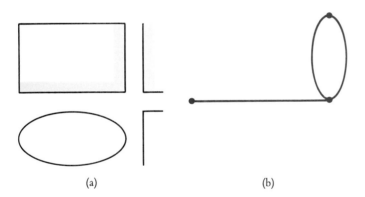

(a)   (b)

Figure 1.5   (a) The three shaded sidewalks cannot be covered by an Euler circuit. (b) The graph of the shaded sidewalks in part (a). The three-valent vertex rules out an Euler circuit.

Finding Euler Circuits

Now that we know the conditions an Euler circuit satisfies, we have two obvious questions to ask about Euler circuits:

1. Is there a way to tell by calculation, not by trial and error, if a particular graph has an Euler circuit?

2. Is there a method, other than trial and error, for finding an Euler circuit when one exists?

Euler answered these questions in 1735, by using the concepts of valence and connectedness. The **valence** of a vertex in a graph is the number of edges meeting at the point. Figure 1.6 illustrates the concept of valence, with vertices A and D having valence 3, vertex B having valence 2, and vertex C having valence 0. (Isolated vertices such as vertex C are an annoyance in Euler circuit theory. Because they don't occur in applications, we henceforth assume that our graphs have no vertices of valence 0.)

A graph is said to be **connected** if for each pair of its vertices there is at least one path of edges connecting the two vertices. The graph in Figure 1.7 is not connected because we are

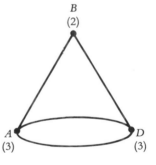

Figure 1.6

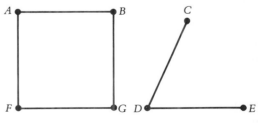

Figure 1.7

unable to join A to D with a path of edges. However, it does consist of two connected components.

Now we can state Euler's theorem, his simple answer to the problem of detecting when a graph G has an Euler circuit:

1. If G is connected and has all valences even, then G has an Euler circuit.

2. Conversely, if G has an Euler circuit, then G must be connected and all its valences must be even numbers.

(For clues about a proof of Euler's theorem, see Box 1.2.)

Once we know there is an Euler circuit in a certain graph, how do we find it? The set of rules Euler gave us is of theoretical interest and could be of practical interest if we wanted to program a computer to find Euler circuits mechanically (see Box 1.3). However, we will not study these rules because it turns out that most human beings, with a little practice, can find Euler circuits by trial and error and a little ingenuity — even in fairly large graphs.

Armed with Euler's theorem, we can now take another look at the parking-meter problem. The key question is this: Is there an Euler circuit in the graph? We can answer this question by checking connectedness and valence, the two conditions in Euler's theorem.

First, we observe that the graph is connected, so the first condition is satisfied. Next, we determine the valence, odd or even, of each vertex. Because each vertex does have even valence, the graph meets the second condition, too. The parking commissioner can now design a route along an Euler circuit. Figure 1.4b is an example of such a circuit.

BOX 1.2 Proving Euler's Theorem

One of the easiest kinds of graphs to deal with is the simple loop, a graph that we can draw completely, starting and ending at the same vertex without revisiting any vertex except the starting point. Here are some examples:

A simple loop generally has the appearance of a polygon, although its edges need not be straight.

It is obvious that a simple loop has an Euler circuit. If we put two simple loops together so that they touch at one or more vertices but not along whole edges, we can build an extended Euler circuit out of the individual Euler circuits. In the figure that follows, we can begin the Euler circuit for the first loop at vertex A, interrupt ourselves when we get to the vertex where the two loops meet, go around the second loop, returning to the common vertex, and then finish the Euler circuit for the first loop at vertex A:

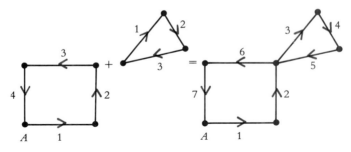

If we hang yet another simple loop on this graph, we can extend the Euler circuit of the two-loop graph to an Euler circuit for the three-loop graph in a similar way. Notice that the four-sided loop is added at two vertices, not along edges:

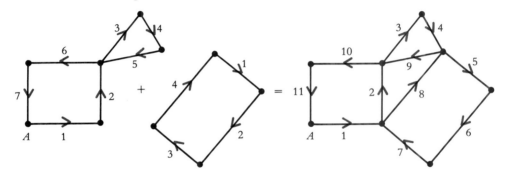

Thus, graphs that can be built out of simple loops by joining the loops at vertices have Euler circuits. This yields a strategy for proving Euler's theorem: it can be shown that the graphs that can be built out of simple loops are precisely the ones that are connected and have all valences even.

BOX 1.3 The Human Aspect of Problem Solving

Thomas Magnanti, professor of operations research and management, heads the department of management science at MIT's Sloan School of Management. Here are some of his observations:

> In applying Euler circuits and other management-science tools and techniques, it's very important to recognize that you need data in order to be effective. In the case of Euler circuits, you must have information on the underlying patterns of the street network. You also need information concerning the demand patterns — what the flow of traffic is like, what the parking situation is, and so on.
>
> But in addition to that, I think one should recognize that the underlying algorithms — the procedures for solving the Euler circuits — are typically only a very small

Thomas Magnanti, department of
management science, Sloan School
of Management, MIT.

part of a management-science approach to
problem solving. We want to be able to collect
information and data, and we want to be able to
use them effectively. In many cases, collecting,
storing, and updating the information is as
much of a challenge to mathematicians and
computer scientists as are the solution methods
themselves. Often, a large-scale data base
underlies the actual application of Euler circuits
or other management-science techniques.

Typically, a management-science approach
has several different ingredients. One is just
structuring the problem — understanding that
the problem is an Euler-circuit problem or a
related management-science problem. After that,
one has to develop the solution methods.

But one should also recognize that you don't
just push a button and get the answer. In using
these underlying mathematical tools, we never
want to losesight of our common sense, of understanding, intuition, and judgment.
The computer provides certain kinds of insights. It deals with some of the combinatorial complexities of these problems very nicely. But a model such as an Euler
circuit can never capture the full essence of a decision-making problem.

Typically, when we solve the mathematical problem, we see that it doesn't quite
correspond to the real problem we want to solve. So we make modifications in the
underlying model. It is an interactive approach, using the best of what computers
and mathematics have to offer and the best of what we, as human beings, with our
own decision-making capabilities, have to offer.

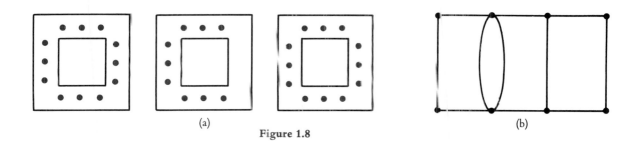

(a)

Figure 1.8

(b)

Circuits with Reused Edges

Now let's see what Euler's theorem tells us
about the three-block neighborhood in Figure
1.8a. Figure 1.8b shows the corresponding
graph. (Notice that the sidewalk with no
meters is not represented by any edge in the
graph. We only use edges to represent side-

walks along which the officer must walk.)
This graph has two odd valences, so Euler's
theorem tells us that there is no Euler circuit
for this graph.

Since we must reuse some edges in this
graph if we are to cover all edges in a circuit,
for efficiency we need to keep the total length
of reused edges to a minimum. This type of

problem is often called the **Chinese postman problem** (like parking-control routes, mail routes need to be efficient). Although the Euler circuit theory doesn't deal directly with reused edges or edges of different lengths, we can extend the theory to help solve the Chinese postman problem.

In a realistic Chinese postman problem, we need to take account of the *lengths* of the sidewalks, streets, or whatever the edges represent because we want to minimize the total length of the reused edges. However, we will simplify things at the start by supposing that all edges represent the same lengths. (This is often called the *simplified* Chinese postman problem.) In this case, we need only count reused edges and need not measure their lengths. Thus, we want to find a circuit that covers each edge and that has the minimal number of reuses of edges already covered.

To illustrate the theory we are going to develop, consider the graph of Figure 1.9a. It has no Euler circuit, but there is a circuit that has only one reuse of an edge (*AD*), namely *ADA-BEDC*. Let's draw this circuit so that when an edge is about to be reused, we install a new, extra edge in the graph for the circuit to use (see Box 1.4). By duplicating edge *AD*, we can avoid reusing the edge. To duplicate an edge we must join the vertices that are already joined by the edge. We have now created the graph of Figure 1.9b. In this graph the original circuit can be traced as an Euler circuit, using the new edge when needed.

Our theory will be based on using this idea in reverse, as follows:

1. Take the given graph and add edges, duplicating existing ones until you arrive at a graph that is connected and even-valent. We call this process **eulerizing** the graph because the graph we produce will have an Euler circuit.

2. Find an Euler circuit on the eulerized graph.

3. Trace out this Euler circuit on the original graph (before eulerization), reusing an edge each time the circuit on the eulerized graph uses an added edge.

Figure 1.10b shows one way to eulerize the graph of Figure 1.10a. After eulerization, each

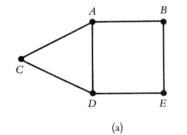

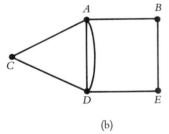

(a) (b)

Figure 1.9

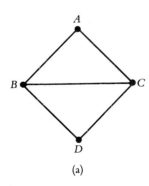

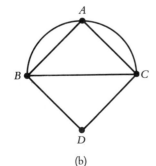

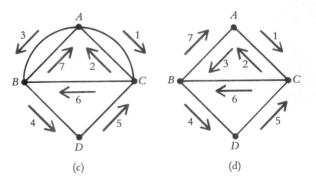

(a) (b) (c) (d)

Figure 1.10

BOX 1.4 How to Add New Edges

When adding new edges to eulerize a graph, it is desirable to add only edges that are duplicates of existing edges. For example, in graph (a) below, we need to make X and Y even. Adding one long edge from X to Y [graph (b)] might seem like an attractive idea, but this would not duplicate only one existing edge. (Remember, to duplicate an edge means to add an edge that joins two vertices that are already joined by the edge to be duplicated.)

If we add this long edge and then try to squeeze an Euler circuit on this new graph back into the old graph, this long edge won't count for just one reused edge. This is because the alternative to using the long edge, which we can't use during squeezing, is not one reused edge but a whole series stretching from X to Y [the heavy line in graph (c)]. If we don't add the long edge but instead follow the rule for duplicating existing edges in graph (d), we'll be able to count the edges that will be reused according to how many duplicate edges we added (three in this case).

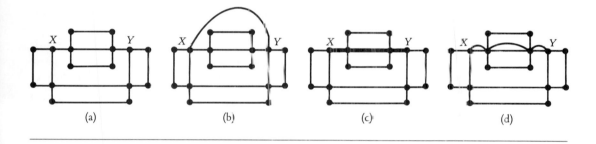

| (a) | (b) | (c) | (d) |

vertex has even valence. Figure 1.10c shows an Euler circuit on the eulerized graph: simply follow the edges in numerical order and in the direction of the arrows, beginning and ending at vertex A. The final step, in Figure 1.10d, is to squeeze our Euler circuit into the original graph. There are two reuses of previously covered edges. Notice that each reuse of an edge corresponds to an added edge. This is generally true in this type of problem: the number of reuses of edges equals the number of edges added during eulerization.

A special caution about eulerizing: be sure that each added edge connects vertices that are already joined by an edge. For example, the procedure shown in Figure 1.11a is not a proper eulerization, whereas the procedure shown in Figure 1.11b is. The added edges must be duplicate versions of already existing ones for a proper eulerization.

Now that we have learned to eulerize, the next step is to try to get the best eulerization we can. It turns out that there are many ways to eulerize a graph.

In Figure 1.12, we begin with the same graph we used in Figure 1.10, but we eulerize it in a different way—by adding only one edge. Figure 1.12c shows an Euler circuit on the eulerized graph, and in Figure 1.12d we see how it is squeezed into the original graph. There is one reuse of an edge, because we added one edge during eulerization.

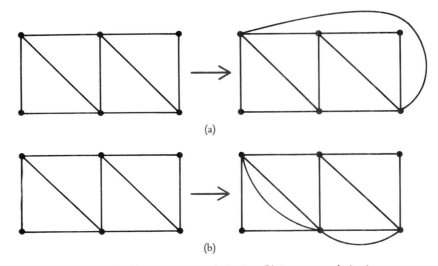

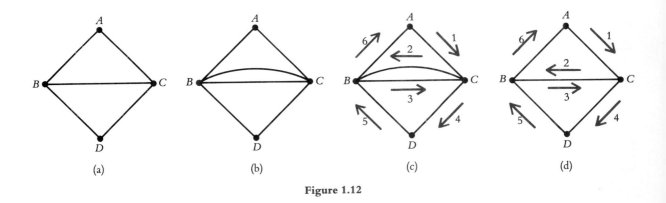

Figure 1.11 (a) An incorrect eulerization. (b) A correct eulerization.

Figure 1.12

The solution in Figure 1.12 is better than the solution in Figure 1.10 because one reuse is better than two. These examples suggest the following addition to our solution procedure: *Try to find the eulerization with the smallest number of added edges.* This extra requirement makes the problem both more interesting and more difficult. For large graphs, the best eulerization may not be obvious. We can try out a few and pick the best among the ones we find,

however, there may be an even better one that our haphazard search does not turn up.

A systematic procedure for finding the best eulerization does exist, but it is complicated, we present only a few hints about it in Box 1.5. With a little practice, most people can find the best or nearly best eulerization, using only trial and error and a little ingenuity. This is especially easy for rectangular street networks (see Box 1.6).

BOX 1.5 Finding Good Eulerizations

Suppose we wanted a perfect procedure for eulerizing a graph. What theoretical ideas and methods could we use to build such a tool?

One building block we could use is a method for finding the shortest path between two given vertices of a graph. For example, suppose we attempt to eulerize vertices X and Y in graph (a) below by connecting them with a pattern of duplicate edges, as in graph (b). The cost of this is the length of the path we duplicated from X to Y. A shorter path from X to Y, such as the one shown in graph (c), would be better. Fortunately, the shortest-path problem has been well studied, and we have many good procedures for solving it exactly, even in large, complex graphs.

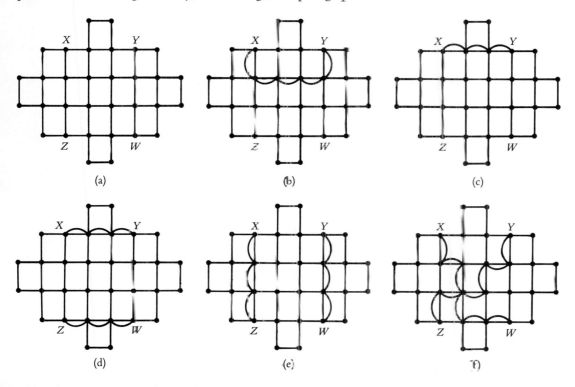

But there is more to eulerizing graph (a) than dealing with X and Y: notice that we have odd valences at Z and W. Should we connect X and Y with a path and then connect Z and W, as in graph (d)? Or should we connect X to Z and Y to W, as in graph (e)? Another alternative is to use diagonal connections X to W and Y to Z, as in graph (f). The problem is how to pair up vertices for connection to get a set of paths whose total length is minimal. This is called a *matching problem*, and it too can be solved by well-known methods.

Detailed treatments of the shortest path and matching problems would take us too far afield but can be found in the following texts:

- Bogart, Kenneth P. *Introductory Combinatorics,* Pitman, 1983.
- Roberts, Fred S. *Applied Combinatorics,* Prentice-Hall, 1984.
- Tucker, Alan. *Applied Combinatorics,* Wiley, 1980.

BOX 1.6 Good Eulerizations for Rectangular Networks

Many street networks are composed of a series of rectangular blocks that form a large rectangle *m* blocks long by *n* blocks wide. For such networks, we can follow simple patterns to get a best eulerization. The figures that follow show three different patterns. Which pattern to use depends on whether, in the large rectangle, the length and width are both an odd number of blocks [graph (a)], whether both are an even number of blocks [graph (c)], or whether one is even and the other is odd [graph (b)].

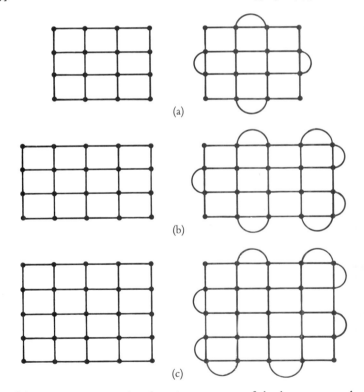

(a)

(b)

(c)

To create one of these patterns, start by choosing a corner of the large rectangle. (For the figures shown we chose the upper left corner in each case.) Now, proceed around the boundary of the large rectangle in both directions simultaneously, duplicating every second edge. Think of this as being done by two separate processors. The procedure ends when the processors meet at the corner diagonally opposite the one at which they started

BOX 1.7 Israel Electric Company Reduces Meter-Reading Task

The Beersheva branch of Israel's major electric company wanted to make the job of meter reading more efficient. When the branch managers decided to minimize the number of people required to read the electric meters in the houses of one particular neighborhood, they set a precedent by applying management science. Formerly, each person's route had been worked out by trial and error and intuition, with no help from mathematics. The whole job required 24 people, each doing a part of the neighborhood in a five-hour shift.

Finding a more efficient way of doing the work sounds a lot like the Chinese postman problem, but there are two important differences. First, the neighborhood was big enough to negate any possibility of having only one route assigned to one person. It was necessary to find a number of routes that, taken together, covered all the edges (sidewalks). Second, a meter reader who was done with a route was allowed to return home directly. Thus, there was no reason for the individual routes to return to their starting points: routes could be paths instead of circuits.

The Beersheva researchers started by doing a partial eulerization in which each odd valence was converted to even valence, except for two vertices that remained odd. In such a partially eulerized graph, it is possible to find a path, not a circuit, that covers every edge and that starts at one odd-valent vertex and ends at the other.

Next, they chopped the path into five-hour chunks by stepping through the path and marking each time five hours' worth of edges were covered. By following this procedure, researchers managed to cover the neighborhood with 15 five-hour routes, a 40% reduction over the original 24 five-hour routes. Altogether, these routes involve a total of 4338 minutes of walking time, of which 41 minutes (less than 1%) is deadheading.

Circuits with More Complications

The theory we have described has many more practical applications than just checking parking meters. Almost any time services need to be delivered along streets or roads, our theory can make the job more efficient. Examples include collecting garbage, salting icy roads, plowing snow, inspecting railroad tracks, and reading electric meters (see Box 1.7).

However, each of these problems has its own special requirements that may call for modifications in the theory. For example, in the case of garbage collection, the edges of our graph will represent streets, not sidewalks. If some of the streets are one-way, we need to put arrows on the corresponding edges, resulting in a directed graph, or **digraph.** The circuits

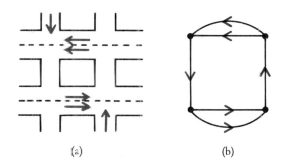

(a)   (b)

Figure 1.13 (a) Salt-spreading route, where each east-west street has two one-way lanes, and (b) an appropriate digraph model.

we seek will have to obey these arrows. In the case of salt spreaders and snow plows, each lane of a street needs to be modeled as a directed edge. (See Figure 1.13.)

(a)

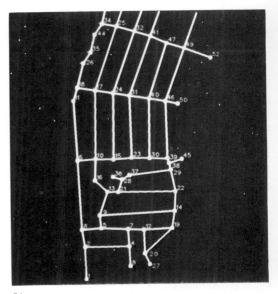

(b)

Figure 1.14 (a) Today, finding optimal routes within complex street networks is often done with sophisticated computer-based color graphics systems that allow for detailed modeling of such networks in real towns and cities. (b) A computer-generated street network. [Courtesy of Sidney B. Bowne & Son, Consulting Engineers.]

Like salt spreaders, street-sweeping trucks can travel in only one lane at a time and need to obey the direction of traffic. Street sweepers, however, have an additional complication: parked cars. It is very difficult to clean the street if cars are parked along the curb. Yet for overall efficiency, those who are responsible for routing street sweepers want to interfere with parking as little as possible. The common solution is to post signs specifying times when parking is prohibited, such as the second Thursday of each month between 8 a.m. and 2 p.m. Because the parking-time factor is a constraint on street sweeping, it is important not only to find an Euler circuit, or a circuit with very few duplications, but a circuit that can be completed in the time available. Once again, the theory can be modified to handle this constraint.

Finally, because towns and cities of any size will have more than one street sweeper, parking officer, or garbage truck, a single best route will not suffice. Instead, they will have to divide the territory into multiple routes. The general goal is to find optimal solutions while taking into account traffic direction, number of lanes, time restrictions, and divided routes. (See Figure 1.14.)

Management science makes all this possible. For example, a pilot study done in the 1970s in New York City showed that applying these techniques to street sweepers in just one district could save about $30,000 per year. Because there are 57 sanitation districts in New York, this comes to a savings of more than $1.5 million in a single year. In addition, the same principles could be extended to garbage collection, parking control, and other services carried out on street networks.

But New York City never adopted this plan. Because city services take place in a political context, several other factors come into play. Unions want to protect the jobs of city workers, bureaucrats want to keep their departmental budgets high, and elected politicians don't want to be accused of cutting jobs.

Thus, political obstacles interfered with the goal of efficiency, and New York lost an opportunity to save $1.5 million.

Despite the complications of real-world problems, management-science principles provide ways to understand them by using graphs as models. We can reason about the graph and then return to the real-world problem with a workable solution. The results we get can have a lasting effect on the efficiency and economic well-being of any organization or community.

REVIEW VOCABULARY

Chinese postman problem The problem of finding a connected sequence of edges on a graph that covers every edge of the graph at least once, returns to its starting vertex, and has the shortest length of all such sequences.

Connected graph A graph is connected if it is possible to reach any vertex from any specified starting vertex by traversing edges.

Digraph A graph in which each edge has an arrow indicating the direction of the relationship. Such directed edges are appropriate when the relationship is "one-sided" rather than symmetric (e.g., one-way streets as opposed to regular streets).

Edge An arc joining two vertices in a graph.

Euler circuit A connected sequence of edges that traverses each edge of the graph exactly once and returns to the starting vertex.

Eulerizing Adding new edges to a graph so as to make a graph that possesses an Euler circuit.

Graph A mathematical structure in which points (called vertices) are used to represent things of interest and where arcs (called edges) are used to connect vertices as a means of displaying that the connected vertices have a certain relationship.

Management science A discipline in which mathematical methods are applied to management problems in pursuit of optimal solutions that cannot readily be obtained by common sense.

Operations research Another name for management science.

Optimal solution When a problem has various solutions that can be ranked in preference order (perhaps according to some numerical measure of "goodness"), the optimal solution is the best-ranking solution.

Valence of a vertex The number of edges touching that vertex.

Vertex A point in a graph where one or more edges end.

EXERCISES

1. What are the valences of the vertices in this graph?

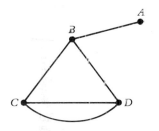

2. Examine the numbered sequences of edges in the figures below. Explain why none is an Euler circuit.

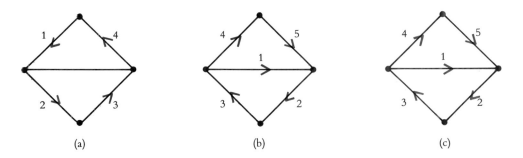

(a)   (b)   (c)

3. Which graphs in the figures below have Euler circuits? In the ones that do, find the Euler circuits. (Show this by numbering the edges in the order the Euler circuit uses them.)

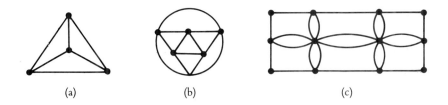

(a)   (b)   (c)

4. In the graph of the figure below, add one or more edges to produce a graph that has an Euler circuit.

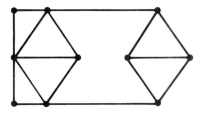

5. Suppose for a certain graph that it is possible to disconnect it by removing one edge. Explain why such a graph (before the edge is removed) must have at least one odd valence. (Hint: show that it cannot have an Euler circuit.)

6. Can you find an eulerization with 7 added edges for a 2-block by 5-block rectangular street network? Can you do better than 7?

7. Can you find an eulerization with 6 added edges for a 3-block by 5-block rectangular street network? Can you do better than 6?

8. Can you find an eulerization with 10 added edges for a 4-block by 6-block rectangular street network? Can you do better than 10?

9. In the figure below, all blocks are 1000 by 1000 feet, except for the middle column of blocks, which are 1000 by 4000 feet. Can you find an eulerization in which the total length of all duplicated edges is 8000?

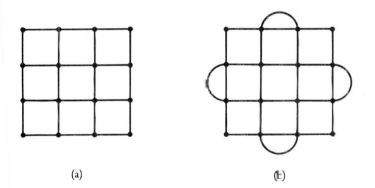

10. Find an Euler circuit on the eulerized graph (b) of the following figure. Use it to find a circuit on the original graph (a) that covers all edges and only reuses edges four times.

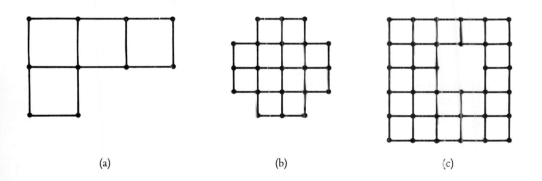

(a)   (b)

11. Find good eulerizations for these graphs:

(a)   (b)   (c)

12. Squeeze the circuit shown in graph (a) on the next page into graph (b). Show your answers by numbered arrows on the edges.

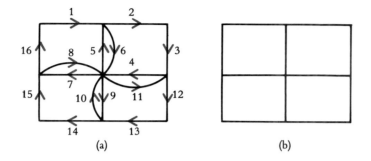

(a)   (b)

13. Find a circuit in the graph below that covers every edge and has as few reuses as possible.

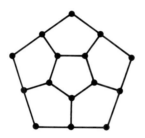

14. a. For the street network below, draw the graph that would be useful for finding an efficient route for checking parking meters. (Hint: notice that not every sidewalk has a meter.)

b. For the same street network, draw the graph that would be useful for routing a garbage truck. Assume that all streets are two-way and that passing once down a street suffices to collect from both sides.

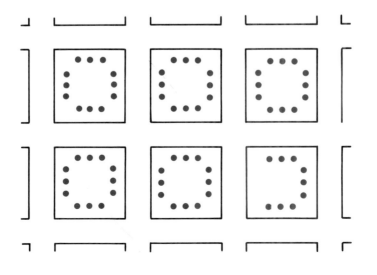

15. In the graph below, we see a territory for a parking-control officer that has no Euler circuit. Which sidewalk (edge) could be dropped in order to enable us to find an Euler circuit?

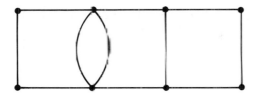

16. Many street networks are composed of a series of rectangular blocks that we can eulerize using one of the three different patterns shown in the right-hand panels of parts (a), (b), and (c) of the figure in Box 1.6. After reviewing Box 1.6, "Good Eulerizations for Rectangular Networks," try to apply the appropriate pattern in each of the following cases:

    a. A 5 × 5 rectangle

    b. A 5 × 4 rectangle

    c. A 6 × 6 rectangle

CHAPTER 2

Visiting Vertices

In the last chapter, we saw that it is relatively easy to determine if there is a route traversing the edges of a graph exactly once — for example, a route for street sweepers that covers the streets in a section of a city exactly once. However, the situation changes radically if we make an apparently innocuous change in the problem: When is it possible to find a route along the edges of a graph that visits each *vertex* once and only once in a loop or simple circuit? For example, the wiggly line in Figure 2.1 shows a circuit we can take to tour that graph, visiting each vertex once and only once. This tour can be written *ABDGIHFECA.*

A tour such as the wiggly edges in Figure 2.1 is a **Hamiltonian circuit,** named for the Irish mathematician William Rowan Hamilton, who was one of the first to study the concept. (We now know that the concept was

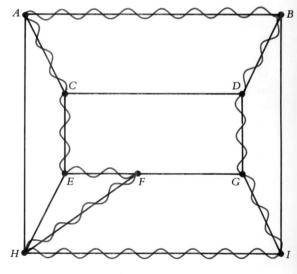

Figure 2.1

discovered somewhat earlier by Thomas Kirkman, a British minister with a penchant for mathematics.) This problem is typical of the many new problems mathematicians create as part of their professional training, here by changing vertices for edges.

The concepts of Euler and Hamiltonian circuits are similar in that both forbid reuse: the Euler circuit of edges, the Hamiltonian circuit of vertices. However, it is far more difficult to determine which connected graphs admit a Hamiltonian circuit than to determine which connected graphs have Euler circuits. As we saw in Chapter 1, looking at the valences of vertices tells us if a connected graph has an Euler circuit, but we have no such simple method for telling whether or not a graph has a Hamiltonian circuit. Some special classes of graphs are known to have Hamiltonian circuits and some special classes of graphs are known to lack them, but a general solution to this problem, as we will later see, is unlikely.

The Hamiltonian-circuit problem and the Euler-circuit problem are both examples of graph-theory problems. Although we posed the Hamiltonian-circuit problem merely as a variant of another graph-theory problem with many applications (that is, the Euler-circuit problem), the Hamiltonian-circuit problem itself has many applications. This is not unusual in mathematics. Often the mathematics used to solve some particular real-world problem leads to new mathematics that suggests applications to other real-world situations.

Suppose inspections or deliveries need to be made at each vertex (rather than along each edge) of a graph. An "efficient" tour of the graph would be a route that passed through all the vertices without reuse, or repetition; that is, the route would be a Hamiltonian circuit. Such routes would be useful for inspecting traffic signals or for delivering mail to drop-off boxes, which hold heavy loads of mail, so that urban postal carriers do not have to carry it

long distances. There are many similar examples, but rather than pursue problems involving applications of Hamiltonian circuits, we will study instead a more important class of related problems.

Let's imagine that you are a college student studying in Chicago. During spring break you and a group of friends have decided to take a car trip to visit other friends in Minneapolis, Cleveland, and St. Louis. There are many choices as to the order of visiting the cities and returning to Chicago, but you want to design a route that minimizes the distance that you have to travel. Presumably, you also want a route that cuts costs and you know that minimizing distance will minimize the cost of gasoline for the trip. (Similar problems with different complications would arise for railroad or airplane trips.)

Imagine now that the local automobile club has provided you with the intercity driving distances between Chicago, Minneapolis, Cleveland, and St. Louis. We can construct a graph model with this information, representing each city by a vertex and the legs of the journey between the cities by edges joining the vertices. To complete the model, we add a number called a **weight** to each edge. The weight represents the distances between the cities, which correspond to the endpoints of the edges in Figure 2.2. We want to find a minimal-cost tour that starts and ends in Chicago and visits each other city once. Using our earlier terminology, what we wish to find is a **minimum-cost Hamiltonian circuit** — a Hamiltonian circuit with the lowest possible sum of the weights of its edges.

How can we determine which Hamiltonian circuit has minimum cost? There is a conceptually easy **algorithm,** or mechanical step by step process, for solving this problem:

1. Generate all Hamiltonian tours (starting from Chicago).

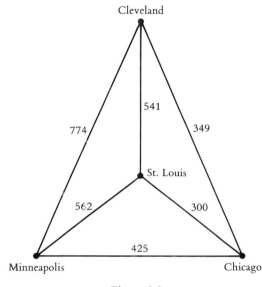

Figure 2.2

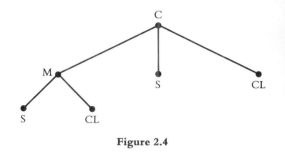

Figure 2.4

2. Add up the distances on the edges of each tour.

3. Choose the tour of minimum distance.

Steps 2 and 3 of the algorithm are straightforward. Thus, we need worry only about step 1, generating the Hamiltonian circuits in a systematic way. To find the Hamiltonian tours, we will use the **method of trees.** Starting from Chicago, we can choose any of the three cities to visit after leaving Chicago. The first stage of the enumeration tree is shown in Figure 2.3. If Minneapolis is chosen as the first city to visit, then there are two possible cities to visit next, namely, Cleveland and St. Louis. The possible branchings of the tree at this

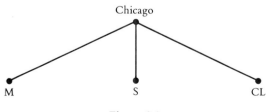

Figure 2.3

stage are shown in Figure 2.4. In this second stage, however, for each choice of first city we visited, there are two choices from this city to the second city visited. This would lead to the diagram in Figure 2.5.

Having chosen the order of the first two cities to visit, and knowing that no revisits (reuses) can occur in a Hamiltonian route, there is only one choice left for the next city. From this city we return to Chicago. The complete tree diagram showing the third and fourth stages for these routes is given in Figure 2.6. Notice, however, that because we can traverse a circular tour in either of two directions, the paths enumerated in the tree diagram of Figure 2.6 do *not* correspond to different Hamiltonian circuits. For example, the left-most path (C–M–S–CL–C) and the right-most path (C–CL–S–M–C) represent the same Hamiltonian circuit. For this problem, therefore, we find three distinct Hamiltonian circuits among the six paths in the tree diagram.

Note that in generating the Hamiltonian circuits we disregard the distances involved. We are concerned only with the different patterns of carrying out the visits. But which route is the optimal route? Adding up the distances on the edges to get each tour's length shows that the optimal tour is Chicago, Minneapolis, St. Louis, Cleveland, Chicago. The length of this tour is 1877 miles.

The method of trees is not always as easy to use as our example suggests. Instead of doing our analysis for four cities, consider the general case of *n* cities. The graph model similar to that

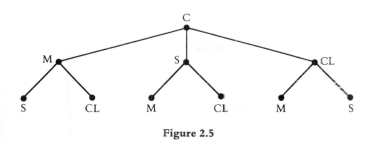

Figure 2.5

in Figure 2.2 consists of a weighted graph with *n* vertices, with every pair of vertices joined by an edge. Such a graph is called **complete** because all possible edges are included in the graph.

How many Hamiltonian circuits are in a complete graph of *n* vertices? We can solve this problem by using the same type of analysis that emerged in the enumeration tree. The method uses the **fundamental principle of counting**, a procedure for counting outcomes in multistage processes (see Box 2.1). The city visited first after the home city can be chosen in $n - 1$ ways, the next city in $n - 2$ ways, and so on until only one choice remains. Using the fundamental principle of counting, there are $(n - 1)! = (n - 1)(n - 2) \cdots 3 \cdot 2 \cdot 1$ routes.

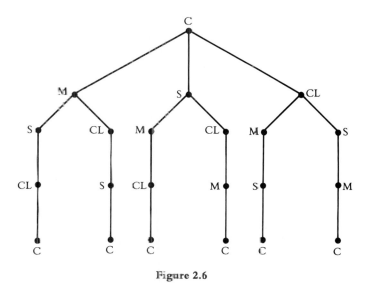

Figure 2.6

BOX 2.1 The Fundamental Principle of Counting

The fundamental principle of counting is a method for counting how many patterns occur in a situation by looking at the number of ways the component parts can occur. For example, if Jack has 9 shirts and 4 pair of trousers, he can wear 36 shirt-pants outfits. Each shirt can be worn with any of the pairs of pants.

In general, if there are *a* ways of choosing one thing, *b* ways of choosing a second after the first is chosen, . . . , and *z* ways of choosing the last item after the earlier choices, then the total number of choice patterns is: $a \times b \times c \times \cdots \times z$.

This multiplication process is used in counting Hamiltonian circuits and also gives us the number of patterns in tossing dice. For example, if one red die (with 6 dots) and one green die (with 6 dots) are thrown, the fundamental principle of counting shows that 36 different outcomes (6×6) are possible. The same 36 patterns occur even if the dice are identical. Starting with these 36 outcomes, a systematic analysis of gambling games such as craps can be made.

[The exclamation mark in "$(n - 1)!$" is read "factorial" and is a shorthand notation for the product $(n - 1)(n - 2) \cdots 3 \cdot 2 \cdot 1$. For example, $5! = 5 \cdot 4 \cdot 3 \cdot 2 \cdot 1 = 120$.]

Pairs of routes correspond to the same Hamiltonian circuit because one route can be obtained from the other by traversing the cities in reverse order. Thus, there are $(n - 1)!/2$ Hamiltonian circuits. Now, if we have only a few cities to visit, $(n - 1)!/2$ routes can be listed and examined in a reasonable amount of time. For example, for 6 cities it would require $(6 - 1)!/2 = 5!/2 = 120/2 = 60$ tours. But for, say, 25 cities, $24!/2$ is approximately 3×10^{23}. Even if these tours could be generated at the rate of one million a second, it would take 10 billion years to generate them all.

If the only benefit were saving money and time in vacation planning, the difficulty of solving the problem for large values of n would not be of great concern. However, the problem we are discussing is one of the most common problems in operations research. We usually call it the **traveling salesman problem** (TSP) because of its early formulation: determine the trip of minimum cost that a salesperson can make to visit the cities in a sales territory, starting and ending the trip in the same city.

There are many situations that require solving a TSP:

1. A lobster fisherman has set out traps at various locations and wishes to pick up his catch.

2. The telephone company wishes to pick up the coins from its pay telephone booths.

3. The electric (or gas) company needs to design a route for its meter readers.

4. In drilling holes in a series of plates, the drill press operator (perhaps a robot!) must drill the holes in a predetermined order.

The meaning of cost can vary from one formulation of TSP to another. We may measure cost as distance, airplane-ticket prices, time, or any other factor that is to be optimized.

In many situations, the TSP arises as a subproblem of a more complicated problem. For example, a supermarket chain may have a very large number of stores to be served from a single large warehouse. If there are fewer trucks than stores, the stores must be grouped into clusters, so that one truck serves each cluster. If we then solve the TSP for every truck, we can minimize total costs for the supermarket chain. Similar vehicle-routing problems for dial-a-ride services and for delivering schoolchildren to their schools often involve solving the TSP as a subproblem.

Strategies for Solving the Traveling Salesman Problem

Because the traveling salesman problem arises so often in situations whose complete graphs would be very large, we must find a better method than the "brute force" method we have described. We need to look at our original problem in Figure 2.2 and try to find an alternative algorithm for solving it. Recall that our goal is to find the minimum-cost Hamiltonian circuit.

Let's try a new approach: Starting from Chicago, first visit the nearest city, then visit the nearest city that has not already been visited. We return to the start city when no other choice is available. This approach is called the **nearest-neighbor algorithm.** Applying it to the TSP in Figure 2.2 quickly leads to the tour of Chicago, St. Louis, Cleveland, Minneapolis, and Chicago, with a length 2040 miles. The nearest-neighbor algorithm is an example of a "greedy" algorithm, because at each stage a best (greedy) choice, based on an appropriate criterion, is made. Unfortunately, as we saw earlier, this is not the optimal tour. Making the best choice at each stage may not yield the best "global" solution. However, even for a large TSP, one can always find a nearest-neighbor route quickly.

Perhaps some other easy method would yield optimal solutions. We might start by sorting or arranging the edges of the complete graph in order of increasing cost (or equivalently, arrange the intercity distances in order of increasing distance). Then we can select at each stage that edge of least cost that (1) never requires that three edges meet at a vertex and that (2) never closes up a circular tour that doesn't include all the vertices.

Applying this sorted-edges algorithm to our problem yields the tour Chicago, St. Louis, Minneapolis, Cleveland, and Chicago, since the edges chosen would be 300, 349, 562, 774. Again, this solution is not optimal because its length is 1985. Note that this algorithm, like nearest neighbor, is "greedy."

Although many "quick and dirty" methods for solving the TSP have been suggested and although some methods give an optimal solution in some cases, none of these methods *guarantees* an optimal solution. Surprisingly, most experts believe that no efficient method that guarantees an optimal solution will ever be found (see Box 2.2).

Recently, mathematical researchers have adopted a somewhat different strategy for dealing with TSP problems. If finding a fast algorithm to generate optimal solutions for large problems is unlikely, perhaps one can show that the "quick and dirty" methods, usually called **heuristic algorithms,** come close enough to giving optimal solutions. For example, suppose one could prove that the nearest-neighbor heuristic never was off by more than 25% in the worst case or by more than 15% in the average case. For a medium-size TSP, one would then have to choose whether to spend a lot of time (or money) to

---

BOX 2.2 NP-Complete Problems

Stephen Cooke, a computer scientist at the University of Toronto, showed that certain hard, frustrating problems are equivalently difficult. This class of problems, now referred to as **NP-complete problems,** have the following characteristic: if a "fast" algorithm for solving *one* of these problems could be found, then a fast method would exist for *all* problems.

In this context, "fast" means that as the size n of the problem grows (the number of cities gives the problem size in the traveling salesman problem), the amount of time needed to solve the problem grows no more rapidly than a polynomial function in n. (A polynomial function has the form $a_n x^n + a_{r-1} x^{n-1} + \cdots + a_1 x - a_0$.) On the other hand, if it could be shown that any problem in the class of NP-complete problems required an exponentially growing amount of time to solve it as the problem size increased, then all problems in the NP-complete class would share this characteristic. If some NP-complete problems had fast solutions, it seems likely that at least one such fast solution would have been found by now. It has been known for some time that the traveling salesman problem is an NP-complete problem; for this reason it is generally thought that no "fast" algorithm for an optimal solution for the TSP will ever be found.

BOX 2.3 Solving the Elusive Traveling Salesman Problem

More than twenty years ago, Shen Lin, a Bell Labs researcher, set out to solve the traveling salesman problem. Although the TSP defies any quick mathematical solution, Shen Lin has discovered a method that works fairly quickly on many practical problems. Today, he applies this knowledge to designing private telephone networks for AT&T's corporate customers (see Box 2.6). Here, Shen Lin recalls his earlier work on the traveling salesman problem:

> I started trying to solve the traveling salesman problem back in 1965 and ended in 1972, collaborating with a colleague, Brian Kernahan, in an algorithm that up to this day is considered to be the most efficient practical method of solving this problem.
>
> In 1965, we knew very little about which kind of problems were hard and which were easy. So naively, I went in and thought I could solve this problem. Of course, I couldn't. I published my first paper on computer approximations by using the so-called *iterative techniques* to solve them.
>
> Today, I essentially use a *heuristic method.* The difference between this and the so-called classical method is that exact-solution methods usually take too long when you are trying to solve a hard problem like the traveling salesman problem. Also, they are formulated for problems that are very well defined. But in real life, we frequently find that no problem is so well defined, so we need some method of solving problems that is more flexible. We want something that can get at approximate solutions, which may turn out to be optimum solutions. The method must be powerful enough to solve a variety of problems. We cannot guarantee that the solution is the absolute and best one, because to find the best solution would take billions of years of computation, even by the world's fastest computer.
>
> Within a few seconds, however, I could give you the solution to the traveling salesman problem, say, for a few hundred [geographical] points. And I'm sure no human being could look at the map, connect those points, and achieve the same result. I can't *prove* that it's the best solution that could ever be found, but it's usually quite close — within 1 to 2 percent of the optimum.

find an optimal solution or to use instead a heuristic algorithm to obtain a fairly good solution. Researchers at AT&T Bell Laboratories have developed many remarkably good heuristic algorithms (see Box 2.3). The best-known guarantee for a heuristic algorithm for a TSP is that it yields a cost that is no worse than 1.5 times the optimal cost. Interestingly, this heuristic algorithm involves solving a Chinese postman problem, for which a "fast" algorithm is known to exist, as part of the algorithm.

Minimum-Cost Spanning Trees

The traveling salesman problem is but one of many graph-theory optimization problems that have grown out of real-world problems in both government and industry. Here is another example:

Imagine that Pictaphone service is to be set up on an experimental basis between five cities (see Figure 2.7). In order to send a Pictaphone message between two cities, a direct communication link is not necessary because it is pos-

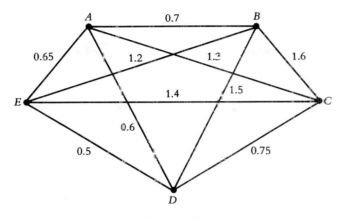

Figure 2.7

sible to send a message indirectly via another city. Thus, in Figure 2.7, sending a message from A to C could be achieved by sending the message from A to B, from B to E, and from E to C, provided the links AB, BE, and EC are part of the network.

The numbers on the edges in Figure 2.7 show, in millions of dollars, the cost of providing Pictaphone service between the cities at the endpoints of the edges. We will assume that the cost of relaying a message, compared with the link cost, is so small that we can neglect this amount. The question that concerns us, therefore, is to provide service between any pair of cities in a way that minimizes the total cost of the links.

Our first guess at a solution is to put in the cheapest possible links between cities first, until all cities could send messages to any other

city. In our example, if the cheapest links are added until all cities are joined, we obtain the connections shown in Figure 2.8a.

The links were added in the order ED, AD, AE, AB, DC. However, because this graph contains the circuit ADEA (wiggly edges in Figure 2.8b), it has redundant edges: we can still send messages between any pair of cities using relays after omitting the most expensive link in a circuit. After deleting an edge of a circuit, a message can still be relayed among the cities of the circuit by sending signals the long way around. After AE is deleted, messages from A to E can be sent via D (Figure 2.8c). This suggests a modified algorithm for our problem: add the links in order of cheapest cost, so that no circuits form and so that every vertex belongs to some link added (Figure 2.8d).

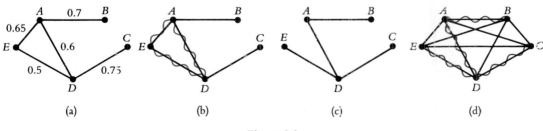

(a)   (b)   (c)   (d)

Figure 2.8

A subgraph formed in this way will be a **tree;** that is, it will consist of one piece and contain no circuits. It will also include all the vertices of the original graph. A subgraph that is a tree and that contains all the vertices of the original graph is called a **spanning tree** of the original graph.

Does the spanning tree found by our modified algorithm link the cities with the minimum possible cost? Although this sounds very plausible, our experience with the traveling salesman problem should suggest caution. Remember that for the TSP, a similar "greedy" algorithm was not optimal! On what basis should we have more faith in our present algorithm?

It turns out that the algorithm described here was first suggested by Joseph Kruskal (AT&T Bell Laboratories) in the mid-1950s to solve a problem in pure mathematics proposed by a Czechoslovakian mathematician. In mathematics it is a surprising but not uncommon finding that ideas used to solve problems with no apparent application often turn out to have many real-world uses. Kruskal's solution to the problem of finding a minimum-cost spanning tree in a graph with weights is a good example of this phenomemon (see Box 2.4). Kruskal showed that the greedy algorithm described does yield the minimal answer, and his work led to applications of these and related ideas in designing minimum-cost computer networks, phone connections, and road and railway systems. For additional discussion of operations research in the communications industry, see Boxes 2.5 and 2.6.

Although we have mentioned many routing problems in graphs, we have not discussed one of the most obvious: finding the path between two specified vertices with the sum of the weights of the edges in the path as small as possible. (Here there is no need to cover all vertices or to cover all edges.) We have seen that the weights on the edges have many possible interpretations, including time, distance,

---

BOX 2.4 Some Reminiscences on Shortest Spanning Trees

Dr. Kruskal has provided the following thoughts on the minimum-cost spanning algorithm, which he refers to as shortest spanning trees.

In 1951, I entered the graduate mathematics program at Princeton. Among those active in combinatorial mathematics at Princeton then were Al Tucker, who soon became chairman of the department, Harold Kuhn, and Roger Lyndon. For a while, there was a joint Princeton–Bell Laboratories combinatorics seminar, which met alternately at two locations. My first introduction to Bell Laboratories, where I subsequently have spent several decades, came in this connection.

One day, someone in the combinatorics community handed me a blurry, carbon-copy typescript on flimsy lightweight paper about shortest spanning trees. It was in German, and appeared to be the foreign-language summary of a paper written in some other language. I had the typescript only for a limited period, and then passed it on to someone else. The typescript contained no date, and at the time I had no idea of whether it had been published or where. To this day, I have no idea where the typescript had come from, and why it was circulating. While writing my own paper stimulated by this material, I must have been told that the typescript was (part of?) a translation of a 1926 paper by Otakar Boruvka, for I give an exact citation in my paper. However, I have no memory of who told me this, nor do I remember ever seeing the full paper.

By the way, I regret that the phrase "minimum-spanning tree" has become dominant. What is actually meant is the minimum *length* spanning tree (MST), which is more concisely rendered by *shortest* spanning tree (SST). That is the phrase I used in 1954, and continue to use. If you wish to stick with MST rather than SST, however, at least do not make the common error of misconstruing MST as minimal, rather that minimum, spanning tree. The thing being minimized, length, has a unique minimum under very general conditions. [In this book, the term "minimum-cost spanning tree" is used rather than MST, as is common in the technical mathematics literature.]

The chief point of Boruvka's paper, I believe, was the uniqueness of the SST of a graph if the edges of the graph all have distinct lengths. His proof was based on a construction of the SST that I had difficulty in fully grasping. Although the construction has a certain spare elegance, I decided that it was difficult to follow because it was unnecessarily complicated. That led me to devise some much simpler methods, which are described in my 1956 paper. In 1935 Graham and Hell wrote a full history of the SST and closely related ideas. Well over 100 papers dating back to 1909 are cited, though the earliest paper specifically on the SST is Boruvka's work in 1926.

It is interesting to note that I was very uncertain about whether the material in my simple paper, which occupies only two and a half printed pages and requires few equations, was worth writing up for publication. I was a graduate student in my second year, and did not have a good sense of what was publishable. I asked a friend for advice (I don't remember whom), and fortunately received encouragement. It was my second published paper.

BOX 2.5 AT&T Manager Explains How Long-Distance Calls Run Smoothly

Although long-distance calls are now routine, it takes great expertise and careful planning for a company like AT&T to handle its vast amounts of telephone traffic. Rich Wetmore is district manager of AT&T's Communications Network Operations Center in Bedminster, New Jersey. Here are his responses to questions about how AT&T handles its huge volumes of long-distance traffic and how it tracks its operations to keep things running smoothly.

How do you make sure that a customer doesn't run into a delayed signal when attempting a long-distance call?

When a customer places a call, it's our job to see that it goes through. We monitor the performance of our AT&T network by displaying data collected from all over the country on a special wallboard. The wallboard is configured to tell us if a customer's call is not going through because the network doesn't have enough capacity to handle it.

That's when we step in and take control to correct the problem. The typical control we use is to reroute the call. Instead of sending the customer's call directly to its destination, we'll route it via a third city — to someplace else in the country that has the capacity to complete the call. When we put the reroute through, people's calls are completed, and everyone is happy.

It would seem that routing via another city would take longer. Is the customer aware of this process?

Routing a call via a third city is entirely transparent to the customer. I'm an expert about the network, and even when I make a phone call, I have no idea how that individual call was routed. It's transparent both in terms of how far away the other person sounds and in how quickly the telephone call gets set up. With the signalling network we use, this process works very, very quickly. It takes milliseconds for switching systems to "talk" to each other to set up a call. So the fact that you are involved in a third switch in some distant city is something you would never know.

There is obviously more traffic between New York and Los Angeles than between many other cities. Does that mean AT&T has to put in more links between those two cities?

The fact that different points of the country are demographically different does impact on how we engineer the AT&T network. It's very similar to the way you would engineer a highway system. Between major cities, you have major highways with 10 to 12 lanes. Between smaller cities, there are 2-lane highways.

That's the way our network looks, in terms of circuits and capacity. But our fundamental principle is that no matter where you live, you receive equitable treatment in terms of the way the network performs, and in terms of the blocking level you get. The traditional guideline we use is that whether you live in Chicago or Horseheads, New York (where I'm from), you should get an average of zero blocking throughout the day. And we try to build just enough capacity both to Chicago and to Horseheads so that you don't get blocked.

You want to be sure to keep costs down while supplying enough service to customers. So how do you balance company benefits with customer benefits?

In terms of making the network efficient, we want to do two things. First, we want our customers to be happy with our service and for all their calls to go through, which means we must build enough capacity in the network to allow that to happen. Second, we want to be efficient for our stockholders and not spend more money than we need to for the network to be at the optimum size.

There are basically two costs in terms of building the network. There is the cost of switching systems and the cost of the circuits that connect the switching systems. Basically, you can use operations-research techniques and mathematics to determine cost trade-offs. It may make sense to build direct routing between two switching systems and use a lot of circuits, or maybe to involve three switching systems, with fewer circuits between the main two, and so on.

BOX 2.6 Made-to-Order Phone Networks Benefit Large-Scale Users

Shen Lin is executive consultant and head of Bell Lab's Network Configuration and
Planning Department, Murray Hill, New Jersey. A scientist-turned-salesman for AT&T,
he devised a method for solving the traveling salesman problem in 1972 (see Box 2.3).
More recently, he developed the "interactive network optimization" for AT&T. Here are
his comments on the network:

> Before our customers will sign a contract to buy a multimillion dollar network, they
> usually want someone to tell them about the methodology used for optimizing their
> networks. So I am frequently asked to present the network design results to our
> customers, not only to explain how it serves their needs, but also to give them an
> idea of how the design was worked out. It takes a lot of time from my research
> work, but I like to do it very much.
>
> Here is how it works: Perhaps hundreds of locations in a large organization need
> to "talk" to each other. The people may need to talk to one another, or their
> terminals may need to communicate with computers. There may be an almost
> infinite variety of choices in configuring their network.
>
> For example, an individual may pick up his phone, make a call, and get charged
> for a regular long-distance call. If the calling volume is rather high, a company can
> use the Wide Area Telecommunications Service, commonly called WATS, and
> reduce the cost by 20% or 30%. Another option is to connect locations via private
> lines. So the problem is how to configure all communications needs into an optimal
> network that will minimize overall costs.
>
> This has always been considered a difficult problem. Let's consider the first step
> that must be taken: because not every location can be connected to every other
> location, a set of locations that are close together will come into what is called a
> "switch," or concentrator. Their calls will come into the switch and be completed
> thereafter.
>
> So one of the first questions that has to be answered is, Where do you put the
> switch? It turns out that this problem is a mathematically difficult problem in itself.
> Before we know where to put the concentrators, we have to estimate the cost of
> calls reaching their destination via the switch, and since we don't know where the
> switches will be located, we must use many different iterations to get at our optimal
> answer.
>
> An efficient telephone routing system saves money for customers, perhaps as
> much as 5% to 10%. To be more specific, let's say all users of a big corporation use
> the regular so-called long-distance service. It would probably cost them $30.00 per
> hour of usage. Now, if they buy WATS service, they will spend something like
> $20.00 an hour. But if a small corporation makes only $10.00 worth of calls per
> month, there's no need for a WATS. The best and most economical way is through
> plain old telephone service. But if their bill is several hundred dollars higher, it may
> be more cost-effective for the corporation to buy some WATS service.
>
> In general, a large corporation may generate about five million calls per month,
> and they may be calling from several hundred — or even two thousand — locations,
> to tens of thousands of destinations. The job of optimization is to group all these
> together in such a way that the total bill is minimized.

INOS (Interactive Network Optimization System) does all the investigation and configuration. It's very computation-intensive. To do a customer network, we run something over 300 to 500 million arithmetical instructions, and that's after months of work collecting the traffic data.

Right now, INOS is moving in the direction of configuring data networks. For example, my terminal has to talk to a computer in Massachussetts at this moment, so I have a data line. People may use a dial-up or data line, and in some cases many, many terminals may be connected to use a common line called a multi drop line. All of this can amount to a great deal of savings to the customer simply by solving the mathematical problem of optimum configuration.

---

and cost. Here are some of the many possible applications:

1. Design routes to be used by a fire engine to get to a fire as quickly as possible.

2. Design delivery routes that minimize gasoline use.

3. Design routes to bring soldiers to the front as quickly as possible.

The need to find shortest paths seems natural. Next we shall investigate a situation where finding a longest path is the right tool.

Critical-Path Analysis

One of the delights of mathematics is its ability to confirm the obvious in certain situations while showing that our intuition is wrong in other circumstances. Consider the common problem of scheduling a series of interrelated subtasks that are part of some major undertaking, such as building a skyscaper, installing a major computer system, or landing a man on the moon.

If the tasks cannot be performed in arbitrary sequence or order, we can specify the order in an **order-requirement digraph.** A typical example of an order-requirement digraph is shown in Figure 2.9. There is a vertex in this

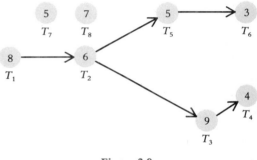

Figure 2.9

digraph for each task. If one task must be done immediately before another, we draw a directed edge, or arrow, from the prerequisite task to the subsequent task. The numbers within the circles representing vertices are the times it takes to complete the tasks. In Figure 2.9 there is no arrow from T_1 to T_5 because task T_2 intervenes. Also, T_1, T_7, and T_8 have no tasks that must precede them. Hence, if there are at least three processors available, tasks T_1, T_7, and T_8 can be worked on simultaneously at the start of the job.

Consider an airplane that carries both freight and passengers. The plane must have its passengers and freight unloaded and new passengers and cargo loaded before it can take off again. Also, the cabin must be cleaned before departure can occur. Thus, the job of "turning the plane around" requires the completion of five tasks:

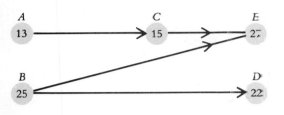

Figure 2.10

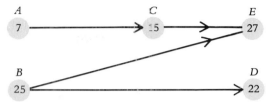

Figure 2.11

Task *A*	Unload passengers	13 minutes
Task *B*	Unload cargo	25 minutes
Task *C*	Clear cabin	15 minutes
Task *D*	Load new cargo	22 minutes
Task *E*	Load new passengers	27 minutes

The order-requirement digraph for the problem of turning an airplane around is shown in Figure 2.10. Because we want to find the earliest completion time, it might seem that finding the shortest path through the digraph (that is, path *BD* with length 25 + 22 = 47) would solve the problem. But this approach shows the danger of ignoring the relationship between the mathematical model (the digraph) and the original problem.

The time required to complete all the tasks must be at least as long as the time necessary to do the tasks on any particular path. Thus, the earliest completion time actually corresponds to the length of the *longest* path. In the airplane example, this earliest completion time is 55 (13 + 15 + 27) minutes, corresponding to the path *ACE*. We call *ACE* the **critical path** because the times of the tasks on this path determine the earliest completion time.

Suppose it were desirable to speed the turnaround of the plane to below 55 minutes. One way to do this might be to build a second jetway to help unload passengers. Then, for example, we could unload passengers (task *A*) in 7 minutes instead of 13. However, reducing task *A* to 7 minutes does not reduce the completion time by 6 minutes because in the new digraph (Figure 2.11) *ACE* is no longer the

critical (that is, longest) path. The longest path is now *BE*, which has a length of 52 minutes. Thus, shortening task *A* by 6 minutes only results in a 3-minute saving in completion time. This may mean that building a new jetway is uneconomical. Note also that shortening the time to complete tasks that are not on the original critical path *ACE* will not shorten the completion time at all. Speeding tasks on the critical path will shorten completion time of the job only up to the point where a new critical path is created.

This example is typical of many scheduling problems that occur in practice (see Box 2.7). Perhaps the most dramatic use of critical-path analysis is in the construction trades. No major new building project is now carried out without first performing a critical-path analysis to ensure that the proper personnel and materials are available at the construction site at the right times, in order to have the project finished as quickly as possible. Of course, many such problems are too large and complicated to be solved without the aid of computers.

The critical-path method was popularized and came into wider use as a consequence of the Apollo project. As we saw in the introduction, this project, which aimed at landing a man on the moon within 10 years of 1960, was one of the most sophisticated projects in planning and scheduling ever attempted. The dramatic success of the project can be attributed partly to the use of critical-path ideas and the related Program Evaluation and Review Technique (PERT), which helped keep the project on schedule.

BOX 2.7 **Every Moment Counts in Rigorous Airline Scheduling**

When people think of airline scheduling, the first thing that comes to mind is how quickly a particular plane can safely reach its destination. But using ground time efficiently is just as important to an airline's timetable as the time spent in flight. Bill Rodenhizer is the manager of control operations for Eastern Airlines in Boston. He is considered to be an expert on airplane turnaround time, the process by which an airplane is prepared for almost immediate take-off once it has landed. He tells us how this well-orchestrated effort works:

> Scheduling, to the airline, is just about the whole ballgame. Everything is scheduled right to the minute. The whole fleet operates on a strict schedule. Each of the departments responsible for turning around an aircraft has an allotted period of time in which to perform its function. Manpower is geared to the amount of ground time scheduled for that aircraft. This would be adjusted during off-weather or bad-weather days or during heavy air-traffic delays.
>
> Most of our aircraft in Boston are scheduled for a 42- to 65-minute ground time. Boston is the end of the line, so it is a "terminating and originating station." In plain talk, that means almost every aircraft that comes in must be fully unloaded, refueled, serviced, and dispatched within roughly an hour's time.
>
> This is how the process works: In the larger aircraft, it takes passengers roughly 20 minutes to load and 20 minutes to unload. During this period, we will have completely cleaned the aircraft and unloaded the cargo, and the caterers will have taken care of the food. The ramp service may take 20 to 30 minutes to unload the baggage, mail, and cargo from underneath the plane, and it will take the same amount of time to load it up again. We double-crew those aircraft with heavier weights so that the work load will fit the time it takes passengers to load and unload upstairs.
>
> While this has been going on, the fueler has fueled the aircraft. As to repairs, most major maintenance is done during the midnight shift, when all but 20 of Eastern's several hundred aircraft are inactive.
>
> We all work under a very strict time frame. There are four functional departments. If any of the four cannot fit its work into its time frame, then it advises us at the control center, and we adjust the departure time or whatever, so that the other departments can coordinate their activities accordingly.

REVIEW VOCABULARY

Algorithm A step-by-step description of how to solve a problem.

Complete graph A graph in which each pair of vertices is joined by an edge.

Critical path The longest path in an order-requirement digraph. The length of this path gives the earliest completion time for all the tasks making up the job consisting of the tasks in the digraph.

Fundamental principle of counting A method for counting outcomes of multistage processes.

Greedy algorithm An approach for solving an optimization problem, where at each stage of the algorithm the best (or cheapest) action is taken. Unfortunately, greedy algorithms do not always lead to optimal solutions.

Hamiltonian circuit A tour visiting each vertex of a graph once and only once.

Heuristic algorithm A method of solving an optimization problem that is "fast" but does not guarantee an optimal answer to the problem.

Method of trees A visual method to carry out the fundamental principle of counting.

Minimum-cost spanning tree The spanning tree of a weighted connected graph having minimal cost. The cost of a tree is the sum of the weights on the edges of the tree.

Nearest-neighbor algorithm An algorithm used to solve the traveling salesman problem that begins at a "home" city and generates a tour that at each stage visits next that city not already visited that can be reached most cheaply. When all other cities have been visited, the tour returns to home.

Order-requirement digraph A directed graph that shows which tasks precede other tasks among the collection of tasks making up a job.

Spanning tree A subgraph of a connected graph that is a tree and includes all the vertices of the original graph.

Traveling salesman problem (TSP) The problem of finding a minimum-cost Hamiltonian circuit in a complete graph where each edge has been assigned a cost (or weight).

Tree A connected graph with no circuits.

Weight A number assigned to an edge of a graph that can be thought of as a cost, distance, or time associated with that edge.

EXERCISES

1. For each of the graphs below, determine if there is a circuit of all the vertices that visits each vertex exactly once.

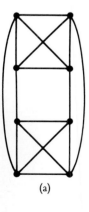

(a)

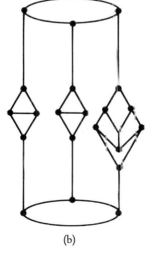

(b)

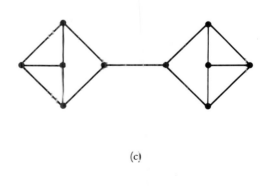

(c)

2. Suppose two Hamiltonian circuits are considered different if the collections of edges that they use are different. How many other tours can you find in the graph in Figure 2.1 that visit all the vertices once and only once (different from the one shown)?

3. Do the following graphs have tours of the vertices, visiting each vertex only once? If not, can you demonstrate why not?

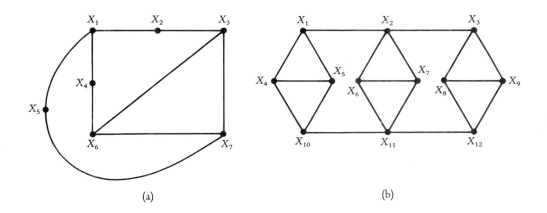

(a)   (b)

4. The graph below is known as a 3×4 grid graph. What conditions on m and n guarantee that an $m \times n$ grid graph has a Hamiltonian circuit?

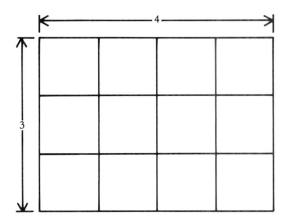

Can you invent a real-world situation in which finding a Hamiltonian circuit in an $m \times n$ grid graph would represent a solution to the problem? If an $m \times n$ grid graph has no Hamiltonian circuit, can you find a tour that repeats a minimum number of vertices?

5. To practice your understanding of the concepts of Euler circuit and Hamiltonian circuit, determine for each graph on the next page if there is an Euler circuit and/or a Hamiltonian circuit. If so, write it down.

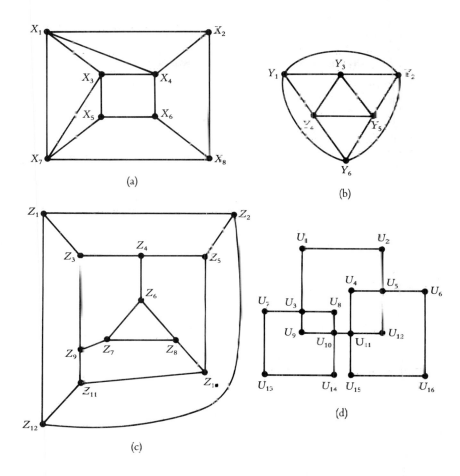

(a)

(b)

(c)

(d)

6. For each of the graphs below, add wiggly edges to indicate a Hamiltonian circuit.

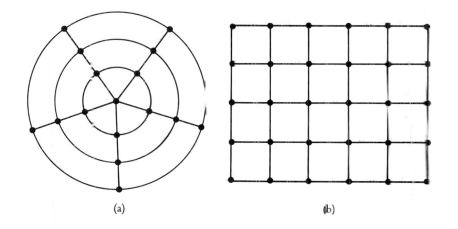

(a)

(b)

Can you generalize your method to *m* spokes and *n* concentric circles (where graph (a) shows *m* = 5, *n* = 3) and to *m* rows and *n* columns (where graph (b) shows *m* = 4 and *n* = 5)?

7. An airport limo must take its 6 passengers to different downtown hotels from the airport. Is this a traveling salesman problem or an Euler-circuit problem?

8. The *n*-dimensional cube is obtained from two copies of an (*n* − 1)-dimensional cube by joining corresponding vertices. (The process is illustrated for the 3-cube and the 4-cube in the following figure.)

Find formulas for the number of vertices and the number of edges of an *n*-cube. Can you show that every *n*-cube has a Hamiltonian circuit? (Hint: Show that if you know how to find a Hamiltonian circuit on an (*n* − 1) cube, then you can use two copies of this to build a Hamiltonian circuit on an *n*-cube.)

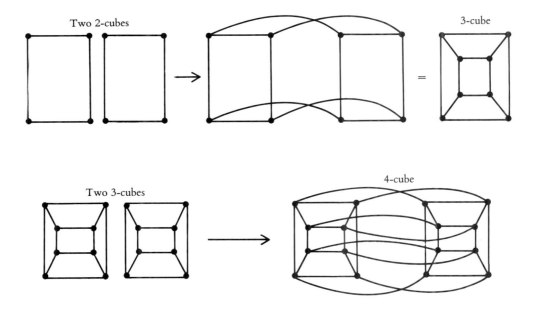

9. Use a calculator to find the values of 5!, 6!, 7!, and 8!. Then find the approximate number of TSP tours in the complete graph with 9 vertices.

10. Draw the complete graph with 4 vertices; 5 vertices; 6 vertices. How many edges do these graphs have? Can you generalize to *n* vertices? How many TSP tours would these graphs have?

11. For each of the complete graphs that follow, find the costs of the nearest-neighbor tour and of the tour generated using the sorted-edges algorithm. Do you think these tours are optimal?

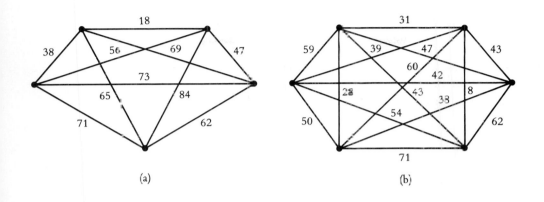

(a)   (b)

How many tours would have to be examined to check your answers by checking all the Hamiltonian circuits in the graph? Can you invent an algorithm different from the sorted-edges and nearest-neighbor algorithms that is easy to apply for finding TSP solutions?

12. The following figure represents a town where there is a sewer located at each corner (where two or more streets meet). After every thunderstorm, the department of public works wishes to have a truck start at its headquarters (at vertex *H*) and make an inspection of sewer drains to be sure that leaves are not clogging them. Can a route start and end at *H* that visits each corner exactly once? (Assume that all the streets are two-way streets.) Does this problem involve finding an Euler circuit or a Hamiltonian circuit?

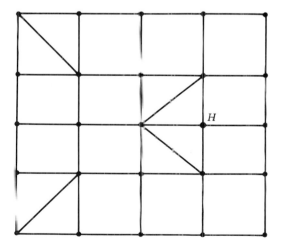

Assume that at equally spaced intervals along the blocks in this graph there are storm sewers that must be inspected after each thunderstorm to see if they are clogged. Is this a Hamiltonian-circuit problem, an Euler-circuit problem, or a Chinese postman problem? Can you find an optimal tour to do this inspection?

13. Construct an example (if possible) of a complete graph on 5 vertices, with weights on the edges for which the nearest-neighbor algorithm and the sorted-edges algorithm yield different solutions for the traveling salesman problem. Can you find a 5-vertex complete graph with weights on the edges in which the optimal solution, the nearest-neighbor solution, and the greedy algorithm solution are all different?

14. Find all the spanning trees in the graphs below:

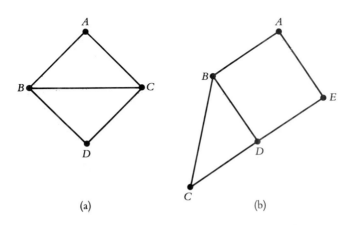

(a)   (b)

15. Use Kruskal's algorithm to find the minimum-cost spanning tree for graphs (a), (b), (c), and (d).

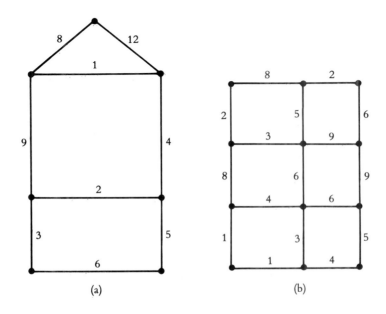

(a)   (b)

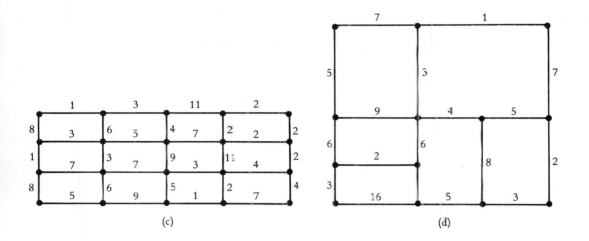

(c)   (d)

16. Can Kruskal's algorithm be modified to find a maximal-weight spanning tree? Can you think of an application for finding a maximal-weight spanning tree?

17. Suppose that Pictaphone service is not possible between 2 towns where there is a hill of more than 800 feet along a straight line that runs between them. In constructing a model for solving a relay network problem, how would you handle the question of putting a link between 2 cities with a 1200-foot hill between them?

18. Find the cost of providing a relay network between the 5 cities with the largest population in your home state, using the road distances between the cities as costs. Does it follow that the same solution would be obtained if air distances were used instead?

19. Let G be a graph with weights assigned to each edge. Consider the following algorithm:

a. Pick any vertex V of G.

b. Select that edge E with a vertex at V that has a minimum weight. Let the other endpoint of E be W.

c. Contract the edge VW so that edge VW disappears and vertices V and W coincide (see the following figures).

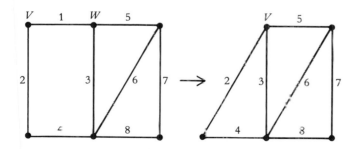

If in the new graph two or more edges join a pair of vertices, delete all but the cheapest. Continue to call the new vertex V.

d. Repeat steps **b** and **c** until a single point is obtained. The edges selected in the course of this algorithm (called Prim's algorithm) form a minimum-cost spanning tree. Apply this algorithm to the following graphs:

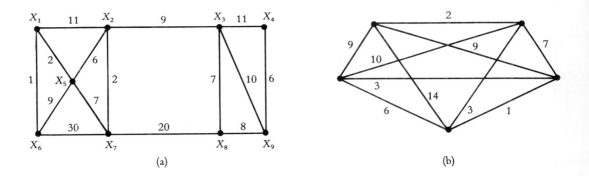

(a)   (b)

20. In the following graphs, the number in the circle for each vertex is the cost of installing equipment at the vertex if *relaying* must be done at the vertex, while the number or an edge indicates the cost of providing service between the endpoints of the edge. In each case, find the minimum-cost spanning tree for sending messages between any pair of vertices.

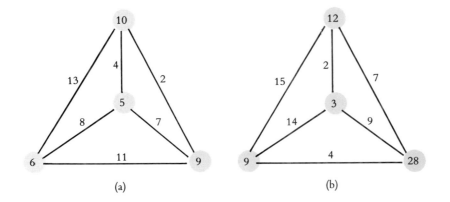

(a)   (b)

21. Determine whether each of the following items is true or false for a minimum-cost spanning tree T for a weighted connected graph G:

a. T contains a cheapest edge in the graph.

b. T cannot contain a most expensive edge in the graph.

c. T contains one fewer edge than there are vertices in G.

d. There is some vertex in T to which all others are joined by edges.

22. Suppose that a letter requires postage of p (positive integer) and that stamps of various denominations are available, say, $d_1, \ldots, d_v$ (positive integers). We are interested in finding the minimum number of stamps to choose with the available denominations to obtain the postage p exactly.

  a. Give an example to show that unless other conditions are put on p and $d_1, \ldots, d_v$, it might happen that no selection of stamps will achieve the desired postage.

  b. Show that even if the desired postage can be obtained for some choice of stamps, it does not follow that a minimum number of stamps can be achieved by using a greedy algorithm, that is, by selecting at each stage the largest denomination possible.

  c. Show that for at least one choice of denominations, a greedy algorithm will produce the fewest stamps for any given postage!

23. Find the earliest completion time and critical paths for the order-requirement digraphs below:

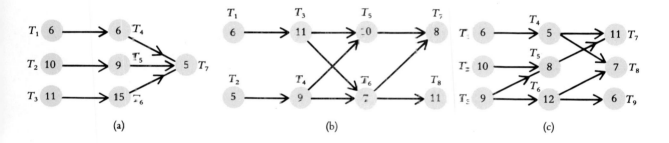

(a)   (b)   (c)

24. Construct an example of an order-requirement digraph with 3 different critical paths.

25. In the order-requirement digraph below, determine which tasks, if shortened, would reduce the earliest completion time and which would not. Then find the earliest completion time if task T_5 is reduced to time length 7. What is the new critical path?

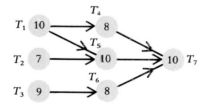

26. In order to build a new adition on a house, the following tasks must be completed. Construct reasonable time estimates for these tasks and a reasonable order-requirement digraph. What is the fastest time in which these tasks can be completed?

  a. Lay foundation

  b. Erect sidewalls

    c. Erect roof

    d. Install plumbing

    e. Install electric wiring

    f. Lay tile flooring

    g. Obtain building permits

    h. Put in door that adjoins new room to existing house

    i. Install track lighting on ceiling

27. For the order-requirement digraph below, find the critical path and the task in the critical path whose time, when reduced the *least*, creates a new critical path.

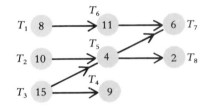

28. Construct an order-requirement digraph with 6 tasks that has 2 critical paths of length 24.

Planning and Scheduling

In a society as complex as ours, everyday problems such as providing services efficiently and on time require accurate planning of both people and machines. Take the example of a hospital in a major city. Around-the-clock scheduling of nurses and doctors must be provided to guarantee that people with particular expertise are available during each shift. The operating rooms must be scheduled in a manner flexible enough to deal with sudden emergencies. Equipment used for x-ray, CAT, or NMR scans must be scheduled for maximal efficiency.

Although we often solve many scheduling problems on an *ad hoc* basis, we can also use mathematical ideas to gain insight into the complications that arise in scheduling. The theory we develop in this chapter has practical value in a relatively narrow range of applications, but it throws light on many characteristics of more realistic and hence more complex scheduling problems.

Scheduling Tasks

Assume that a certain number of identical **processors** (machines, humans, or robots) work on a series of tasks that make up a job. Associated with each task is a specified amount of time required to complete the task. For simplicity, we assume that any of the processors can work on any of the tasks. Our problem is to decide how the tasks should be scheduled.

Even with these simplifying assumptions, complications in scheduling will arise. Some tasks may be more important than others and perhaps should be scheduled first. When "ties" occur, they must be resolved by special rules. As an example, suppose we are scheduling patients to be seen in the emergency room at a hospital staffed by one doctor. If two patients arrive simultaneously, one with a bleeding foot, the other with a bleeding arm, which patient should be processed first? Suppose the doctor treats the arm patient first, and while treatment is going on, a person in cardiac arrest arrives. Scheduling rules must establish appropriate priorities for cases such as these.

Another common complication arises with jobs that cannot be done in an arbitrary order. For example, as we saw in Chapter 2, if the job of putting up a new house is treated as a scheduling problem, the task of laying the foundation must precede the task of putting up the walls, which in turn must be completed before work on the roof can begin.

To simplify our analysis, we need to make additional assumptions:

1. If a processor starts work on a task, the work on that task will continue without interruption until the task is completed.

2. No processor stays voluntarily idle. In other words, if there is a processor free and a task available to be worked on, then that processor will immediately begin work on that task.

3. The requirements for ordering the tasks are given by an order-requirement digraph. (A typical example is shown in Figure 3.1, with task times circled within each vertex.)

4. The tasks are arranged in a priority list that is independent of the order requirements.

When considering a scheduling problem, there are various goals one might wish to achieve. Among these are

1. minimizing the completion time of the job

2. minimizing the total time that machines are idle

3. finding the minimum number of processors necessary to finish the job by a specified time

For the moment, we will concentrate on goal (1), finishing all the tasks at the earliest possible time. Note, however, that optimizing with respect to one criterion or goal may not optimize with respect to another.

The scheduling problem we have described sounds more complicated than the traveling salesman problem (TSP). Indeed, like the TSP, it is known to be NP-complete. This means that it is unlikely that anyone will ever find a computationally fast algorithm that can find an optimal solution. Thus, we will be content to seek a solution method that is computationally fast and gives only approximately optimal answers.

The algorithm we will use to schedule tasks is the **list-processing algorithm.** In describing it, we will call a task *ready* if all its predecessors have already been completed. The algorithm works as follows: at a given time, assign to the lowest-numbered free processor the first ready task on the priority list that hasn't already been assigned to a processor.

Let's apply this algorithm to the priority list $T_8, T_7, T_6, \ldots, T_1$ using two processors, and the order-requirement digraph in Figure 3.1. The result is the schedule shown in Figure 3.2, where idle processor time is indicated by color shading. How does the list-processing algorithm generate this schedule?

Because T_8 (task 8) is first on the priority list and ready at time 0, it is assigned to the lowest-numbered free processor, processor 1. Task 7, next on the priority list is also ready at time 0 and thus is assigned to processor 2. The first processor to become free is processor 2 at time 5. However, task T_6, the next unassigned task

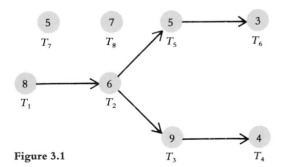

Figure 3.1

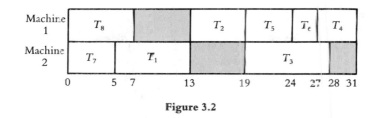

Figure 3.2

on the list, is not ready at time 5. In fact, at time 5, the only ready task on the list is T_1, so that task is assigned to processor 2. At time 7, processor 1 becomes free, but no task becomes ready until time 13. Thus, processor 1 stays idle from time 7 to time 13. At this time, because T_2 is the first ready task on the list not already scheduled, it is assigned to processor 1. Processor 2, however, stays idle because no other ready task is ready at this time. The remainder of the scheduling is completed in a similar manner.

The schedule in Figure 3.2 has a lot of idle time, so it may not be optimal. Indeed, if we apply the list-processing algorithm for two processors to the priority list $T_1, \ldots, T_8$, using the digraph in Figure 3.1, the resulting schedule (Figure 3.3) is optimal because the path T_1, T_2, T_3, T_4, with length 27, is the critical path in the order-requirement digraph. As we saw in Chapter 2, the earliest completion time for the job made up of all the tasks is the length of the longest path in the order-requirement digraph. Of course, different priority lists may yield different schedules with the same completion time.

The list-processing algorithm shows us four factors that affect the final schedule. The answer we get depends on

1. the times of the tasks

2. the order-requirement digraph

3. the number of processors

4. the ordering of the tasks on the list

To see the interplay of these four factors, consider another scheduling problem associated with the order-requirement digraph shown in Figure 3.4 (the circled numbers are task time lengths).

The schedule generated by the list-processing algorithm applied to the list $T_1, T_2, \ldots, T_9$, using three processors, is given in Figure 3.5.

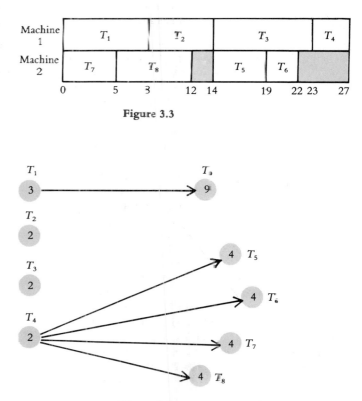

Figure 3.3

Figure 3.4

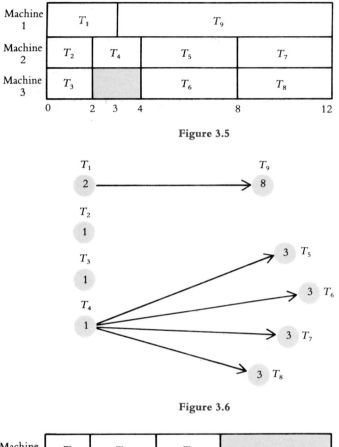

Figure 3.5

Figure 3.6

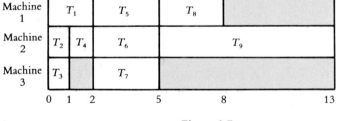

Figure 3.7

How might we make the completion time earlier? Our alternatives are to

1. reduce task times

2. use more processors

3. "loosen" the constraints of the order-requirement digraph

Let's consider each alternative in turn, changing one feature of the original problem at a time, and see what happens to the resulting schedule. If we adopt strategy 1, reducing the time of each task by one unit, it seems intuitively clear that the completion time would go down. Figure 3.6 shows the new order-requirement digraph, and Figure 3.7 shows the schedule produced for this problem, using the list-processing algorithm with three processors applied to the list $T_1, \ldots, T_9$. The completion time is now 13, longer than the completion time of 12 for the case (Figure 3.5) with the *longer* task times. Here is something unexpected! Let's explore further and see what happens.

Next we consider strategy 2, increasing the number of machines. Surely this should speed matters up. When we apply the list-processing algorithm to the original graph in Figure 3.4, using the list $T_1, \ldots, T_9$ and four machines, we get the schedule shown in Figure 3.8. The completion time is now 15! Here is a shock, an even later completion time than the previous alteration.

Finally, we consider strategy 3, trying to shorten completion time by erasing *all* constraints in Figure 3.4. By increasing flexibility of the ordering of the tasks, we might guess we could finish our tasks more quickly. Figure 3.9 shows the schedule using the list $T_1, \ldots, T_9$: now it takes 16 units! This is the worst of our three strategies to reduce completion time.

The failures we have seen here appear paradoxical at first glance, but they are typical of what can happen when a situation is too complex to analyze with naive intuition. Sometimes our common sense leads us astray. The value of using mathematics rather than intuition or trial and error to study scheduling and other problems is that it points up flaws that can occur in unguarded intuitive reasoning.

The paradoxical behavior we see here is a consequence of the rules we set up for generating schedules. The list-processing algorithm has many nice features, including the fact that it is easy to understand and fast to implement. However, the results of the model in some cases can appear strange. Since we have been explicit about our assumptions, we could go back and make changes in these assumptions in hopes of eliminating the strange behavior. But the price we may pay is more time spent in constructing schedules and perhaps even new types of strange behavior.

Mathematicians suspect that no computationally efficient algorithm for solving scheduling problems optimally will ever be found, even for the special case in which the order-requirement digraph has no arrows, that is, where the tasks are independent of each other and can be performed in any order.

Independent Tasks

To further simplify our discussion we now consider scheduling only independent tasks. There are two approaches we can consider. First, we might ask, How close on the average do we usually get to an optimal solution when we use the list-processing algorithm? Problems of this sort are amenable to mathematical solution but require methods of great sophistication. Second, we can ask, How far from optimal is a schedule we find by using the list-processing algorithm. This problem can be answered using a surprisingly simple argument developed by Ronald Graham of AT&T Bell Laboratories (see Box 3.1). The idea is that if the tasks are independent, no processor can be idle at a given time and then busy on a task at a later time. We will return to Graham's idea after exploring the problem of independent tasks in more detail.

Geometrically, we can think of the independent tasks as rectangles of height 1 whose lengths are equal to the time length of the task. Finding an optimal schedule amounts to packing the task rectangles into a longer rectangle

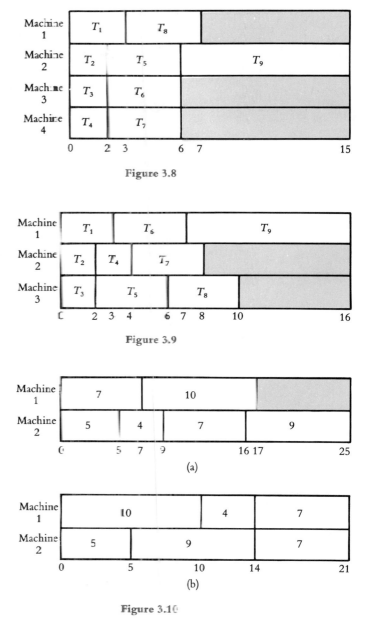

Figure 3.8

Figure 3.9

(a)

(b)

Figure 3.10

whose height equals the number of machines. For example, Figure 3.10 shows two different ways to schedule tasks of length 10, 4, 5, 9, 7, 7 on two machines. Scheduling basically means efficiently packing the task rectangles into the

BOX 3.1 Ronald Graham on Mathematics and Mathematicians

Ronald Graham is Director of the Mathematics Sciences Research Department of AT&T Bell Labs, Murray Hill, New Jersey. In addition to his own research, he supervises some of the country's most distinguished mathematicians as they investigate the puzzles of management science. Here are some of his comments:

On Scheduling

Scheduling is a very interesting area. You have some number of processors that you can think of as computers or as people, and you have a number of tasks or jobs that you would like to get done. One might think that adding more processors to the system would guarantee that you would be done sooner. In fact, just the opposite can happen. It's an example of the old adage, "Too many cooks can spoil the broth." You can add more machines or decrease the amount of time it takes for each job, or relax the constraints between the various jobs, but all of these can actually end up taking longer. To try to understand why that is and how to avoid it is the arena of mathematics.

One of the earliest applications of scheduling came in looking at the Apollo moon shot. In that case, there were three different processors, namely, the three astronauts, who had a large collection of tasks that they were supposed to do—various experiments to run on the way to and from the moon. And of course, there were various constraints. You want the astronauts to sleep a little every day. Other constraints you might not think of. For example, you have to rotate the capsule so that the same side isn't facing the sun all the time, because of heating problems. Then, within these constraints, you would next like to know: If you sequence the experiments in different orders, how much time could you save over sequencing them in orders that aren't so good?

It turns out that in fact NASA had developed very good sequencing already. All I could show them was that they couldn't improve on it very much, no matter what they might try. And of course this is one of the problems in management science where it's just impossible to enumerate the possibilities. So that's where the mathematics comes in: it enables you to say that no matter what you do, you'll never be able to "do any better than this" or "do it any faster."

You might well ask, How can you know about every one of the possibilities even though you didn't try them all? That's an interesting question, and it really goes to the root of how mathematics is used to analyze the real world. There are several well-known steps in going from a real-world problem to a model of the problem—a model in which you try to capture the essential aspects of the problem, but in a way that can be dealt with mathematically. Once you convert the problem into the world of mathematics, you can say something about all the mathematical possibilities. Then you translate it back to the real-world situation. How well it works depends crucially on how accurate the translation of the model was.

Now it's always good to check a few of the things that you've predicted are going to happen. If the translation is good and you've captured the essence of the problem, then there is a good chance that you can say something sensible about the thing you've studied. If it isn't, you can get some pretty bizarre conclusions. There are famous examples of this. One that comes to mind is the case where the people who first analyzed how bees fly were able to prove mathematically that bees *couldn't* fly. That didn't bother the bees, of course, and the model was eventually modified to show that bees could fly after all.

On the Spirit of Investigation

I think in order to do the best work in any area, you have to enjoy it and be intrigued by it. You have to wonder, "Why is this happening?" There are many examples where something slightly out of the ordinary occurs and 99 people notice it yet go on to something else. But one in a hundred—or fewer—is fascinated by that slightly anomalous or even bizarre behavior in one mathematical area or another. They start probing it more deeply and soon find that it's really the tip of a giant iceberg. Once you start to melt it, you have a much deeper understanding about what is really going on. There are those who feel mathematics is a giant game, albeit one that is amazingly relevant to everything that is going on in the scientific world.

Often, half the battle in trying to solve a difficult problem is knowing the right question to ask. If you know what it is you are trying to look for, you can often be very far along in finding the solution. It's useful for me, and for many others who are working on a difficult problem, to look at a special case first. You can work your way up to the full problem by trying easier special cases that you still can't do. You may be bouncing the idea around a bit and not forcing it, and it's amazing how often it happens that the next day or next week something will seem obvious that was very hard to imagine even a week earlier.

On Managing the Mathematicians

One of the rules or axioms that we have here [at Bell Labs] is that each person is best equipped to know how he or she functions optimally. Some people are night people, for example, and they just don't function before 12 o'clock.

In our mathematics group, there are roughly 65 professionals, and quite a few are leading-edge, world-class researchers. People of this caliber you don't so much manage as try to keep up with what they are doing. In many cases, you act as a sounding board, because one of the most useful things you can do with a colleague who has some mathematical ideas is to act as a sympathetic listener. As the person explains the ideas or where he is stuck, he will more often than not gain a much clearer insight into what he's doing and where he's going.

Many times there is much more interactive collaboration. There is quite a lot of joint work that goes on here, not just from within the mathematics area, but among mathematicians, computer scientists, physicists, and chemists. It is a very interactive environment which, I think, works to everyone's benefit.

---

machine rectangle. Finding the optimal answer among all possible ways to pack these rectangles is like looking for a needle in a haystack. The list-processing algorithm produces a packing, but it may not be a good one.

What Graham showed for independent tasks was that no matter which list L one uses, if the optimal schedule requires time T, then the completion time for the schedule produced by the list-processing algorithm applied to list L with m processes is less than or equal to $(2 - 1/m)T$. For example, for two machines ($m = 2$), if an optimal schedule yields completion at time 30, then no list would ever yield a completion later than $(2 - \frac{1}{2})(30) = 45$. Although it is of great theoretical interest, Graham's result does not provide much comfort to those who are trying to find good schedules for independent tasks.

Is there some way of choosing a priority list that consistently yields relatively good schedules? The surprising answer is yes! The idea is that when long tasks appear toward the end of the list, they often seem to "stick out" on the right end, as in Figure 3.10a. This suggests that before trying to schedule a collection of tasks, the tasks should be placed in a list, so that longest tasks are listed first. The list-process-

ing algorithm applied to a list arranged in this fashion is called the **decreasing-time-list algorithm.** If we apply it to the set of tasks listed previously, we obtain the times 10, 9, 7, 7, 5, 4 and the schedule (packing) shown in Figure 3.11. This packing is again optimal, but it is different from the optimal scheduling in Figure 3.10b.

It is important to remember that the decreasing-time-list algorithm does not guarantee optimal solutions. This can be seen by scheduling the tasks with times 11, 10, 9, 6, 4 (Figure 3.12). However, the rearranged list 9, 4, 6, 11, 10 yields the schedule in Figure 3.13, which finishes at time 20. This solution is obviously optimal. Note that when tasks are independent, if there are m machines available, the completion time cannot be less than the sum of the task times divided by m.

The problems we have encountered in scheduling independent tasks seem to have taken us a bit far from our goal of *applying* the mathematics we have developed. Sometimes mathematicians will pursue their mathematical ideas even though they have reached a point where there appear to be no applications. Fortunately, however, it is very common to be able to find applications for the "abstract" extensions. This is the case in the current instance.

Imagine a photocopy shop with three photocopiers. Photocopying tasks that must be completed overnight are accepted until 5 p.m. The tasks are to be done in any manner that minimizes the finish time for all the work. Because this problem involves scheduling machines for independent tasks, the decreasing-time-list algorithm would be a good heuristic to apply.

For another example, consider a typing pool at a large corporation or college, where individual typing tasks can be assigned to any typist. In this setting, however, the assumption that the processors (typists) are identical in skill is less likely to be true. Hence, the tasks

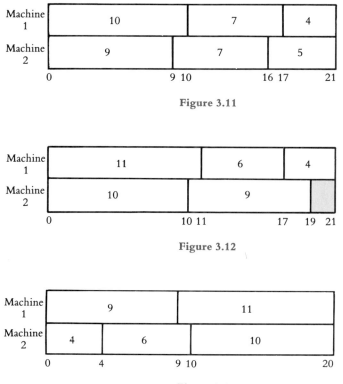

Figure 3.11

Figure 3.12

Figure 3.13

might have different times with different processors. This phenomenon, which occurs in real-world scheduling problems, violates one of the assumptions of our mathematical model.

Graham's result for the list-processing algorithm—that the finishing time is never more than $(2 - 1/m)T$—offers us the small comfort of knowing that even the worst choice of priority list will not yield a completion time worse than twice the optimal time. Compared with the list-processing algorithm the decreasing-time-list algorithm seems to improve completion times. Thus, it is not surprising that Graham was able to provide an improved *bound*, or time estimate, for this case: the decreasing-list algorithm gives a completion time no more than $\frac{4}{3}T - \frac{1}{3}m$ where m is the number of processors and T is the optimal time in which the tasks can be completed. In particular, when the number of processors is 2, the schedule produced by the decreasing-time-list algorithm is never off by more than 17%! Usually, the error is much less. This result is a remarkable instance of the value of mathematical research into applied problems.

Critical-Path Schedules

The scheduling methods developed so far are not the only approaches to constructing schedules from an order-requirement digraph. Recall from our discussion of critical-path analysis in Chapter 2 that no matter how a schedule is constructed, the finish time cannot be earlier than the length of the longest path in the order-requirement digraph. This suggests that we should try to schedule first those tasks that occur early in long paths, because they can bottleneck the flow of tasks to be completed.

To illustrate this method, consider the order-requirement digraph in Figure 3.14. Suppose we wish to schedule these tasks on two processors. Initially, there are two critical

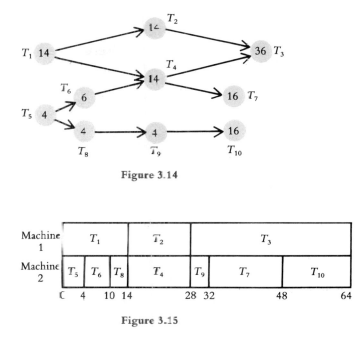

Figure 3.14

Figure 3.15

paths of length 64: T_1, T_2, T_3 and T_1, T_4, T_3. Thus, we schedule T_1 on the first machine. Once T_1 is scheduled, there is a new critical path of length 60 that starts with T_5 and that is put on machine 2. We can continue in this fashion, obtaining the schedule in Figure 3.15. (In case of ties, we schedule a lower-number task on a lower-number machine first.)

This example shows that **critical-path scheduling**, as this method is called, can sometimes yield optimal solutions. Unfortunately, this algorithm does not always perform well. For example, the critical-path method on four processors applied to the order-requirement digraph shown in Figure 3.16 yields the schedule in Figure 3.17. In fact, there can be no *worse* schedule than this one. An optimal schedule is shown in Figure 3.18.

Many of the results we have examined so far are negative because we are dealing with a general class of problems that defy our using computationally efficient algorithms to find an optimal schedule. But we can close on a more

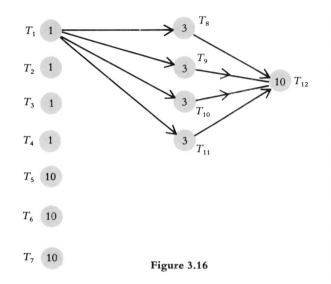

Figure 3.16

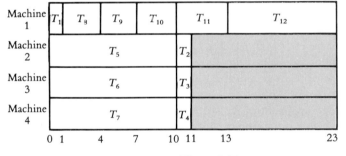

Figure 3.17

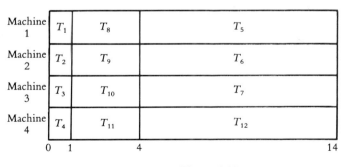

Figure 3.18

positive note. Consider an arbitrary order-requirement digraph but assume all the tasks have equal time. It turns out that we can always construct an optimal schedule using two processors in this situation. Ironically, we can choose among many algorithms to produce these optimal schedules. The algorithms are easy to understand (though not easy to prove optimal) and have all been discovered since 1969! Many people think that mathematics is an already known subject, that all its ideas and methods were discovered hundreds of years ago. They assume mathematics consists of nothing more than the arithmetic, algebra, and trigonometry taught in high school. This stereotype is easily broken down, as we have just seen, and it is well to consider that scheduling problems account for only a narrow portion of mathematics in general.

Bin Packing

Suppose you plan to build a wall system for your books, records, and stereo set. It requires 24 wooden shelves of various lengths: 6, 6, 5, 5, 5, 4, 4, 4, 4, 2, 2, 2, 2, 3, 3, 7, 7, 5, 5, 8, 8, 4, 4, 5. The lumber yard, however, sells wood only in boards of length 9 feet. If each board is $8, what is the minimum cost to buy sufficient wood for this wall system?

Because all shelves required for the wall system are shorter than the boards sold at the lumber yard, the largest number of boards needed is 24, the precise number of shelves needed for the wall system. Buying 24 boards would of course be a waste of wood and money because several of the shelves you need could be cut from one board. For example, pieces of length 2, 2, 2, and 3 feet can be cut from one 9-foot board.

To be more efficient, we think of the boards as bins of capacity W (9 feet in this case) into which we will pack (without overlap) weights (in this case, lengths) $w_1, \ldots , w_n$, where each $w_i \leq W$. We wish to find the minimum number of bins into which the weights can be packed. In this formulation, the problem is known as the **bin-packing problem**. At first glance, it may appear unrelated to the machine-scheduling problems we have been studying; however, there is a connection.

Let's suppose we want to schedule independent tasks so that each machine working on the tasks finishes its work by time W. Instead of fixing the number of machines and trying to find the earliest completion time, we must find the minimum number of machines that will guarantee completion by the fixed completion time (W). Despite this similarity between the machine-scheduling problem and the bin-packing problem, the discussion that follows will use the traditional terminology of bin-packing.

By now, it should come as no surprise to learn that no one knows a fast algorithm that always picks the optimal (smallest) number of bins (boards). In fact, the bin-packing problem belongs to the class of NP-complete problems, which means that most experts think it unlikely that any fast optimal algorithm will ever be found.

We will think of the items to be packed, in any particular order, as constituting a list. In what follows we will use the list of 24 shelf lengths given for the wall system. We will consider various heuristic methods. Probably the easiest approach is simply to put the weights into the first bin until the next weight won't fit, and then start a new bin. (Once you open a new bin, don't use leftover space in an earlier, partially filled bin.) Continue in the same way until as many bins as necessary are used. The resulting solution is shown in Figure 3.19. This algorithm, called **next fit,** has the advantage of not requiring knowledge of all the weights in advance; only the remaining space in the bin currently being packed must be remembered. The disadvantage of this heuristic is that a bin packed early on may have had room for small items that come later in the list.

Our wish to avoid permanently closing a bin too early suggests a different heuristic — **first fit:** put the next weight into the *first* bin already opened that has room for this weight; if no such bin exists, start a new bin. Note that a computer program to carry out first fit would have to keep track of how much room was left in *all* the previously opened bins. For the 24 wall-system shelves, first fit would generate a

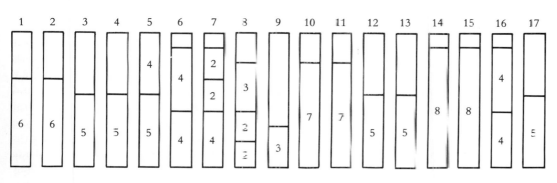

Figure 3.19

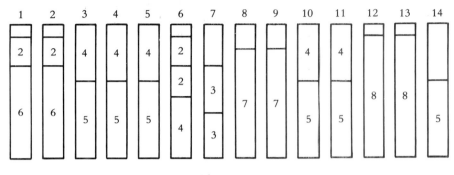

Figure 3.20

solution that uses only 14 bins (see Figure 3.20) instead of the 17 bins generated by next fit.

If we are keeping track of how much room remains in each unfilled bin, we can put the next item to be packed into the bin that currently has the most room available. This heuristic is called **best fit.** The solution generated by this approach is the same as that shown in Figure 3.20. Although this heuristic also leads to 14 bins, the items are packed in a different order. For example, the first item of size 2, the tenth item in the list, is put into bin 6 in best fit, into bin 1 in first fit.

One difficulty with all three of these heuristics is that large weights that appear late in the list can't be packed efficiently. Therefore, we should first sort the items to be packed in order of decreasing size, assuming that all items are known in advance. We can then pack large items first and pack the smaller items into leftover spaces. This approach yields three new heuristics: **next-fit decreasing, first-fit decreasing,** and **best-fit decreasing.** Here is the original list sorted: 8, 8, 7, 7, 6, 6, 5, 5, 5, 5, 5, 5, 4, 4, 4, 4, 4, 4, 3, 3, 2, 2, 2, 2. Packing using first-fit decreasing order yields the solution in Figure 3.21. This solution uses only 13 bins.

Is there any packing that uses only 12 bins? No. In Figure 3.21, there are only 2 free units

(1 unit in each of bin 1 and bin 2) of space in the first 12 bins but 4 occupied units (two 2s) in bin 13. We could have predicted this by dividing the total length of the shelves (110) by the capacity of each bin (board): $\frac{110}{9} = 12\frac{2}{9}$. Thus, no packing could squeeze these shelves into 12 bins — there would always be at least 2 units left over for the 13th bin. (In Figure 3.21, there are 4 units in bin 13 because of the 2 wasted empty spaces in bins 1 and 2.) Even if this division had created a zero remainder, there would still be no guarantee that the items could be packed to fill each bin without wasted space. For example, if the bin capacity is 10 and there are weights of 6, 6, 6, 6, 6 the total weight is 30; dividing by 10, we get 3 bins as the minimum requirement. Clearly, however, 5 bins are needed to pack the five 6s.

None of the six heuristic methods shown will necessarily find the optimal number of bins for an arbitrary problem. How, then, can we decide which heuristic to use? One approach is to see how far from the optimal solution each method might stray. Various formulas have been discovered to calculate the maximum discrepancy between what a bin-packing algorithm actually produces and the best possible result. For example, in situations where a large number of bins are to be packed, first fit (FF) can be off as much as 70%, but first-fit decreasing (FFD) is never off by more

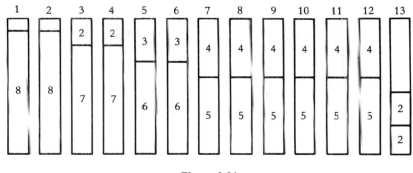

Figure 3.21

than 22%. Of course, FFD doesn't give an answer as quickly as FF, because extra time for sorting a large collection of weights may be considerable. Also, FFD requires knowing the whole list of weights in advance, whereas FF does not. It is important to emphasize that 22% is a worst-case figure. In many cases, FFD will perform much better. Recent results obtained by computer simulation (experiment) indicate excellent average-case performance for this algorithm.

When modeling real-world problems, we always have to look at the interface of the mathematics with the real world. Thus, first-fit decreasing usually results in fewer bins than next fit, but next fit can be used even when all the weights are not known in advance. Next fit also requires much less computer storage than first fit, because once a bin is packed, it need never be looked at again. Fine tuning of the conditions of the actual problem often results in better practical solutions and in interesting new mathematics as well. (See Box 3.2 for a discussion of some of the tools mathematicians use to verify and even extend mathematical truths.)

BOX 3.2 Using Mathematical Tools

The tools of a carpenter include the saw, T-square, level, and hammer. A mathematician also requires tools of the trade. Some of these tools are the proof techniques that enable verification of mathematical truths. Another set of tools consists of strategies to sharpen or extend the mathematical truths already known. For example, suppose that if A and B hold, then C is true. What happens if only A holds? Will C still be true? Similarly, if only B holds, will C still be true?

This type of thinking is of value because such questions will result either in more general cases where C holds or in examples showing that B alone can't imply C. For example, we saw that if a graph G is connected (hypothesis A) and even-valent (hypothesis B), then G has a tour of its edges using each edge only once (conclusion C). If either hypothesis is omitted the conclusion fails to hold. Examples of this are shown in the accompanying figure on the next page. On the left you see an even-valent but nonconnected graph, and on the right, a connected graph with two odd-valent vertices; neither graph has an Euler circuit.

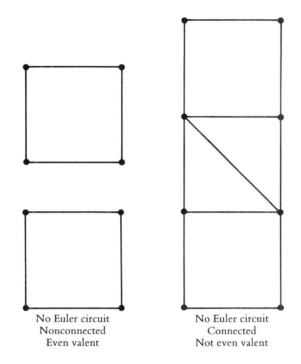

No Euler circuit
Nonconnected
Even valent

No Euler circuit
Connected
Not even valent

Here is another way that a mathematician might approach extending mathematical knowledge. If A and B imply C, will A and B imply both C and D, where D extends the conclusion of C? For example, not only can we prove that a connected, even-valent (hypotheses A and B) graph has an Euler-circuit, but we can also show that the first edge of the tour can be chosen arbitrarily (conclusions C and D). It turns out that being able to specify the first two edges of the tour may not always be possible. Mathematicians are trained to vary the hypotheses and conclusions of results they prove in an attempt to clarify and sharpen the range of applicability of the results.

We have seen that machine scheduling and bin packing are probably computationally difficult to solve because they are NP-complete. A mathematician could then try to find the simplest version of a bin-packing problem that would still be NP-complete: What if the items to be packed can have only eight weights? What if the weights are only one and two? Asking questions like these is part of the mathematician's craft. Such questions help to extend the domain of mathematics and hence the applications of mathematics.

Cryptography

We conclude with a variant of the bin-packing problem that is of interest to those who make and break secret codes. Given numbers (weights) $w_i, \ldots, w_n$ and a number S, find an algorithm that selects a subcollection of the weights whose sum is exactly S. This problem, also known to be NP-complete, is a special case of what is called the **knapsack problem** because S can be thought of as the capacity of a knapsack to be filled by the items selected from the list of weights. This particular knapsack problem is called the **subset sum problem.** As tame as it may sound, it is related to a revolutionary new proposal for ensuring the security of data stored in computers such as bank records and classified military information.

The security of data has become an important new branch of **cryptography,** the study of codes and ciphers and how to break them. Traditional cryptography is based on a single key that is used to transform or encrypt a message (plaintext) into garbled form (ciphertext). The same key used to encrypt the message is used to decipher the message. However, in the new scheme, called *public-key cryptography*, there are two keys. One key, made public, is used by anyone wishing to send a secure message to X. The other key is known only to X, the receiver of the message. The keys are designed so that knowledge of the public key does not compromise the private key.

The idea behind public-key cryptography is that certain processes are very easy to carry out in one direction but very hard to reverse. For example, one can easily add $708 + 259 + 871 + 1836 + 82$ to get 3756, but it is not so easy to find a subcollection of the numbers 1036, 708, 82, 259, 589, 871, and 1836 that add to $S = 3756$.

Because knapsack problems are thought to be computationally hard to solve, schemes have been suggested for applying these ideas to public-key systems. Although recent work has shown that knapsack systems are faulty as a base for public-key systems, other public-key systems appear to be viable. Furthermore, the idea of NP-completeness, which arose as a concept in the abstract analysis of the complexity of algorithms, turned out to be related to the mundane issue of how to protect money being transferred between banks! We can see once more how the use of clever (though not necessarily impenetrable) mathematics can affect and enrich our lives.

REVIEW VOCABULARY

Best fit A heuristic algorithm for bin packing in which the next weight to be packed is placed into the open bin with the largest amount of room remaining. If the weight fits in no open bin, a new bin is opened.

Best-fit decreasing A heuristic algorithm for bin packing where the best-fit algorithm is applied to the list of weights sorted so that they appear in decreasing order.

Bin-packing problem The problem of determining the minimum number of containers of capacity W into which objects of size w_1, . . . , w_n ($w_i \leq W$) can be packed.

Cryptography The study of how to make and break codes. These codes are now used primarily for data and computer security rather than for national security.

Decreasing-time-list algorithm The heuristic algorithm that applies the list-processing algorithm to the priority list obtained by listing the tasks in decreasing order of their time length.

First fit A heuristic algorithm for bin packing in which the next weight to be packed is placed in the lowest-numbered bin already opened into which it will fit. If it fits in no open bin, a new bin is opened.

Heuristic algorithm An algorithm that is fast to carry out but that doesn't necessarily give an optimal solution to an optimization problem.

Knapsack problem Given a knapsack of size W and a collection of weights w_1, . . . , w_n, find a largest collection of weights that will fit in the knapsack.

List-processing algorithm A heuristic algorithm for assigning tasks to processors: assign the first ready task on the priority list that has not already been assigned to the lowest-numbered processor that is not working on a task.

Machine scheduling The problem of assigning tasks to processors so as to complete the tasks by the earliest time possible.

Next fit A heuristic algorithm for bin packing in which a new bin is opened if the weight to be packed next will not fit in the bin that is currently being filled; the current bin is then closed.

Next-fit decreasing A heuristic algorithm for bin packing where the next-fit algorithm is applied to the list of weights sorted so that they appear in decreasing order.

Processor A person, machine, robot, operating room, or runway whose time must be scheduled.

Ready task A task whose predecessors have already been finished.

Subset sum problem Given an integer W and integer numbers $w_1, \ldots, w_n$, find a subcollection of the numbers whose sum is W.

EXERCISES

1. Use the list-processing algorithm to schedule the tasks in the following order-requirement digraph on three processors, using the list $T_1, \ldots, T_{11}$. From the schedule so constructed, for each task list the start and finish time for that task.

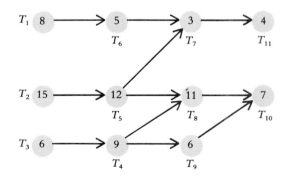

2. For the order-requirement digraph at the top of the next page, apply the list-processing algorithm, using three processors for lists **a** and **b**. How do the completion times obtained compare with the length of the critical path?

    a. $T_1, T_2, T_3, T_4, T_5, T_6, T_7, T_8$

    b. $T_2, T_4, T_6, T_8, T_1, T_3, T_7, T_5$

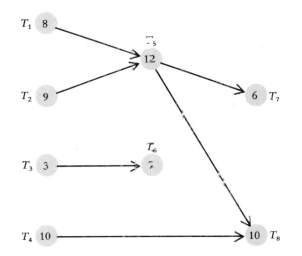

3. List as many scheduling situations as you can for the following environments:

  a. Hospital

  b. Railroad system

  c. Airplane

  d. Automobile repair garage

  e. Machine shop

  f. Your home

4. Can you think of situations other than those mentioned in the text where scheduling independent tasks on processors occurs?

5. Find the completion time for independent tasks of length 8, 11, 17, 14, 16, 9, 2, 1, 18, 5, 3, 7, 6, 2, 1 on three processors, using the list-processing algorithm. Then find the completion time for these tasks on three processors, using the decreasing-time-list algorithm. Does either solution give rise to an optimal schedule? Repeat for tasks of lengths 19, 19, 20, 20, 1, 1, 2, 2, 3, 3, 5, 5, 11, 11, 17, 18, 18, 17, 2, 16, 16, 2.

6. Could the following schedule have arisen from the list-processing algorithm? Could it have arisen from the application of the list-processing algorithm to a collection of independent tasks?

Machine 1	T_1	T_5		T_8
Machine 2	T_2			T_9
Machine 3	T_3	T_6		T_{10}
Machine 4	T_4	T_7		T_{11}

7. For the following schedules, can you produce a list so that the list-processing algorithm produces the schedule shown when the tasks are independent? What are the task times for each task?

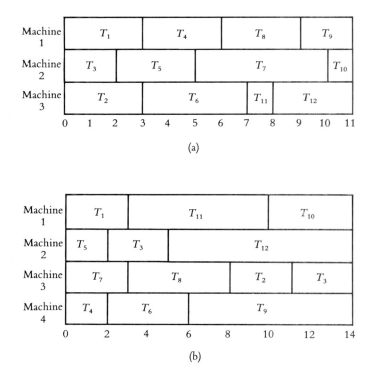

(a)

(b)

8. Discuss different criteria that might be used to construct a priority list for a scheduling problem.

9. Give examples of scheduling problems in which

    a. processors available can be treated as if they are identical

    b. processors available cannot be treated as if they are identical

10. Give examples of scheduling problems for which it is *not* reasonable to assume that once a processor starts a task, it would always complete that task before it works on any other task. Give examples for which this approach would be reasonable.

11. Can you find a schedule (use the order-requirement digraph in Figure 3.4) with a completion time earlier than 12 as shown in the following figure, using a list other than $T_1, \ldots, T_9$? If not, why not?

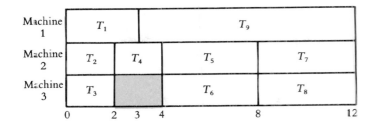

12. Can you give examples of scheduling problems for which it seems reasonable to assume that all the task times are the same?

13. Describe possible modifications for the list-processing algorithm that would allow for

a. a machine to be "voluntarily" idle

b. a machine to interrupt work on a task once it has begun work on the task

14. Can you think of real-world scheduling situations in which all the tasks have the same time and are independent? Can you find an algorithm for solving this problem optimally? (If there are n independent tasks of time length k, when will all the tasks be finished?)

15. Show that when tasks to be scheduled are independent, the critical-path method and the decreasing-time-list method are identical.

16. Find a list that produces the following optimal schedule when the list-processing algorithm is applied to this list.

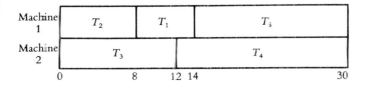

What completion time and schedule are obtained when the decreasing-time-list algorithm is applied to this list?

17. A radio station's policy allows advertising breaks of no longer than 2 minutes, 15 seconds. Using (a) first fit and (b) first-fit decreasing, determine the minimum number of breaks into which the following ads will fit (lengths given in seconds): 80, 90, 130, 50, 60, 20, 90, 30, 30, 40. Can you find the optimum solution? Do the same for these ads: 60, 50, 40, 40, 60, 90, 90, 50, 20, 30, 30, 50.

18. It takes 4 seconds to photocopy one page. Manuscripts of lengths 10, 8, 15, 24, 22, 24, 20, 14, 19, 12, 16, 30, 15, and 16 pages are to be photocopied. How many photocopy machines would be required, using the best-fit-decreasing algorithm to guarantee that all manuscripts are photocopied in 2 minutes or less?

19. How many bins are necessary to pack the numbers 1, 1, 2, 2, . . . , 20, 20 into bins of size 25? Repeat for 1, 1, 1, 2, 2, 2, . . . , 20, 20, 20. Generalize!

20. Two-dimensional bin packing refers to the problem of packing rectangles of various sizes into a minimum number of $m \times n$ rectangles, with the sides of the packed rectangles parallel to those of the containing rectangle.

   a. Suggest some possible real-world applications of this problem.

   b. Devise a heuristic algorithm for this problem.

   c. Give an argument to show that the problem is at least as hard to solve as the usual bin-packing problem.

   d. If you have $1 \times m$ rectangles with total area W to be packed into a single rectangle of area $p \times q = W$, can the packing always be accomplished?

21. Can you find an example of weights that, when packed into bins using first fit, use fewer bins than the number of bins used when the first-fit algorithm is used with the first weight on the list removed?

22. Can you formulate "paradoxical" situations for bin packing that are analogous to those we found for scheduling processors?

23. In what situations would packing bins of different capacities be the appropriate model for real-world situations? Suggest some possible algorithms for this type of problem.

Linear Programming

A corporate manager's job often involves a series of very complicated decisions. In a typical decision problem, managers have objectives to meet and resources to deploy to meet them. They also have obstacles to overcome and constraints that must be respected. All too often, the alternative solutions are so many that it is impossible to evaluate them all individually. Despite this, millions of dollars may ride on the manager's decision.

When we have decisions to make in our personal lives, most of us use seat-of-the-pants methods that defy description. Traditionally, large organizations such as corporations and governments did the same — and often made bad decisions as a result. Shortly after World War II, however, an organized method of mathematical decision making called **linear programming** changed all that. Computer calculations supplemented the boss's intuition,

and departments engaged in linear programming became almost as common as accounting departments. The payoff from linear programming is large, often many millions of dollars, and the method has wide applications.

Mixture Problems

We will focus on a class of linear-programming problems called **mixture problems.** (Box 4.1 explains how linear programming applies to other types of problems.) In mixture problems, limited resources need to be allocated in an appropriate mixture to maximize profits on the products produced. We now examine an actual example of a mixture problem that confronted a lumber company.

Plywood Ponderosa of Mexico produces various plywood products by gluing together

---

BOX 4.1 Case Studies in Linear Programming

Linear programming is not limited to mixture problems. Here are two case studies that do not involve mixture problems, yet where applying linear-programming techniques produced impressive savings:

• The Exxon Corporation spends several million dollars per day running refineries in the United States. Because running a refinery takes a lot of energy, energy-saving measures can have a large effect. Managers at Exxon's Baton Rouge plant had over 600 energy-saving projects under consideration. They couldn't implement them all because some conflicted with others, and there were so many ways of making a selection from the 600 that it was impossible to evaluate all selections individually.

Exxon used linear programming to select an optimal configuration of about 200 projects. The savings are expected to be about $100 million over a period of years.

• American Edwards Laboratories uses heart valves from pigs to produce artificial heart valves for human beings. Pig heart valves come in different sizes. Shipments of pig heart valves often contain many of some sizes and too few of others; however, each supplier tends to ship roughly the same imbalance of valve sizes on every order, so that the company can expect consistently different imbalances from the different suppliers. Thus, if they order shipments from all the suppliers, the imbalances could cancel each other out in a fairly predictable way. The amount of cancellation, of course, will depend on the sizes of the individual shipments. Unfortunately, there are too many combinations of shipment sizes to consider all combinations individually.

American Edwards used linear programming to figure out which combination of shipment sizes would give the best cancellation effect. This reduced their annual cost by $1.5 million.

---

thin sheets (veneers) of lumber. Some of these products are more profitable than others, yet it is not sensible to produce only the highest-profit product because a lot of lumber would be left over in some grades. How much of each product — that is, which product mix — should be made in order to maximize total profit? Because Plywood Ponderosa had too many possible product mixes to consider individually, the company used linear programming to find the product mix that would give the highest overall profit. This turned out to be different from the mix they had been making. In fact, changing to the new mix increased profits by 20%.

Real-world mixture problems, like Plywood Ponderosa's problem, deal with very large numbers of resources and products — too many for hand calculation. We will therefore simplify the plywood mixture problem in our discussion.

Suppose the company produces plywood by using a press to glue veneers together. The veneers come in two grades, A and B, and two kinds of plywood are made: exterior and interior. One panel of exterior plywood requires two panels of grade A veneer, two panels of grade B veneer, and 10 minutes on a press. One panel of interior plywood requires four panels of grade B veneer, no grade A veneers, and 5 minutes on the press. On a certain day there are 1000 panels of grade A veneer available and 3000 panels of grade B. There are 12 presses, each of which can press four veneer panels into a finished product. Each press can be run for 500 minutes per day, yielding a total of 6000 machine-minutes. The profit per panel is $5 for interior plywood and $6 for exterior plywood. How much of each product should be produced that day to maximize profit?

Let's list the features or conditions of this problem that make it a mixture problem:

• *Resources.* Definite resources are available in limited, known quantities for the time period in question. The resources include the two grades of veneers and the amount of time available on the presses.

• *Products.* Definite products can be made by combining, or mixing, the resources. In this example, the products are exterior and interior plywood.

• *Recipes.* A recipe for each product specifies how many units of each resource are needed to make one unit of that product.

• *Profits.* Each product earns a known profit per unit. (It is assumed that every unit produced can be sold.)

• *Objective.* The objective in a mixture problem is to find how much of each product to produce in order to maximize the profit without exceeding any of the resource limitations.

Mixture problems are widespread because nearly every product produced by our economy is created by combining resources. For example, various kinds of steel alloys can be produced by combining iron with different amounts of other elements such as carbon and tungsten. How much of each alloy should be made if one has limited quantities of raw materials on hand? Consider another example: Ce-

real manufacturers use oats, corn, wheat, and rice to produce breakfast cereals that contain either one type of grain or a mix of grains. How many boxes of each product should be made?

One of the most important skills required to solve a mixture problem is the ability to understand its underlying structure. (This is at least as important as being able to do the subsequent algebra and arithmetic.) Understanding the underlying structure means being able to answer these questions:

1. What are the resources?

2. What quantity of each resource is available?

3. What are the products?

4. What are the recipes for creating the products from the resources?

5. What are the unknowns?

6. What is the profit formula?

We will display the answers to these questions in a diagram called a mixture chart.

Let's examine this underlying structure for a fruit juice manufacturer that produces two kinds of juice for sale: appleberry and cranapple. These products are made by combining pure cranberry juice and pure apple juice in the following proportions: 3 quarts of cranberry juice and 1 quart of apple juice make 1 gallon of cranapple; 2 quarts of apple juice and 2 quarts of cranberry juice are combined to make 1 gallon of appleberry. There are 200 quarts of cranberry juice and 100 quarts of apple juice available. The profit on a gallon of cranapple is $0.03 and the profit on a gallon of appleberry is $0.04. How many gallons of cranapple and how many gallons of appleberry should be produced to obtain the highest profit without exceeding available supplies?

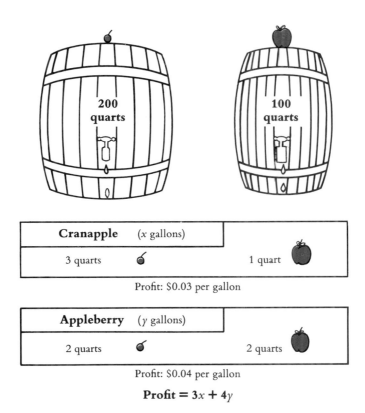

Figure 4.1 A mixture chart for fruit juice production.

Our answers to questions 1 through 6 are shown on the mixture chart on Figure 4.1. The resources (question 1) are cranberry juice and apple juice, and the quantities (question 2) are 200 quarts of cranberry juice and 100 quarts of apple juice (shown on the top of the mixture chart as two resource barrels with spigots). Questions 3 and 4 on products and recipes are answered in the boxes under the resource barrels, one box for each product.

The boxes also show the unknowns (question 5): how many units (gallons) of cranapple juice and how many units of appleberry to make. We call these x and y:

x = the number of gallons of cranapple to make

y = the number of gallons of appleberry to make

The profits for each product (question 6) are shown under the box for that product. The total profit we obtain will depend on x and y, according to the formula

$$\text{Profit} = \underset{\substack{\textbf{Profit from} \\ \textbf{making } x \textbf{ units} \\ \textbf{of cranapple}}}{3x} + \underset{\substack{\textbf{Profit from} \\ \textbf{making } y \textbf{ units} \\ \textbf{of appleberry}}}{4y}$$

In this formula, $3x$ stands for 3 times the (presently unknown) quantity x, whereas $4y$ stands for 4 times the (presently unknown) quantity y.

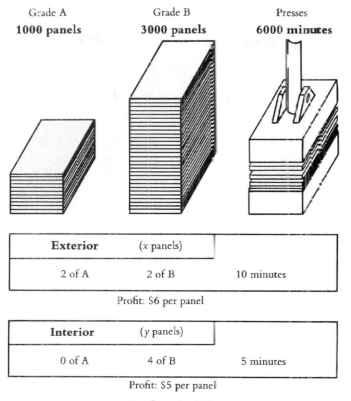

Grade A	Grade B	Presses
1000 panels	**3000 panels**	**6000 minutes**

Exterior	(*x* panels)	
2 of A	2 of B	10 minutes

Profit: $6 per panel

Interior	(*y* panels)	
0 of A	4 of B	5 minutes

Profit: $5 per panel

Profit = 6x + 5y

Figure 4.2   A mixture chart for plywood production.

The fruit juice and plywood problems deal with different industries and different numbers, but to a mathematician they are similar because both fit into the framework of a mixture problem and can be represented by mixture charts. (Figure 4.2 shows the mixture chart for the plywood problem.) The task of recognizing a mixture problem (that is, making a mixture chart for it) is mostly a matter of careful reading and logical analysis. Some might be fooled into thinking that it is not mathematics at all. But actually, it is an extremely important skill in the practice of applied mathematics.

Graphing the Feasible Region

The reader with some exposure to mathematics may sense that there must be some way to obtain the optimal *x* and *y* using a formula or procedure that uses the profit formula and the various numbers on the mixture chart. There are, indeed, such procedures. (One of them, the simplex method, will be discussed later in the chapter.)

However, the details of these procedures are technical and soon forgotten, so we shall not go into them here. Instead, we will present a pictorial view of linear programming that has been valuable to theorists and practitioners alike and that is readily understandable by nonmathematicians.

The basic notion is that a pair of numerical values, such as (5, 3) or (−1, 2) or more generally (x, y), can be interpreted as a point in the plane. In a mixture problem involving two products, we are looking for a pair of values x and y that gives the most profit. This can be interpreted as a search for the best point on the plane.

We can't locate this best point directly, so we begin with another question: Which points are the reasonable candidates for this paragon of points? A candidate for best point must satisfy two feasibility constraints:

1. Both x and y must be nonnegative because they represent quantities of products. (For example, you can't make −6 cans of apple juice.) These constraints restrict feasible points to the shaded (first quadrant) region in Figure 4.3.

2. The combination (x, y) must not require using more of some resource than we have available.

To illustrate a constraint of type 2, suppose the fruit juice manufacturer decides to make 50 gallons each of cranapple and appleberry. Using the recipes recorded on the mixture chart, we can compute that this would require 250 quarts of cranberry juice and 150 quarts of apple juice, which exceeds the supply of both resources. We are not permitted to exceed the supply of even a single resource.

Points that do not violate either type of constraint are called **feasible points.** All others are infeasible points. The collection of feasible points is called the **feasible set,** or **feasible region.**

What does the feasible set look like? Are the feasible points interspersed with infeasible points in a salt and pepper pattern? Or do the feasible points clump together in some type of connected shape? Techniques of algebra and analytic geometry allow us to answer this question precisely, using the data in the mixture chart to deduce the following facts:

1. For any mixture problem of the kind we are discussing, the feasible region is a polygon in the first quadrant, with one corner at the origin and one on each coordinate axis. There may be additional corners off the axes.

2. The polygon has neither dents (as in Figure 4.4a) nor the holes (as in Figure 4.4b). Figure 4.4c is a typical example. Figure 4.4d shows the feasible region for the fruit juice example.

The precise shape of the feasible region, including the coordinates of the corner points, can be computed from the data on the mixture chart.

The Corner Principle

To evaluate the profit that goes with a given point (x, y), use the profit formula recorded on the mixture chart. For example, to find the profit associated with the point (40, 10), substitute x = 40 and y = 10 into the formula for profit:

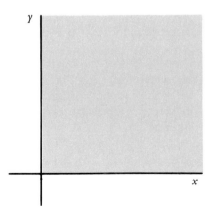

Figure 4.3 Feasible points for x and y occur only in the first quadrant (shaded) of a graph.

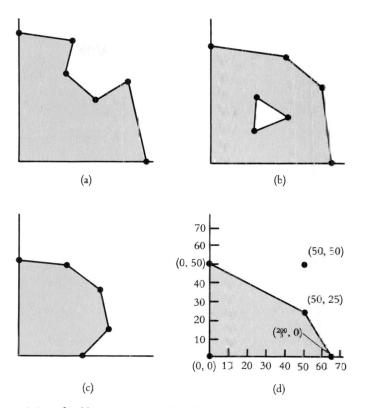

Figure 4.4   A feasible region may not have dents (a) or holes (b). Graph (c) shows a typical feasible region. Graph (d) shows the feasible region for the fruit juice example.

$$\text{Profit} = 3x + 4y$$
$$= 3(40) + 4(10)$$
$$= 160$$

If we had some way of applying this formula to each point of the feasible region, we could then pick out the point with the highest profit. Unfortunately, the feasible region has infinitely many points, making it impossible to compute the profit for each point. Luckily, we can narrow our search because of the **corner principle:** *The highest profit value on a polygonal feasible region can always be found at a corner point.*

The corner principle is probably the most important insight in the theory of linear pro-

gramming. It is the geometric nature of this principle that explains the value of creating a geometric model from the data in a mixture chart.

The corner principle allows us to use the following method to solve a mixture problem:

1. Compute the corner points of the feasible region.

2. Evaluate the profit at each corner of the feasible region.

3. Choose the corner with the highest profit.

The corners of the feasible region for the juice mixture problem are shown in Figure 4.4d, and the following table shows the profit at each corner:

Corner	Profit (= $3x + 4y$)
(0, 0)	$3(0) + 4(0) = 0$
(0, 50)	$3(0) + 4(50) = 200$
(50, 25)	$3(50) + 4(25) = 250$
($\frac{200}{3}$, 0)	$3(\frac{200}{3}) + 4(0) = 200$

The highest profit is 250 and occurs at the corner point (50, 25). Therefore, to achieve this profit, we should make 50 gallons of cranapple and 25 gallons of appleberry.

The situation here is reminiscent of Alexander the Great's approach to the problem of the Gordian knot, a legendary knot so large and tight and tangled that no one had been able to open it. Alexander's solution was to slice the knot open with his sword. Mixture problems and other linear-programming problems are like Gordian knots because there are infinitely many feasible points — we can't calculate the profit for all of them. The corner principle functions like Alexander's sword and cuts through the problem.

You can visualize a mathematical proof of the corner principle by imagining that each point of the plane is a tiny light bulb that is capable of lighting up. For the example in Figure 4.4d, imagine what would happen if we ask this question: Will all points with profit = 360 please light up? What figure do these lit-up points form?

In algebraic terms, we can restate the profit question in this way: Will all points (x, y) with $3x + 4y = 360$ please light up? As it happens, this version of the profit question is one mathematicians learned to answer hundreds of years before linear programming was born. The answer is a straight line. Furthermore, it is a routine matter to determine the exact position of the line. We call this line the **profit line** for

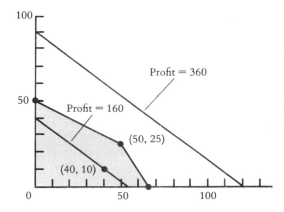

Figure 4.5 The profit line for 360 lies outside the feasible region, whereas the profit line for 160 passes through the region.

360; it is shown in Figure 4.5. For numbers other than 360, we would get different profit lines.

Unfortunately, there are no points on the profit line for 360 that are feasible, that is, that lie in the feasible region. Therefore, the profit of 360 is unachievable. *If the profit line corresponding to a certain profit doesn't touch the feasible region, then that profit can't be achieved.*

Because 360 is too big, perhaps we should ask the profit line for a more modest amount, say 160, to light up. You can see that the new profit line of 160 in Figure 4.5 is parallel to the first profit line and closer to the origin. This is no accident: well-known elementary theorems prove it must be so.

The most important feature of the profit line for 160 is that it has points in common with the feasible region. For example, (40, 10) is on that profit line because $3(40) + 4(10) = 160$ and in addition is a feasible point. This means that it is possible to make 40 gallons of cranapple and 10 gallons of appleberry and that if we do so, we will have a profit of 160.

Can we do better than a 160 profit? As we slowly increase our desired profit from 160

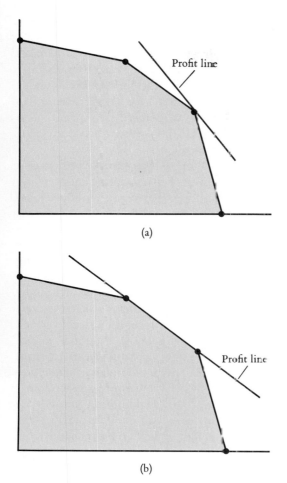

(a)

(b)

Figure 4.6 The highest profit will occur when the profit line is just touching the feasible region, either at a corner point (a) or along a line segment (b).

toward 360, the location of the profit line that lights up shifts smoothly upward away from the origin. As long as the line continues to cross the feasible region, we are happy to see it move away from the origin because the more it moves, the higher will be the profit represented by the line. We would like to stop the movement of the line at the last possible instant, while the line still has one or more points in common with the feasible region. It

should be obvious that this will occur when the line is just touching the feasible region either at a corner point (Figure 4.6a) or along a line segment joining two corners (Figure 4.6b). This is just what the corner principle says.

Suppose now that each bulb (point of the plane) has a color determined by the profit associated with that point. All points with the same profit (that is, points on a profit line) have the same color. Furthermore, suppose the colors range continuously from violet to blue to green to yellow to red, just as they do in a rainbow. The cool colors represent low profits, the hot colors higher ones: the higher the profit, the hotter the color. In effect, we are superimposing a straightened-out rainbow on the picture containing our feasible region. Finding the highest profit point can be thought of as finding a hottest-colored point in the feasible region (Figure 4.7).

It would be convenient if all linear-programming problems yielded simple feasible regions in two-dimensional space. However, two complications arise for practical problems:

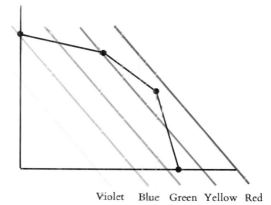

Violet   Blue   Green   Yellow   Red

Figure 4.7 If profit lines are colored according to the hues of the rainbow, the highest profit point will be the hottest-colored point in the feasible region.

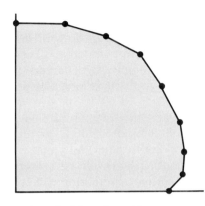

Figure 4.8 A feasible region with many corners.

1. Sometimes, as in Figure 4.8, we have a great many corners. Naturally, the more corners there are, the more calculations we need to do to calculate the coordinates of all of them and the profit at each one.

2. It will not be possible to visualize the feasible region as a part of two-dimensional space when there are more than two products. Each product is represented by an unknown, and each unknown is represented by a dimension of space. If we had 50 products, we would need 50 dimensions, and we couldn't visualize the feasible region.

Most practical linear-programming problems present us with both cases, that is, with many corners and more than two products. The upshot is that the number of corners literally can exceed the number of grains of sand on the earth. Even with the fastest computer, computing the profit of every corner is impossible.

The Simplex Method

Two main methods have been proposed to get around this difficulty. The older method is the **simplex method,** which is still the most

commonly used. Discovered by the American mathematician George Dantzig (see Box 4.2), this ingenious mathematical invention makes it possible to find the best corner by evaluating only a tiny fraction of all corners. In this way, a problem that might have taken centuries to solve on a supercomputer (if each corner had to be checked) can be solved in seconds on a home computer.

The operation of the simplex method may be likened to the behavior of an ant crawling on the edges of a polyhedron (a solid with flat sides), looking for one particular target vertex (Figure 4.9). The ant cannot see where the target vertex is. As a result, if it were to wander along the edges randomly, it might take a long time to reach the target. The ant will do much better if it has a temperature clue to let it know it is getting warmer (closer to the target vertex) or colder (farther from the target vertex).

Think of the simplex method as a way of calculating these temperature hints (it turns out there are fairly simple algebraic manipulations to do this). We begin at any randomly chosen vertex. All neighboring vertices are evaluated to see which ones are warmer and which are colder. A new vertex is chosen from

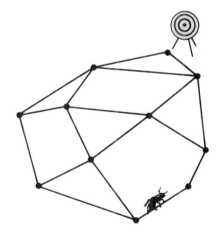

Figure 4.9 The simplex method can be compared to an ant crawling along the edges of a polyhedren, looking for one particular target vertex.

BOX 4.2 Father of Linear Programming Recalls Its Origins

George Dantzig is professor of operations research and computer science at Stanford University. He is credited with discovering the linear-programming technique called the simplex method. For the past 40 years, the simplex method has provided solutions to linear-programming problems that have saved both industry and the military time and money. Dantzig talks about the background of his famous technique:

Initially, all the work we did had to do with military planning. During World War II, we were planning on a very extensive scale. The civilian population and the military were all performing scheduling and planning tasks, perhaps on a larger scale than at any time in history. And this was the case up until about 1950. From 1950 on, the whole emphasis shifted from military planning to practical planning for the civilian population, and industry picked it up.

The first areas of industry to use linear programming were the petroleum refineries. They used it for blending gasoline. Nowadays, they run all of the refineries in the world (except for one) using linear-programming methods. They are one of the biggest users of it, and it's been picked up by every other industry you can think of—the forestry industry, the steel industry—you could fill up a book with all the different places it's used.

The question of why linear programming wasn't invented before World War II is an interesting one. In the postwar period, various technologies just evolved that had never been there before. Computers were one example. These technologies were talked about before; you can go back in history and you'll find isolated papers on them. For example, the very famous French mathematician Joseph Fourier, who invented Fourier series, had a paper on problems similar to linear programming, as did the Belgian mathematician de la Valee Pousson. But these were isolated cases that never went anywhere.

In the immediate postwar period, everything just fermented and began to happen, and one of the things that began to happen was linear programming. Mathematicians as well as economists and others who do practical planning and scheduling saw the possibilities. Things then began to happen very rapidly.

In the immediate postwar period, the whole idea of using computers to mechanize the process was an obvious one. The question was then asked, How could you formulate the process as a sort of mathematical system, and how could computers be used to make this happen?

The problems we solve nowadays have thousands of equations, sometimes a million variables. One of the things that still amazes me is to see a program run on the computer—and to see the answer come out. If we think of the number of combinations of different solutions that we're trying to choose the best of, it's akin to the stars in the heavens. Yet we solve them in a matter of moments. This, to me, is staggering. Not that we can solve them—but that we can solve them so rapidly and efficiently.

The simplex method has been used now for close to 40 years. There has been steady work going on trying to use different versions of the simplex method, nonlinear methods, and interior methods. It has been recognized that certain classes of problems can be solved much more rapidly by special algorithms than by using the simplex method. If I were to say what my field of specialty is, it is in looking at these different methods and seeing which are more promising than others.

For example, when the system is too big, we use something called the "decomposition principle" to break it into parts and solve the parts. This has been very efficient for certain classes of problems. Certain methods that have been reported in the news that are quite well known in nonlinear programming, called "interior techniques," have been experimented on very successfully with some kinds of problems, and not so successfully with others.

In my classes at Stanford, we review these different methods, and it's surprising how good some of them are. There's a lot of promise in this—there's always something new to be looked at.

---

among the warmer ones, and the evaluation of neighbors is repeated—this time checking neighbors of the new vertex. The process ends when we arrive at the target vertex.

Part of what the simplex method has going for it in the speed sweepstakes is that it works faster in practice than its worst-case behavior would lead us to believe. Although mathematicians have devised artificial cases for which the simplex method bogs down in unacceptable amounts of arithmetic, the examples arising from real applications are never like that. This may be the world's most impressive counterexample to Murphy's law, that if something can go wrong, it will.

Although the simplex method usually avoids visiting every vertex, it may require visiting many intermediate vertices as it moves from the starting vertex to the optimal one. The simplex method has to search along edges on the boundary of the polyhedron. If it happens that there are a great many small edges lying between the starting point and the optimal vertex, the simplex method must operate like a slow-moving bus that stops at every street corner.

An Alternative to the Simplex Method

In 1984, Narendra Karmarkar, a mathematician working at Bell Laboratories, devised an alternative method for linear programming that avoids such slowdowns by making use of search routes through the interior of the feasible region (see Box 4.3). The potential applications of Karmarkar's algorithm are important to a lot of industries, including telephone communications and the airlines (see Box 4.4). Routing millions of long-distance calls, for example, means deciding how to use the long-distance landlines, repeater amplifiers, and satellite terminals to best advantage. The problem is similar to the juice company's need to find the best use of its stocks of juice to create the most profitable products. We find another example in the airline business. American Airlines is working with Dr. Karmarkar to see if his algorithm can cut fuel costs. According to Thomas Cook, director of operations research for American Airlines, "It's big dollars. We're hoping we can solve harder problems faster, and we think there's definite potential."

Scientists at Bell Labs have recently applied Karmarkar's algorithm to a problem of unprecedented complexity: deciding how to economically build telephone links between cities so that calls can get from any city to any other, possibly being relayed through intermediate cities. Figure 4.10 shows two different linkings. The number of possible linkings is unimaginably large, so picking the most economical one is difficult. For any given linking that one has built, or contemplates building,

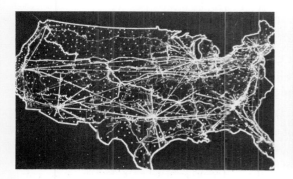

Figure 4.10 A map of the United States showing one conceivable network of major communication lines connecting major cities. Routing millions of calls over this immense network requires sophisticated linear programming techniques and high-speed computers. [Courtesy of AT&T Bell Laboratories.]

Figure 4.11 Narendra Karmarkar, a researcher at AT&T Bell Laboratories, has discovered a powerful new linear programming algorithm that may solve many complex linear programming problems faster and more efficiently than any previous method.

there is also the problem of deciding how to economically route calls through the network to reach their destinations.

Although difficult, these problems are definitely worth solving. Nat Levine, director of the transmission facilities planning center at Bell Labs speculates that if one finds the best solution, "the savings could be in the hundreds of millions of dollars."

Recent work on these problems at Bell Labs involved a linear-programming problem with about 800,000 variables, which Karmarkar's algorithm solved in 10 hours of computer time. The scientists involved in this project believe that the problem might have taken weeks to solve if the simplex method had been used.

Karmarkar's algorithm has been incorporated into a software product that will be sold or licensed for use around the world. Consequently, it won't be long before lots of data on its performance are available and we will be able to tell whether it is as big an improvement over the simplex method as early experiments suggest.

BOX 4.3 Linear Programming and the Cold War

In 1979, mathematics found itself on the front page of the *New York Times* as a result of a linear-programming technique discovered by the Russian mathematician L. G. Khachian. This algorithm, called the *ellipsoid method*, seemed at first to be much better than the simplex method for large problems. Perhaps because so much is at stake in this area of applied mathematics, the story spread like wildfire, with little regard for accuracy. In the pages of the popular press it appeared as though the Russians had achieved a breakthrough that might imperil our national security. In particular, it was suggested that the method might be applied to break cryptographic codes that had previously been considered secure.

As it turned out, there was no real reason for alarm. The ellipsoid method doesn't apply to the variant of linear programming that occurs in cryptography. Furthermore, the kinds of linear-programming problems for which it does work better than the simplex method are those which are far larger than any that will ever occur in practice. Best of all, following a tradition of international communication in the mathematics community that dates back thousands of years, the Russian mathematicians made no attempt to keep the ellipsoid algorithm a secret.

In fact, the impact of ellipsoid algorithm on American mathematics was undoubtably positive. Although it does not seem to be a practical algorithm, it did break a certain theoretical barrier, thus creating an incentive for other researchers. Narendra Karmarkar, an American mathematician working at Bell Laboratories, broke through this barrier in another way, finding what seems to be a truly practical method for linear programming.

BOX 4.4 Finding Fast Algorithms Means Better Airline Service

Thomas Cook is Director of Operations at American Airlines. Linear-programming techniques have a direct impact on the efficiency and profitability of major airlines, and Cook shares his ideas on why optimal solutions are essential to his business:

> Finding an optimal solution means finding the best solution. Let's say you are trying to minimize a cost function of some kind. For example, we may want to minimize the excess costs related to scheduling crews, hotels, and other costs that are not associated with flight time. So we try to minimize that excess cost, subject to a lot of constraints, such as the amount of time a pilot can fly, how much rest time is needed, and so forth.
>
> An optimal solution, then, is either a minimum-cost solution or a maximizing solution. For example, we might want to maximize the profit associated with assigning aircrafts to the schedule; so we assign large aircraft to high-need segments and small aircraft to low-load segments. Whether it's a minimum or maximum solution depends on what function we are trying to optimize.
>
> Finding fast solutions to linear-programming problems is also essential. If we can get an algorithm that's 50 to 100 times faster, we could do a lot of things that we can't do today. For example, some applications could be real-time applications, as opposed to batch applications. So instead of running a job overnight and getting an answer the next morning, we could actually key in the data or access the data base, generate the matrix, and come up with a solution that could be implemented a few minutes after keying in the data.
>
> A good example of this kind of application is what we call a major weather disruption. If we get a major weather disruption at one of the hubs, such as Dallas or Chicago, then a lot of flights may get cancelled, which means we have a lot of crews and airplanes in the wrong places. What we need is a way to put that whole operation back together again, so that the crews and airplanes are in the right places. That way, we minimize the cost of the disruption and minimize the passenger inconvenience.

We're working on that problem today, but in order to solve it in an optimal fashion, we need something as fast as Narendra Karmarkar's algorithm. In the absence of that, we'll have to come up with some heuristic ways of solving it that won't be optimal.

The simplex method, which was developed some 40 years ago by George Dantzig, has been very useful at American Airlines and, indeed, at a lot of large businesses. The difference between his solution and Karmarkar's is that if we can get an algorithm that comes up with basically the same optimal answer 50 to 100 times faster, then we can apply that technology to new problems, and even to problems that we wouldn't have tried using the simplex method. I think that's the primary reason for the excitement.

REVIEW VOCABULARY

Corner principle   The principle that states that there is a corner point of the feasible region that yields the optimal solution.

Feasible point   A possible solution (but not necessarily the best) to a linear-programming problem.

Feasible region   A representation of the feasible set as a portion of *n*-dimensional space.

Feasible set   The set of possible solutions to a linear-programming problem.

Linear programming   A set of organized methods of decision making in which the mathematical formulation of the problem involves only linear equations and inequalities.

Mixture problems are usually solved by some type of linear programming.

Mixture problem   A problem in which a variety of resources are available in limited quantities and in which these resources can be combined in various ways to make different products. It is usually desired to find the way of combining them that produces the most profit.

Profit line   The set of all feasible points that yield the same profit (for two-dimensional linear-programming problems).

Simplex method   One of a number of solution methods for linear-programming problems.

EXERCISES

1. Make a mixture chart for the following problem. A cereal manufacturer has on hand 800 lb of wheat and 40 lb of sugar. He can make a box of Hefties from $\frac{1}{2}$ lb of wheat and no sugar and he can make a box of Sweeties from $\frac{4}{9}$ lb of wheat and $\frac{1}{10}$ lb of sugar. The profit on a box of Hefties is \$0.10 while the profit on a box of Sweeties is \$0.13. How much of each product should he make to earn the maximum profit without exceeding the materials on hand?

2. Make a mixture chart for the following problem. A paper recycling company uses scrap paper and scrap cloth to make two different grades of recycled paper. A single batch of grade A recycled paper is made from 40 lb of scrap cloth and 180 lb of scrap paper, whereas one batch of grade B recycled paper is made from 10 lbs of scrap cloth and 150 lb of scrap paper. The company has 100 lb of scrap cloth and 660 lb of scrap paper on hand. A batch of grade A paper brings a profit of $500, whereas a batch of grade B paper brings a profit of $250. What amounts of each grade should be made?

3. In the example of the juice manufacturer, how does the profit formula change if the profit per gallon doubles for both products? Do you think this change will affect the mixture that gives the maximum profit?

4. Sketch the general appearance of the mixture chart for a firm that uses 6 resources to make 4 products.

5. For each of the following points, determine whether it is feasible or infeasible in connection with the example in Figure 4.4d: (30, 30), (−10, 60), (70, 90), and (10, 10).

6. For each of the following points, all of which are feasible for the example of Figure 4.4d, compute the profit using the formula on the mixture chart of Figure 4.1: (20, 25), (0, 50), and (30, 10).

7. Here is the feasible region for the plywood manufacturing process (see Figure 4.2 for the mixture chart). For each corner point shown in the following, say how many panels of each type that point represents. Find how many sheets of each type of plywood must be made in order to get the most profit.

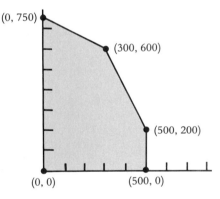

8. What role does the mixture chart play in relation to the feasible region? Does one determine the other?

9. Suppose the available amounts of apple juice and cranberry juice are doubled. Using algebra, we can show that doubling resources causes the two boundary lines of the feasible region that don't contain (0, 0) to move to new positions parallel to the original positions but farther from the origin. Can you predict what effect this will have on the best profit?

10. Discuss the validity of the corner principle if the solution (x, y) to the mixture problem is required not only to lie in the feasible region but also to have integer coordinates.

11. In our colored-light-bulb visualization, we superimposed a straightened-out rainbow on the feasible region. Suppose it were not straightened out, as in the following figure, and we still wanted to find the hottest-colored point. Would the corner principle still hold? (This is an example of nonlinear programming.)

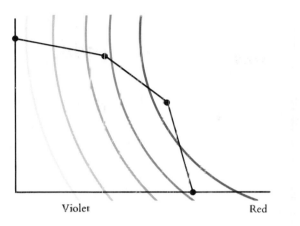

Violet   Red

12. In a mixture problem, resources must be limited during the time period for which one seeks the ideal production levels for the products. Discuss whether this is or is not a reasonable assumption for the examples in this chapter.

13. In the real world, firms often give discounts for large-volume purchases. Does this necessarily contradict the assumption of a fixed (constant) profit on each unit sold?

14. We learn in economics that prices are determined by the interplay of supply and demand. For example, prices may be lowered if much is produced. However, in mixture problems we assume a fixed (constant) profit regardless of how much is produced. Is there a contradiction here?

15. In economics, it is often useful to distinguish between a firm that has a monopoly (for example, is the only supplier of a product) and firms that supply only a small share of the market. How would the presence of a monopoly affect the relation between production and price, and what effect would this have on the fixed-profit assumption of linear programming?

16. Devise and describe a mixture problem for the production of cakes. State exactly what the products, resources, and profits are. You need not specify the recipes numerically or the amounts of the resources available.

PART II

STATISTICS: THE SCIENCE OF DATA

Statistics provides quantitative insights into the similarities and differences exhibited by populations, whether they be composed of people, animals, or things. [Photo by Mark Antman.]

Numerical facts, or data, make up an increasing part of the information we need in order to understand our world. We are told that a car gets 29 miles per gallon, or we read that the rate of inflation dropped two-tenths of a percent last July. This kind of information allows us to make informed, intelligent decisions. How we will manage our own lives depends on knowing which car yields the best overall mileage for the price or whether the state of the economy is conducive to starting a new business. Opinion polls, market research, and government efforts to collect economic and social information all produce data. *Statistics* is the science of data — of gathering data, of putting them into clear and usable form, and of interpreting them to draw conclusions about the world around us.

The information researchers gather may be as vital as the fact that 7.2% of the American labor force is out of work, or as superficial as the fact that the average U.S. household watches television for exactly seven hours and two minutes each day. But in each case we must know how to collect accurate numbers, how to describe the resulting data briefly and clearly, and how to draw conclusions. In fact, collecting data, finding ways to display them, and making intelligent predictions based on them are the building blocks of the science called statistics.

Collecting Data

The news media often present information in the form of numbers. Headlines announce that the unemployment rate has dropped to 7.2%. The Gallup poll claims that 45% of Americans are afraid to go out at night because of crime. Where do these numbers come from? Most people aren't personally interviewed to determine their employment status. And the Gallup poll asks only a few of us if fear of being mugged or assaulted keeps us indoors at night.

The unemployment rate and other data about the labor force are estimated by the Bureau of Labor Statistics from data gathered by the Current Population Survey. The Census Bureau conducts this survey, interviewing 60,000 households each month. Considering that the country has over 80 million households, it seems remarkable that data about such a small group — less than 1 in 1000 households — could represent the unemployment rate of the whole nation. How can infor-mation collected from a small number of people justify conclusions about a much larger group?

Even when numbers are not in the headlines, they often affect our lives. When a popular antiarthritis drug is recalled because of doubts about its safety or when the Food and Drug Administration approves a powerful new antiulcer medication for prescription use, these decisions are based on clinical trials using relatively few patients. Public confidence about the safety and effectiveness of new drugs relies on statistical conclusions, or *inferences,* drawn from this clinical data.

Any use of numbers as evidence depends on accurate observations and on the proper collection of data. The straightforward way to calculate the nation's unemployment rate, to screen for reactions to drugs, or to gauge public opinion would be to interview or observe each member of the population separately.

However, the cost and logistics of such a large-scale operation would be unwieldy and obviously impractical. Instead, researchers must take a sample of the population and analyze it.

We often draw conclusions about a whole on the basis of a sample. Everyone has sipped a spoonful of soup and judged the entire bowl on the basis of that taste. But a bowl of soup is homogeneous; the taste of a single spoonful represents the whole. Many other factors, whether fear of crime or reaction to a drug, are not so easily gauged. They vary greatly from person to person and from region to region.

Dealing with this kind of variability is the central task of statistics. The first step is to collect data that represent a large group of individuals. To do this, we must state carefully which group we want information about. Statisticians call this group the **population.**

For example, if we want to measure unemployment, we must first define the population we want to describe. Which age groups will we include? Will we include illegal aliens or people in prisons? The Bureau of Labor Statistics must answer these questions in order to collect monthly unemployment information (see Box 5.1). In fact, the Current Population Survey defines its population as all U.S. residents (whether citizens or not) 16 years of age and over who are civilians and are not institutionalized. The unemployment rate published

---

BOX 5.1 Statistician Heads Agency That Measures U.S. Work Force

Dr. Janet Norwood,
U.S. commissioner of labor statistics.

Dr. Janet L. Norwood's position as commissioner of labor statistics makes her one of the nation's most influential statisticians. As head of the Bureau of Labor Statistics, she supervises the collection and interpretation of data on employment, unemployment, prices, wages, and productivity. These data have a large impact on the U.S. economy. For example, price levels collected by the bureau are used to adjust many federal and private payments for the effect of inflation, including social security payments and union wage scales.

Data on unemployment and prices represent politically sensitive issues. For this reason, the agency collecting and interpreting the data must remain independent and objective. To safeguard the independence of the Bureau of Labor Statistics, the commissioner is appointed by the President and confirmed by the Senate for a fixed term of four years. "That is tremendously important in protecting the basic integrity of the Bureau and the office," Norwood says. "There have been times in the past when commissioners have been in open disagreement with the Secretary of Labor or, in some cases, with the President. We have guarded our professionalism with great care."

Norwood has served as commissioner of labor statistics since 1979 and was reappointed for a second term in 1983. Her tasks require statistical skill, administrative ability, and a facility for working with both Congress and the President.

in newspapers and in other reports always refers to this specific population.

The clinical testing of a new drug for treating ulcers would also be targeted for a particular population. This population could consist of all individuals who have ulcers, but it might be limited to patients with specific types of ulcers or to those requiring hospitalization. The detailed plan, or *protocol,* of the clinical trial must specify exactly the eligible population.

Sampling

A population need not consist of a group of people; it may also consist of a group of objects. Let's say we want to test a population of newly produced fuses to see if they blow under excessive current, as they should. In this case, gathering information about every member of the population would destroy the entire population. In other cases, such as taking inventory of all items in a large warehouse, boredom and fatigue could prevent an accurate accounting. In such situations, as well as in population surveys and clinical trials, information is gathered about only a few items. The part of the population used to draw conclusions about the whole is called a **sample.**

How can we choose a sample that is truly representative of the population? The easiest way to select a sample is to earmark individuals close at hand. If we are interested in finding out how many people have jobs, for example, we might choose a busy intersection and ask people passing by if they are employed.

But there are many problems inherent in such a method. When we exercise *personal choice* in selecting people, we might feel safer approaching well-dressed, middle-income subjects and therefore avoid dangerous, unfriendly, or brusque individuals. It is also likely that people who are currently employed would be at work, not shopping or sightseeing

in the middle of the day. In short, our intersection interviews would not accurately reflect the nation's rate of unemployment.

If we sample individuals on the basis of friendliness, safety, or income level — whether consciously or unconsciously — we will leave a certain segment of the population unrepresented. In our street-corner survey, for example, we would overrepresent the opinions of middle-class or affluent people and underrepresent those of blue-collar or poorer workers. Such a systematic difference between the results obtained by sampling and the truth about the whole population is called **bias.** To collect accurate data, we must follow procedures expressly designed to eliminate bias. In particular, statisticians go to great lengths to eliminate the role of personal choice in the selection of the sample to be measured.

Other sampling methods are biased because they rely on *voluntary response.* For example, some television stations poll the public daily, asking a specific question on the 6 o'clock news and reporting the responses on the 11 o'clock news. Viewers are urged to call special 900-prefix telephone numbers with their responses: dialing one number indicates a "yes" reply; dialing the other, "no." The system works so that talking is not necessary. Completing the telephone call registers a viewer's answer.

It is not difficult to see the potential sources of bias in such a poll. Households without telephones are automatically excluded. Whereas 96% of U.S. households had telephones in 1980, about one-quarter of those in Alaska and one-fifth of those in Mississippi did not. Because dialing the special 900-prefix number incurs a small charge, people in lower-income brackets may be reluctant to pay a fee in order to telephone the station.

The more-serious potential bias in call-in surveys is that the respondents select themselves. Only those who go to the time, trouble, and expense of calling are counted. Any sample chosen by voluntary response draws people with strong feelings, usually negative feelings. So when the newscaster asks the audience if they are afraid to go out at night because of crime, people angry about crime are more likely to call in than those who are not.

It is also possible that the poll could be manipulated. If crime is a local issue with political impact, one political party could arrange for its workers to spend the evening dialing in. Or, because talking isn't needed for registering a response, a computer could be programmed to dial the 900-prefix number repeatedly.

Personal choice is a common cause of bias in sampling. Whether it is expressed in the way an interviewer or study team selects its subjects or in the respondent's voluntary reply, personal choice can distort the results. To reduce bias, it is vital to reduce the influence of personal choice. The way a statistician eliminates personal choice is to select the sample by *chance*.

Figure 5.1   A sampling demonstrator.

Random Sampling

Imagine that we have a glass box containing thousands of beads. All these beads are identical except that most of them are light and some of them are dark. The beads form a population. We can even say that the box represents working-age Americans and that the dark beads represent those who are unemployed (see Figure 5.1).

Our task is to figure out what percentage of the population of beads is dark without examining each bead individually. Using a scoop with 50 recesses in it, we can thoroughly mix the beads in the box and then draw out a sample. Each bead then has the same chance as every other bead of being selected, and each group of 50 beads also has the same chance as any other group of 50 to be selected.

The first time we draw out a sample, we get 12 dark beads. Because 12 out of 50 is equiva-

lent to the fraction $\frac{24}{100}$, this represents 24%. This is an example of a **simple random sample** — a sample drawn in such a way that every possible sample from the population has the same chance to be the sample actually chosen. Because every bead has an equal chance to be chosen, the simple random sample has eliminated bias in selecting the sample. In a survey of employment, a simple random sample gives everyone — rich or poor, male or female, employed or unemployed — an equal chance to be selected.

Because the results of simple random sampling are free of bias, we can use the sample percentage to estimate the truth about the entire population. If 24% of the beads in the sample are dark, we can estimate that 24% of the beads in the box are dark. Here is an important statistical technique: to estimate a characteristic of a population, take a simple random sample and use the sample characteristic as the estimate.

In principle, we now know how to estimate the nation's rate of unemployment. Put everyone's name in a hat, mix the names well, and draw out 60,000 names. Because this is a simple random sample, the percent unemployed among those 60,000 persons is an unbiased estimate of the percent unemployed in the entire labor force.

In practice, however, when a population is too large to fit into a hat, other methods of sampling are needed. For example, each member of the population can be tagged with a numerical label, and then a sample can be chosen from the set of numerical labels, as long as every label has an equal chance of being selected. If the population is relatively small, tagging and drawing samples can be done easily with a table of random digits such as Table 5.1. For larger populations, a computer can be programmed to generate random numbers and then to print out the sample chosen. Box 5.2 explains how statisticians use tables of random digits.

TABLE 5.1 Random digits

101	03918	86495	47372	21870	28522	99445	38783	83307
102	10041	35095	66357	64569	08993	20429	28569	63809
103	43537	58268	80237	17407	89680	04655	24678	61932
104	64301	47201	31905	60410	80101	33382	95255	10353
105	43857	42186	77011	93839	28380	49296	63311	49713
106	91823	39794	47046	78563	89328	39478	04123	19287
107	34017	87878	35674	39212	98246	29735	09924	27893
108	49105	00755	39242	50472	39581	44036	54518	46865
109	72479	02741	75732	99808	02382	77201	44932	88978
110	84281	45650	28016	77753	39495	41847	19634	82681
111	61589	35486	59500	20060	89769	54870	75586	07853
112	25318	01995	87789	41212	74907	90734	31946	24921
113	40113	37395	51406	98099	43023	70195	07013	72306
114	58420	43526	15539	24845	15582	16780	95286	69021
115	18075	45894	09875	42869	20618	07699	80671	54287
116	52754	73124	93276	71521	59618	44966	37502	15570
117	05255	53579	03239	99174	75548	95776	42314	13093
118	76032	35569	23738	38092	74669	00749	17832	64855
119	97050	31553	32350	51491	53659	89336	36912	05292
120	29020	43074	84602	95131	22769	44680	68492	33987
121	28124	29686	63745	12313	15745	11570	20953	17149
122	97469	41277	90524	36459	22178	63785	20466	67130
123	91754	40784	38916	12949	76104	20556	34001	59133
124	84599	29798	57707	57392	91757	76994	43827	69089
125	06490	42228	94940	10668	62072	58983	10263	08832
126	30666	02218	89355	76117	75167	69005	42479	79865
127	87228	15736	08506	29759	74257	85594	75154	48664
128	45133	49229	32502	99698	68202	44704	39191	73740
129	55713	98670	57794	64795	27102	83420	26630	95009
130	20390	38266	30138	61250	07527	02014	43972	49370
131	13400	68249	32459	41627	56194	93075	50520	96784
132	08900	87788	73717	19287	69954	45917	80026	55598
133	86757	47905	16890	99047	78249	73739	97076	00525
134	19862	54700	18777	22218	25414	13151	54954	80615

continued

TABLE 5.1 *continued*

135	96282	11576	59837	27429	60015	40338	39435	94021
136	17463	26715	71680	04853	55725	87792	99907	67156
137	44880	55285	95472	57551	24602	98311	63293	58110
138	61911	78152	96341	31473	58398	61602	38143	93833
139	07769	22819	58373	88466	71341	32772	93643	92855
140	73063	63623	29388	89507	78553	62792	89343	27401
141	24187	60720	74055	36902	22047	09091	79368	35408
142	06875	53335	91274	87824	04137	77579	54266	38762
143	23393	37710	46457	03553	58275	11138	18521	59667
144	00980	73632	88008	10060	48563	31874	90785	78923
145	46611	39359	98036	25351	88031	72020	13837	03121
146	56644	79453	49072	30594	73185	81691	29225	70495
147	98350	36891	04873	71321	29929	37145	95906	41005
148	17444	61728	86112	76261	92519	61569	65672	95772
149	45785	21301	89563	23018	60423	50801	70564	45398
150	54369	08513	36838	19805	67827	74938	66946	01206

BOX 5.2 Random Numbers and Random Sampling

Imagine a hat containing 10 identical tags, each bearing a different digit: 0, 1, 2, 3, 4, 5, 6, 7, 8, 9. Mix the tags thoroughly and draw one. Each of the 10 digits is equally likely to be selected, so the probability of each is $\frac{1}{10}$.

Return the tag to the hat, mix thoroughly, and draw again. Again, each digit has probability $\frac{1}{10}$. And on the two draws together, each of the 100 possible outcomes 00, 01, . . . , 99 is equally likely and thus has probability $\frac{1}{100}$. Repeat this process many times, and your results form a table of random digits (see Table 5.1). A table of random digits is a string of digits, each chosen independently of all the others by a chance mechanism that gives equal probability to each outcome. Such a table offers a shortcut to doing our own mixing and drawing in choosing a simple random sample.

Example. The last 50 cars produced on an assembly line need to be sampled as part of a program to improve quality. Six of the 50 will be pulled off the line for a detailed inspection. To avoid bias, a simple random sample will be chosen.

STEP 1. Give each car a numerical label of the same length. Because two digits are needed to label 50 cars, *all labels* will have two digits.

STEP 2. Look at successive two-digit groups in Table 5.1. All pairs of digits are equally likely to occur in the table, so we obtain a simple random sample. A pair of digits that was not used as a label or that duplicates a label already in the sample is simply ignored. Starting at line 140 (any line will do), we read:

$$73063 \quad 63623 \quad 29388$$

The cars labeled 06, 36, 23, 29, and 38 are chosen. The initial 73 is ignored because it is not a label, and the second 36 is ignored as a duplicate.

Clearly, the Bureau of Labor Statistics doesn't use beads in a box or names in a hat to determine who will be interviewed for the Current Population Survey. In fact, the bureau doesn't even use simple random samples. Both the Gallup poll and the Bureau of Labor Statistics must use more complex versions of random sampling. Chance still determines the sample, but the process of selecting from the entire nation is done in stages instead of all at once. The Census Bureau, under contract with the Bureau of Labor Statistics, does the job. The entire country is first divided into primary sampling units, or PSU's (see Figure 5.2). Each PSU is a group of contiguous counties. The Census Bureau selects a random sample of PSU's. Within each PSU selected, smaller areas of about 500 inhabitants, called census enumeration districts, are chosen. Finally, the Census Bureau selects, also at random, individual households within each district.

These *multistage random samples* offer several practical advantages over simple random samples. For one, they do not require listing every household in the nation. Moreover, the households to be interviewed are clustered together in relatively few locations, so that the travel costs for the interviewers are reduced. The price paid for practicality, however, is complexity in actually choosing the sample and in interpreting the results (see Box 5.3). Because simple random sampling is the essential principle behind all random sampling and because

Figure 5.2   A primary sampling unit for the Current Population Survey.

it is also the main building block of more complex designs, we will focus our study on simple random sampling.

Sampling Variability

Our first sample from the box of beads contained 12 dark beads. If we draw out another sample, we will probably not get 12 dark beads

BOX 5.3 Sampling and Surveys: Taking the National Pulse

Dr. Janet L. Norwood, commissioner of labor statistics, shares a number of observations on her agency and on the task of sampling the nation at large through the Current Population Survey.

How We Go About the Business of Sampling

The household survey is a basic labor-force survey that provides us with an enormous amount of data. Specifically, it provides information on the demographic characteristics of people and about their labor-force status: their employment, the length of time they are employed, whether they are job seekers, whether they are job losers, whether they are employed part-time or full-time, and so on. That survey, which we call the Current Population Survey, is conducted for us on contract by the Bureau of the Census. Interviewers go out to a sample of 60,000 households spread throughout the United States in more than 400 areas of the country. The reason that the sample is so large — and it is a very large sample for a household survey — is that so much data comes out of it. We do not set out simply to collect data on how many people are employed or unemployed. We also must have information on men versus women, on blacks, on whites, on Hispanics, and on trends in the youth population. We need to have some information by region and by state. And as you get down to smaller and smaller groups, it is necessary to have a sample size that is sufficient to provide adequate data.

The first thing we do is develop a basic list for sampling purposes, the universe from which the sample will be selected. In the case of the Current Population Survey, the basic universe is the decennial census. Between the census years this list is updated by new building that occurs throughout the country. We work quite hard to be sure that this is accurate as the basic universe. Then from that universe, we select our samples.

On the Impact of Our Agency

Our price data are probably among our best-known indicators. This is because the consumer price index (CPI) — the best measure we have in this country of inflation — is used in a number of government programs as an escalator to keep income up with inflation. In fact, if you take dependence into account, we estimate that more than half the population of this country has income in some way affected by the CPI. And in 1985, income tax brackets will have begun to be adjusted by the CPI for purposes of calculating income tax.

Dealing with Uncertainty

Let's face it. Everyone wants to live in a world of certainty. Even the statistician wants to live in a world of certainty. Our job is to explain to people that there is no such thing as absolute certainty. Every word we use is looked at to be sure that we understand it and that someone looking at it isn't going to read more into it than we intended.

What we *can* do is to provide a set of statistics that have some error surrounding them, something that we often call sampling error or variance, which will tell what the basic tolerances are. If the unemployment rate, for example, moves by one-tenth of a point, we in the Bureau of Labor Statistics say that it was about the same. If it goes up or down two-tenths of a point, we will say that there has been a change, because the sampling error surrounding that number is such that a two-tenths change is outside of that limit.

If somebody wants to know something about BLS data, they pick up a phone and they call me or they call someone on my staff who is an expert in the area. We have made it our business to provide information as rapidly as we can and as openly as we can. I happen to believe that a statistical agency can only be a good one if it is completely open.

---

again. In fact, when we remix the beads and take a second sample, we find that only 8 out of 50 or 16% are dark, a figure very different from the 24% we got earlier.

Suppose that the Gallup poll conducted its weekly survey of public opinion twice, selecting two random samples, sending out interviewers, and asking the same question of the two samples. Two random samples, each selecting 1500 of the 165 million U.S. residents age 18 and over, will certainly contain different people. And two distinct groups of people will, of course, have somewhat different opinions.

Gallup's announcement, then, that 45% of Americans hesitate to go out at night because they are afraid of crime refers to the 1500 specific individuals in a specific sample. A second sample will no doubt produce a different result. If the Current Population Survey were conducted twice this month, the two samples would differ, producing two different officially announced unemployment rates. Random sampling eliminates bias, but it does not eliminate *variability*. The variation from sample to sample in repeated random samples is an inevitable consequence of variability in the population.

So how can we trust the result of a random sample, knowing that a second sample will give a different result? How can we base economic and political decisions on the unemployment rate, knowing that the rate will vary if sampling is repeated? In fact, random sampling *can* be trusted. To see why, we need to look more closely at sampling variability.

As we have seen, there are different kinds of variability. The answers obtained by sending an interviewer to a busy intersection vary in a haphazard and unanalyzable way. However, repeated random samples vary according to the way the mechanism of chance operates in random selection, and the long-run results are not haphazard. We see such variability in the results of repeated coin tosses or of successive spins of a roulette wheel. Tossing a balanced coin 1500 times is much like choosing a simple random sample of 1500 from a large population, if we imagine that this population is evenly divided over a particular issue, so that heads represents "yes" and tails represents "no."

Let's look more closely at the outcomes of coin tossing. It is unlikely that a person tossing a coin 1500 times will score exactly 750 heads (a 50% "yes" response). The first sample might produce 732 heads (49% in favor), a second, 781 heads (52% in favor), and so on. But coin tossing is subject to the laws of *probability*, which dictate how frequently each outcome will occur in the long run (see Chapter 7). In particular, 1500 tosses of a fair, or balanced, coin will almost never (only 1 chance in 1000) produce less than 46% or more than 54% heads.

Therefore, 1500 tosses of a balanced coin can be trusted to give a result close to 50% heads. Similarly, a Gallup poll of 1500 Americans can be trusted to give a result close to the result obtained by polling all 165 million adult Americans. Probability deals with phenomena that are variable but that nonetheless show a regular pattern in the long run. Tossing a coin and choosing a random sample are such phenomena.

In the case of choosing a simple random sample of 50 beads from a box, we can easily take repeated samples and observe the pattern of outcomes. In 30 samples, the first had 24% dark beads, the second had 16%, the third 26%, and so on. After scooping out the 30 samples, we find that 6 samples each contained 8 beads, or 16%, and that 7 samples each contained 12 dark beads, or 24%. We can display the outcomes of all 30 samples in a **histogram** (Figure 5.3). The height of each bar in the histogram shows how often the outcome marked at the base of the bar occurred.

The histogram for our 30 samples in Figure 5.3 shows a lot of scatter, but most of the samples cluster around the 20% mark. In fact, 20% is the actual proportion of dark beads in the box from which these samples were drawn. Only 7 of the samples we took contained exactly 20%, but 20 of the samples had between 16% and 24% dark beads.

The results of random sampling usually do come quite close to the truth about a population. And as we'll discover in Chapter 7, the laws of probability make possible exact statements about the accuracy of simple random sampling. We will see, for example, that the outcome of a sample of size 1500 is much more likely than that of a sample of size 50 to come close to the truth.

The deliberate use of chance in collecting data is one of the fundamental ideas of statistics. It is important as a way of removing bias. More important, the proper use of random samples enables us to draw confident conclusions from data, and the mathematics of probability provides support for these conclusions.

As we saw earlier, simple random sampling is the building block of all sampling designs. Thus, even though the Current Population Survey and the Gallup poll use more complex random-sampling procedures, the behavior of their multistage random sampling is still described by the laws of probability. Consequently, both simple and multistage random samples usually give results close to the truth about the population. A careful statement of a sample result expresses "usually" in terms of probability and "close to the truth" in terms of a margin of error.

For example, the Gallup poll takes a sample of about 1500 people each week. The probability is 0.95 that a sample percentage is within ±3% of the true population value. If the poll reveals that 45% of the sample fear to go out at night, we can be quite confident that between 42% and 48% of the entire population share that fear.

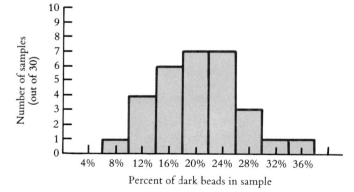

Figure 5.3 Histogram of the outcomes of a sampling demonstration.

The variability in sampling has not vanished, but probability allows us to describe the variability by announcing a margin of error that most samples will meet. In the case of the Gallup poll, 95% of their many sample results fall within a margin of error of ±3 percentage points about the truth. Gallup and other polling organizations report these facts in their press releases, but news editors often cut that part of the story.

What about variability in the unemployment rate? Here are the facts in this case. The Census Bureau prefers to announce the margin of error that 90% of Current Population Survey samples will meet: it is about ±0.2%, or two-tenths of one percent. Thus, when the unemployment rate drops from 8.4% to 8.3%, sampling variability may well account for the change. But a change from 8.4% to 8.0% very probably reflects a real drop in the percent of the American labor force who are without a job.

The smaller margin of error in the unemployment rate is due to the larger sample size. An error of ±3% is acceptable in polling public opinion, but an unemployment rate of 8% ± 3% would be of little value. Because data on employment and unemployment are important for economic planning, the federal government is willing to undertake the expense of a large monthly sample. The margin of error attained with any desired probability (such as 0.90 or 0.95) depends on the size of the sample and on the exact sampling design. The mathematics of probability describes these relationships exactly.

Experimentation

We have seen that sample surveys gather information on part of the population in order to draw conclusions about the whole. When the goal is to describe the population, as the Current Population Survey's aim is to describe employment and unemployment in the United States, statistical sampling is the right tool to use.

Suppose, however, that we want to study the response to a stimulus, to see how one variable affects another when we change existing conditions. Will a new mathematics curriculum improve the scores of sixth graders on a standard test of mathematics achievement? Will taking small amounts of aspirin daily reduce the risk of suffering a heart attack? Does smoking increase the risk of lung cancer? Observational studies, such as sample surveys, are ineffective tools for answering these questions. Instead, we prefer to carry out experiments.

An **experiment** differs from observation in that the experimenter intervenes actively by imposing a *treatment* on the subjects. In sampling, on the other hand, we observe or measure the state of the subjects without trying to change that state by a treatment.

Experiments are the preferred method for examining the effect of one variable on another. By imposing the specific treatment of interest and controlling other influences, we can pin down cause and effect. A survey, in contrast, may show that two variables are related, but it cannot demonstrate that one causes the other. Statistics has something to say about how to arrange experiments, just as it specifies methods for sampling.

Suppose a local school system, concerned about the poor mathematics preparation of American children, adopts an ambitious new mathematics curriculum. After three years of the new curriculum, students completing sixth grade have an average achievement score 10% higher than they had before the treatment. The school pronounces the curriculum a success, and other systems adopt it.

This example describes an experiment with a very simple design. A group of subjects (the schoolchildren) were exposed to a treatment (the new curriculum), and the outcome (achievement-test scores) was observed. Here is the design:

$$\begin{array}{ccc} \text{Treatment} & \longrightarrow & \text{Observation} \\ \text{New curriculum} & \longrightarrow & \text{Improved test scores} \end{array}$$

Most laboratory experiments use a design like this one: apply a treatment and measure the response. In the controlled environment of the laboratory, simple designs are often adequate. But field experiments and experiments with human subjects are exposed to more variable conditions and deal with more variable subjects. They require control of outside factors that can influence the outcome. With greater variability comes a greater need for statistical design.

In this particular case, an atmosphere of concern for education brought about a number of simultaneous changes that could influence the students' achievement-test scores. Elementary teachers were given additional training in mathematics. A parent group began to provide classroom tutors to give children individual help with mathematics. Public concern led parents to pay more attention to their children's progress, and teachers to assign more homework.

In these circumstances, mathematics achievement would have increased without a new curriculum; in fact, the new curriculum could be even *less* effective than the old. In this simple experiment, we cannot distinguish the effects of the changes in parents and teachers from the effects of the new curriculum. Variables, whether part of a study or not, are said to be **confounded** when their effects on the outcome cannot be distinguished from each other.

The remedy for confounding is to do a *comparative experiment* in which some children are taught from the new curriculum and others from the old. Changes in parents' attitudes and involvement, teacher retraining, and other

such variables now operate equally on both groups of students, so that direct comparison of the two curricula is possible.

Once we decide to do a comparative experiment, we need to find a way to assign the students to the two groups. If the groups differ markedly when the experiment begins, bias will result. For example, if we allow students to volunteer for the new curriculum, only adventurous children who are interested in math are likely to sign up for our experimental treatment, and these students are likely to perform well. Personal choice will bias our results in the same way that volunteers bias the results of call-in television surveys. The solution to the problem of bias is the same for experiments and for samples: use impersonal chance to select the groups.

Let's say the school system decides to compare the progress of 100 students taught under the new mathematics curriculum, with that of 100 students taught under the old curriculum. In this case, the treatment group is selected by taking a simple random sample of size 100 from the 200 available subjects. The remaining 100 students form the *control group;* they will continue in the old curriculum.

The selection procedure is exactly the same as it is for sampling: all 200 members of the population are tagged with numerical labels, beginning with 000 and ending with 199. Next, we can use a table of random digits, inspecting only three-digit groups: the first 100 labels encountered will represent the treatment group, to be taught from the new curriculum. Repeated labels and groups of digits not used as labels are ignored. For example, if we begin at line 125 in Table 5.1, the first few students chosen are those labeled 064, 106, 102, 022, 188. (Be sure to inspect numbers according to the directions in Box 5.2.) The remaining 100 students form the control group, to be taught from the old curriculum.

The result is a **randomized comparative experiment** with two groups. The new design is more complex but is also more useful (see Figure 5.4).

Randomized comparative experiments are used whenever environmental variables, such as behavioral change in parents or teachers, threaten to confound the results. The results of such experiments are as reliable as the results of random samples, and for the same reasons: random selection is governed by the laws of probability. Randomized comparative experiments are common tools of industrial and academic research. They are also widely used in medical research. For example, federal regulations require that the safety and effectiveness of new drugs be demonstrated by randomized comparative experiments. Let's look at a typical medical experiment.

There is some evidence that taking low regular doses of aspirin will reduce the risk of

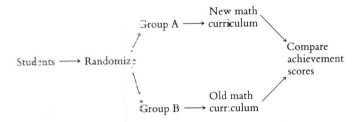

Figure 5.4   The design of a randomized comparative experiment for evaluating a new mathematics curriculum.

heart attacks. Some physicians also suspect that regular doses of beta-carotene (vitamin A) will help prevent some types of cancer. The Physicians' Health Study is a large experiment designed to test these claims (see Box 5.5). The subjects of this study are 22,000 male physicians over 40 years of age. Each physician takes

---

BOX 5.4 Sir Ronald A. Fisher, 1890–1962

Sir Ronald A. Fisher.

While employed at the Rothamsted agricultural experiment station in the 1920s, British statistician and geneticist R. A. Fisher revolutionized the strategy of experimentation. Experimenters there were comparing the effects of several treatments, such as different fertilizers, on field crops. Because fertility and other variables can change in any direction across the planted field, the experimenters used elaborate checkerboard planting arrangements to avoid bias. Fisher realized that random assignment of treatments to growing plots was simpler and better. He introduced randomization, described more complex random arrangements, such as blocks and Latin squares, and worked out the mathematics of the *analysis of variance* to analyze data from randomized comparative experiments.

Fisher contributed many other ideas, both mathematical and practical, to the new science of statistics. His influential books organized the field. Fisher was both opinionated and combative. From the 1930s until his death, he was engaged in sometimes vitriolic debates over the appropriate use of statistical reasoning in scientific inference.

BOX 5.5 Using Statistics to Study Disease

Dr. Julie Buring, associate director of the Physicians' Health Study, discusses the ways in which epidemiologists gather data on disease.

As an epidemiologist, I look at factors that are involved in the distribution and disease frequency in human populations. What is it about what we do, what we eat, what our environment is, what our occupations are, our history — our medical history, our family history — that leads one group of people to be more or less likely to develop a disease than another group of people? It is these factors that we are trying to identify.

We go at it from a couple of different angles. One is called *descriptive epidemiology*, or looking at the trends of diseases over time, trends of diseases in one population relative to another population.

Another is called *observational epidemiology*, in which we observe what people do. We take a group of people who have a disease and a group of people who don't have a disease. We look at their patterns of eating or drinking, medical history, and what their exposures may have been. The other way to go about it is to take a group of people who have been exposed to something such as smoking and a group of people who haven't and follow them up over time to see whether they develop the disease or not. Whether we do one design or another, both are observational. That means we don't interfere in the process. We just observe it.

A second type of epidemiology, of which the Physicians' Health Study is an example, is *experimental epidemiology*, sometimes called an intervention study. We take a group of people who have the treatment and a group of people who do not. The difference between the experimental and observational approach is that in experimental epidemiology, the investigators determine who will be in what treatment group, who will receive the treatment and the placebo, who will be exposed, and who won't be exposed.

From these different approaches — descriptive epidemiology and observational epidemiology — we can judge whether a particular factor causes or prevents the disease that we are looking at.

---

a pill each day over a period of several years. There are four treatments: aspirin alone, beta-carotene alone, both, and neither. The subjects are randomly assigned to one of these treatments at the beginning of the experiment.

The Physicians' Health Study introduces several new ideas important to the proper design of experiments. The first is the importance of counteracting the **placebo effect.** A placebo is a fake treatment, such as a salt pill, that looks and tastes like the real thing. The

placebo effect is the tendency of subjects to respond favorably to any treatment, even a placebo. If subjects given aspirin, for example, are compared with subjects who receive no treatment, the first group gets the benefit of both aspirin and the placebo effect. Any beneficial effect that aspirin may have is confounded with the placebo effect. To prevent confounding, it is therefore important that some treatment be given to *all* subjects in any medical experiment.

In the Physicians' Health Study, all subjects take pills that are identical in appearance, but the pills may contain either a drug or a placebo. The study was designed as a *double-blind experiment*: the subjects do not know which treatment they are receiving, because this knowledge might influence their reaction. In addition, knowing the subjects' treatments might influence the researchers who interview and examine the subjects; therefore, experimental workers are also kept 'blind." Only the study's statistician knows which group each subject belongs to.

The Physicians' Health Study (see Figure 5.6) is a more elaborate experiment than the curriculum study. Not only are four rather than two treatments compared, but two distinct experimental variables are present: aspirin or not, and beta-carotene or not. A *two-factor experiment*, or two-variable experiment, allows us to study the interaction, or joint effect, of the two drugs as well as the separate effects of each. For example, beta-carotene may reinforce (or counteract) the effect of aspirin on future heart attacks. By comparing these four groups, we can study all these possible interactions.

Statistical Evidence

A properly designed experiment, in the eyes of a statistician, is an experiment embodying the principles of *comparison* and *randomization:* comparison of several treatments and randomization in assigning subjects to the treatments. As we saw in the math-curriculum example, comparison eliminates confounding by environmental variables. The immediate appeal of randomization is the elimination of bias by creating groups equivalent in all respects save the treatment they receive.

The future health of the subjects of the Physicians' Health Study may depend on age, past medical history, emotional status, smoking

Figure 5.5   Because the Physicians' Health Study was designed as a double-blind experiment, even the experimental workers don't know who was assigned to which treatment group.

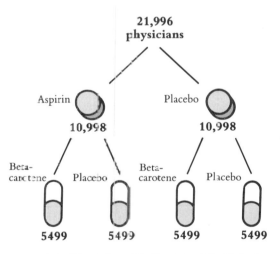

Figure 5.6   The design of the Physicians' Health Study.

habits, and many other variables — known and unknown. Randomization will, on the average, balance the groups simultaneously in all such variables. Because the groups are exposed to exactly the same environmental variables, except for the actual content of the pills, differences among the groups can be attributed to the effect of the medication. That is the logic of randomized comparative experiments.

Let's be a bit more specific: any difference among the groups is due *either* to the medication *or* to the accident of chance in the random assignment of subjects. It could happen, for example, that all the smokers were, by chance, placed in the same group. Once again, statistics calls on probability. Because chance was deliberately used in making assignments, the laws of probability tell us the likely difference among the four groups if nothing but chance were operating. As in sampling, larger numbers of subjects increase our confidence in the results. The Physicians' Health Study follows 22,000 subjects in order to be quite certain that any medically important differences among the groups will be detected and that these differences will be attributable to aspirin or to vitamin A.

The logic of experimentation, the statistical design of experiments, and the mathematics of probability combine to give compelling evidence of cause and effect. Only experimentation can produce fully convincing evidence of causation.

By way of contrast, consider the statistical evidence linking cigarette smoking to lung cancer. This evidence is based on observation rather than experiment. The most careful studies have selected samples of smokers and nonsmokers, then followed them for many years, eventually recording the cause of death. These *prospective* studies are comparative, but they are not properly randomized experiments, because the subjects themselves choose whether or not to smoke. A large prospective study of British doctors found that the lung-cancer death rate among cigarette smokers was 20 times that of nonsmokers; another study of American men aged 40 to 79 found that smokers died of lung cancer 11 times more often. Thus, the observed connection between smoking and lung cancer is high.

This connection is statistically significant: it is far larger than would occur by chance. We can be confident that something other than chance links smoking to cancer. But observation of samples cannot tell us *which* factors other than chance are at work. Perhaps there is something in the genetic makeup of some people that predisposes them both to nicotine addiction and to lung cancer. In that case, a strong link would be observed even if smoking itself had no effect on the lungs.

The statistical evidence that points to cigarette smoking as a cause of lung cancer is about as strong as nonexperimental evidence can be. First, the connection has been observed in many studies in many countries. This eliminates factors peculiar to one group of people, or to one specific study design. Second, specific ways in which smoking could cause cancer have been identified. Cigarette smoke contains tars that can be shown by experiment to cause tumors in animals. Third, no really plausible alternative explanation is available. For example, the genetic hypothesis cannot explain the rise in lung-cancer rates among women that occurred as more and more women became smokers. In 1987, lung cancer, which has long been the leading cause of cancer deaths in men, surpassed breast cancer as the most fatal cancer for women. Moreover, genetics cannot explain why lung-cancer death rates drop among smokers who quit.

This evidence is convincing to most people, and almost all physicians accept it. But it is not quite as strong as the conclusive statistical evidence we get from randomized comparative experiments.

Latin Squares

Most experiments are more complex than a simple comparative randomized design, just as most samples are not really simple random samples. If we anticipate that male and female patients will respond differently to aspirin and beta-carotene, we can first divide the pool of patients into two **blocks** (males and females) and then randomly assign patients to treatments separately within each block. The division into blocks controls one influential variable, the patient's sex, by including it in the design. Other influences are still averaged out by randomization. (The Physicians' Health Study was restricted to male subjects because there are not enough female physicians over 40 to form a second block.)

Even more common than blocking is the simultaneous study of several *factors*, or experimental variables. We saw earlier that the Physicians' Health Study was a two-factor experiment. Industry provides us with another example: a chemical engineer trying to determine the most efficient temperature-pressure setting for a production process would be foolish to rely on experiments taking into account only one variable at a time, because the most productive temperature will change according to the pressure and vice versa. Thus, an effective experiment must change both variables from treatment to treatment. An experiment can combine study of several factors with use of blocks. Because different batches of raw material may not behave identically, the chemical engineer might treat each batch as a block. The two factors temperature and pressure, are then studied within each block.

The combination of several experimental variables with the use of blocks can make experiments prohibitively large and expensive. However, *combinatorics*, the mathematical study of arrangements, can often provide clever arrangements of treatments that help hold down the size and cost of experiments. We can illustrate this by a comparison of motor oils

Some makers of motor oils claim that using their product improves gasoline mileage in cars. To test this claim, we might compare the effect of 4 different oils on mileage. However, because the car model and the driver's habits greatly influence the mileage obtained, the effect of an oil may vary from car to car and from driver to driver. To obtain results of general interest, then, we must compare the oils in several different cars and with several different drivers. If we choose 4 car models and 4 drivers, we will have 16 car-driver combinations. The type of car and driving habits are so influential that we consider each of these 16 combinations a block. If we then complete test drives with 4 oils (in random order) in each block, we need 4 × 16, or 64, test drives.

However, combinatorics gives us a way to test the effects of cars, drivers, and oils that requires only 16 test drives. Call the oils *A, B, C,* and *D* and assign one oil to each car-driver combination in the arrangement shown in Figure 5.7. If we study the arrangement, we

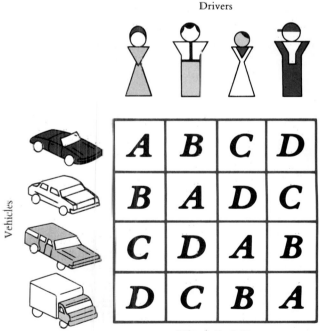

Figure 5.7   The layout of a Latin-square design to compare four motor oils in four types of car with four drivers.

find that each oil appears exactly once in each row (for each car) and also exactly once in each column (for each driver). This setup can, in fact, show how each oil performs with each car and with each driver.

An arrangement like this is called a **Latin square.** Mathematicians studied Latin squares and catalogued many of them long before the statistical design of experiments was invented. The modern use of Latin squares to design experiments illustrates how the study of pure mathematics, originally pursued simply out of curiosity, often turns out to have scientific importance.

The Latin square is a clever arrangement of comparisons. It is almost the opposite of a random assignment. However, the randomization principle of statistical design still governs the application of Latin squares: it is essential to assign the four drivers at random to the columns, the four cars at random to the rows, and the four motor oils at random to the labels *A, B, C, D.* Then the laws of probability will once more dictate whether the gas mileage figures obtained using various oils differ by a greater amount than can be accounted for by chance.

Statistics in Practice

There is more to the wise use of statistics than a knowledge of such statistical techniques as Latin-square designs. A statistician must know which techniques are applicable and appropriate. In the motor-oil experiment, we knew that cars and drivers would influence gas mileage. A Latin square allowed us to compare different motor oils without getting confused by different kinds of cars and different drivers; we

had three known variables. However, in a study of heart disease or cancer, we don't even know all the important influences; in such cases the Latin square does not apply.

Moreover, good data collection requires more than a well-designed sample or experiment. We must also measure the variables of interest. That is easy in the case of gas consumption, but measuring "intelligence" or "attitude toward abortion" is much harder. We must decide exactly what we want to measure and design a procedure for measuring it.

Even with a good design and careful procedures for measurement, other sources of error can creep in. Some of the subjects in a random sample of people may not be at home. Others may misunderstand the questions, refuse to cooperate, or not tell the truth. In experiments with human subjects, it is often hard to apply realistic treatments. A psychologist who studies stress by exposing student volunteers to an artificial situation must ask if a few hours of laboratory stress can simulate months or years of hard living and elicit the same responses.

When we are planning a statistical study, we must also answer some ethical questions. Does the knowledge gained from an experiment or study justify the possible risk to the subjects? In the Physicians' Health Study, doctors gave their informed consent to take either aspirin, beta-carotene, or placebos, in any combination, as prescribed by the study designers. Box 5.6 explains why randomized comparative experiments are the mainstay of medical, agricultural, and many other kinds of research and when such clinical trials are justified.

This chapter has isolated out some statistical ideas and simple techniques, but practical and ethical problems are never far from the surface when statistics is applied to the real world.

BOX 5.6 Experiments and Ethics

Dr. Charles Hennekens.

Dr. Charles Hennekens, director of the Physicians' Health Study, must concern himself with the goals, design, and implementation of his large-scale study. But other questions also arise in the course of such an experiment. Recently, Dr. Hennekens was asked about the ethics of experimenting on human health:

> Much has been made of the ethical concerns about randomized trials. There are instances where it would not be ethical to do a randomized trial. When penicillin was introduced for the treatment of pneumococcal pneumonia, which was virtually 100% fatal, the mortality rate plummeted significantly. Certainly it would have been unethical to do a randomized trial, to withhold effective treatment from people who need it.
>
> There's a delicate balance between when to do or not to do a randomized trial. On the one hand, there must be sufficient belief in the agent's potential to justify exposing half the subjects to it. On the other hand, there must be sufficient doubt about its efficacy to justify withholding it from the other half of subjects who might be assigned to the placebos, the pills with inert ingredients. It was just these circumstances that we felt existed with regard to the aspirin and the beta-carotene hypotheses.
>
> A randomized trial done on a newer therapy or drug is best done when the procedure is first introduced. It becomes very difficult, with regard to both feasibility and ethics, to do such trials after a long period of time has elapsed. Treatments begin to be so accepted by the population that it is difficult to find people willing to have the treatment withheld. Others might feel ethically that it would be difficult to withhold treatment.
>
> One example in contemporary times regards the treatment of breast cancer. William Halsted of John Hopkins, the father of American surgery, invented the radical mastectomy in the early 1900s as a therapy for breast cancer. It's been used widely for more than half a century. However, after all this time, it became apparent to some investigators that less extensive procedures might accomplish the same results, that is, to keep the age-specific mortality rate from breast cancer in affected women at a low level. It's only been in recent years that randomized trials of less extensive forms of treatment for breast cancer have been done. It is important that such research be done. It is optimal to do it early so that we get a clear answer at the beginning of the development of new procedures and new drugs.

Julie Buring, associate director of the Physicians' Health Study, adds:

> Sometimes, in the case where certain procedures are not tested in a trial immediately, the reason is that the procedures seem to make sense intuitively. For example, it

Numerical summaries are important in exploring data, and we will meet several in this chapter. But our focus in this chapter will be on description rather than inference, and the most powerful descriptive tool is not a number but a picture.

Pictorial display of data is not a new idea. In 1861, Charles Minard, a French engineer, used graphs to show the impact of harsh winter conditions on Napoleon's troops during the unsuccessful campaign of 1812. Some 422,000 French soldiers entered Russia, but only 10,000 struggled back. The black band across the map of Eastern Europe in Minard's graph is Napoleon's Grand Army (see Figure 6.1). The width of the band is proportional to the number of surviving troops. As the number of survivors dwindles, the band narrows down to a trickle. Below this graph, Minard drew another showing the temperature during the winter retreat from Moscow. The falling temperatures and shrinking army march together in that famous defeat.

Although visual representations of data abound in books and magazines, not all examples are as imaginative — or as convincing — as Minard's. Recently, the rise of computer graphics has given a new emphasis to pictorial display of data. The computing power of machines can now be combined with the unique ability of the human eye and brain to help us recognize hidden patterns in masses of data.

For example, in order to discover where oil reservoirs lie, geologists have to probe the structure of the earth with explosions. The shock waves of these explosions bounce off underlying rock strata and are reflected up to seismic sensors that are scattered around an exploration site and record the data. A single explosion can produce half a million separate data points, and a skilled seismic crew — operating by helicopter — can set off up to 20

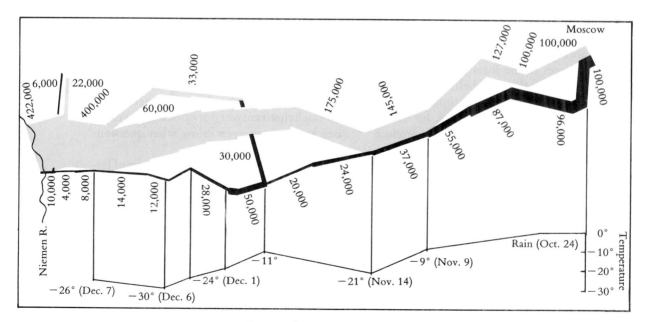

Figure 6.1   Redrawing of Minard's 1861 graph of Napoleon's Russian campaign.

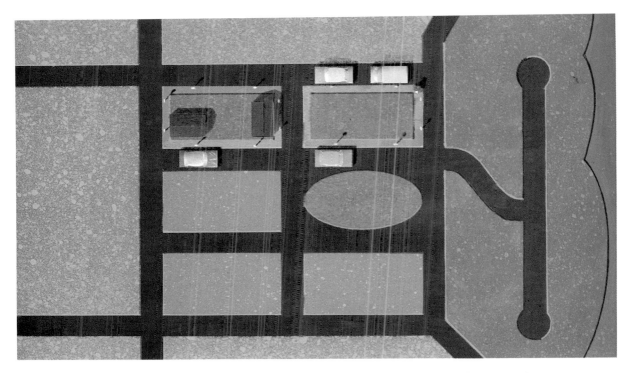

Color Plate 1a A model town showing an aerial view of its street network. (See pp. 5–8.)

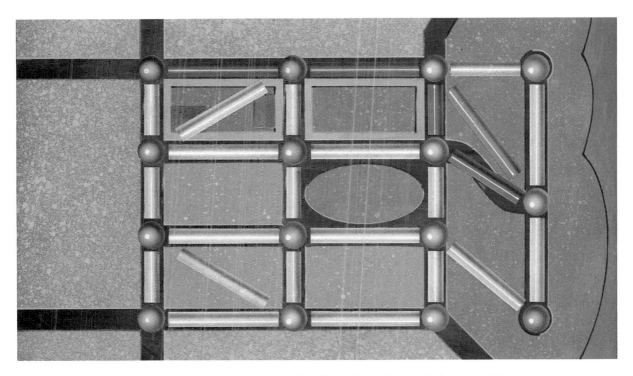

Color Plate 1b The street network in (a) overlaid with a graph. (See pp. 5–8.)

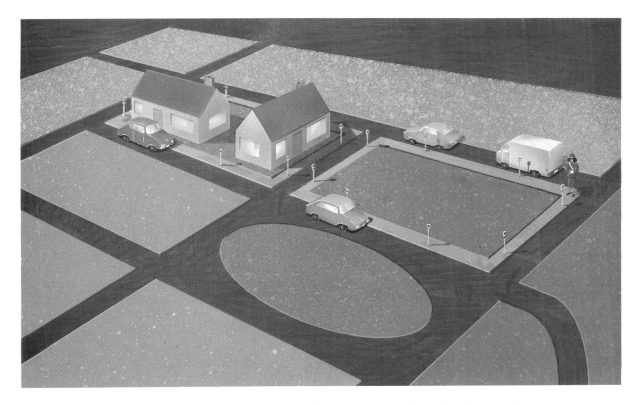

Color Plate 2 A perspective view of the model town street network showing the route of a meter maid. (See pp. 5–8.)

(a)   (b)

Color Plate 3 (a) Computer-generated street network, showing an overview of part of a city. (b) A similar computer-generated street network, but showing more detail in a zoomed in view. [Sidney Bowne & Son, Consulting Engineers.] (See p. 18.)

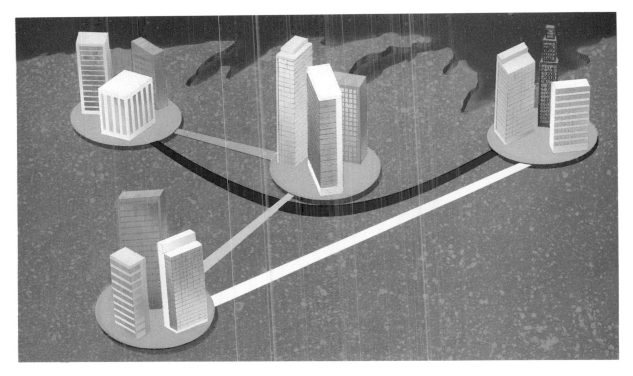

Color Plate 4a The traveling salesman problem solved by using the nearest-neighbor algorithm. (See p. 28–30.)

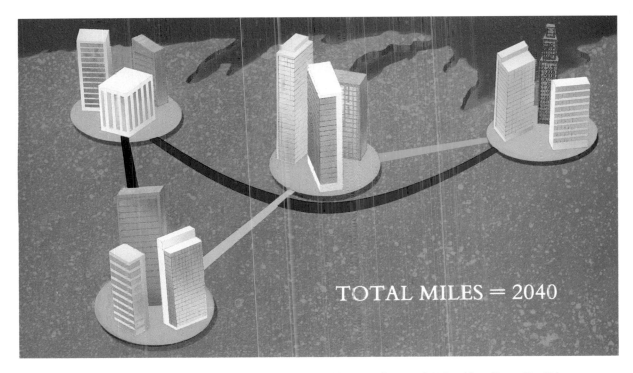

TOTAL MILES = 2040

Color Plate 4b The traveling salesman problem solved by using the "greedy" algorithm. (See p. 28–30.)

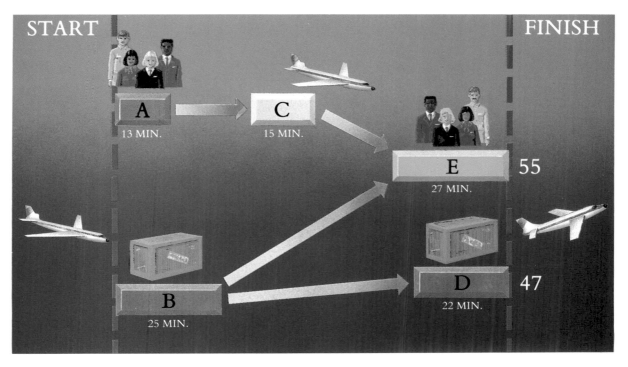

Color Plate 5a Critical path scheduling for servicing an airplane. (See pp. 36–38.)

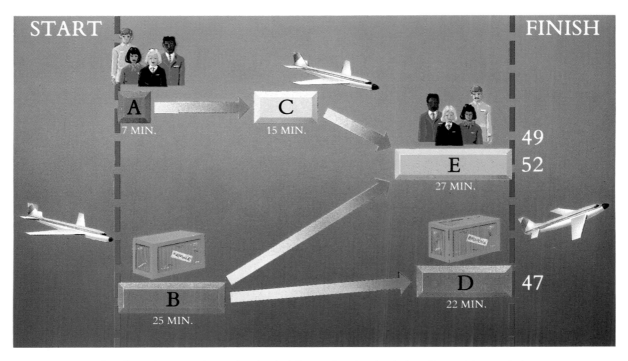

Color Plate 5b Revised critical path scheduling with reduced task times for servicing an airplane. (See pp. 36–38.)

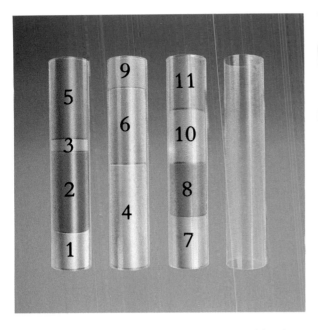

Color Plate 6a Bin packing using the method of first-fit decreasing. (See pp. 58–61.)

Color Plate 6b Bin packing using the method of first fit. (See pp. 58–61.)

Color Plate 7 Sequence of polyhedrons showing graphically how the simplex method of linear programming finds a solution. (See pp. 78–80.)

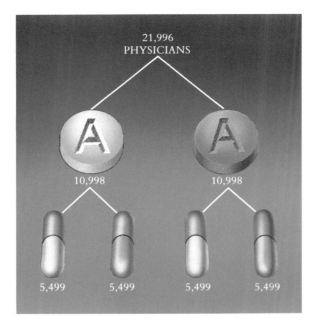

Color Plate 8 Two-part tree diagram showing the Physicians' Health Study two-factor experiment involving aspirin and beta-carotene. (See pp. 101–104.)

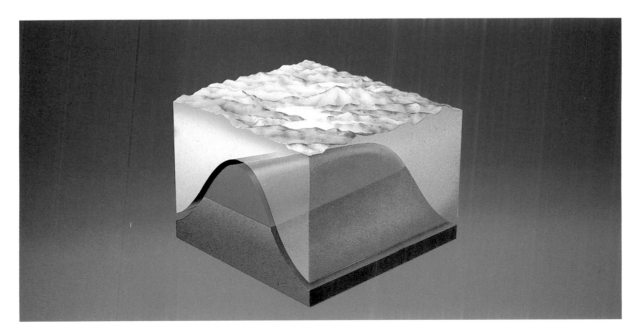

Color Plate 9 A computer graphics representation of geologic seismic data plotted in three dimensions that shows a promising oil reservoir site. (See pp. 114–115.)

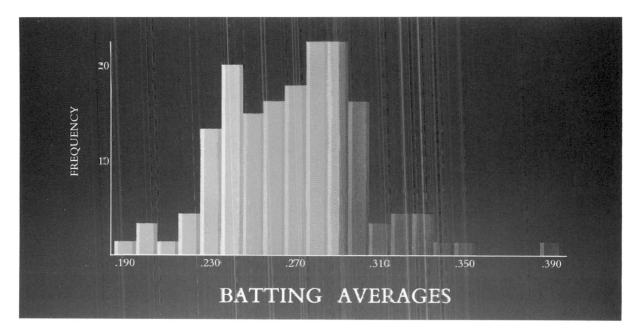

Color Plate 10 Histogram of 1980 American league batting averages. Note the outlier at .390, which was George Brett's batting average that year. (See pp. 115–117.)

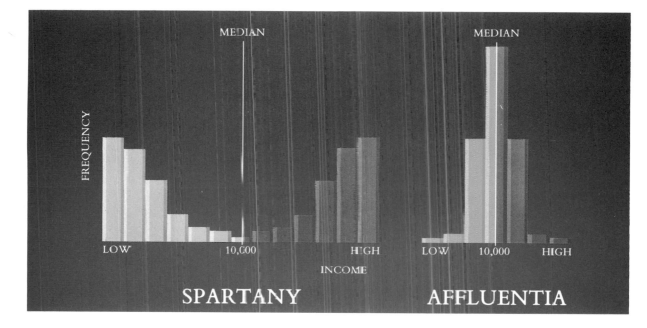

Color Plate 11 Histogram of the wage spread in Spartany and Affluentia. (See pp. 118–119.)

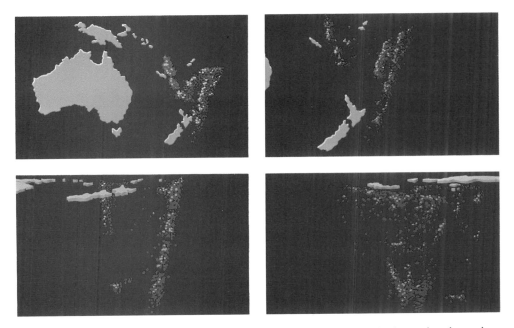

Color Plate 12 A series of computer-generated graphics displays of earthquake data gathered near the Fiji Islands. The first image shows a traditional plotting of earthquake epicenters on a two-dimensional map in which location is given in latitude and longitude. The remaining images show how the computer can manipulate the data to give a three-dimensional view of the epicenters, that is, their depth within the earth. (See pp. 126–129.)

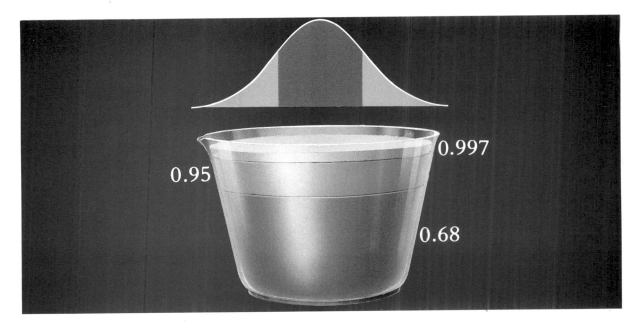

Color Plate 13 A normal distribution curve showing three standard deviations of probability 0.68, 0.95, and 0.997, respectively. This property of any normal curve is known as the 68-95-99.7 rule. The area under the curve of three standard deviations equals 99.7% of the total area. This is also shown in the nearly filled cup. (See p. 146.)

such tests a day. This abundance of data has to be interpreted into meaningful and easily understood patterns.

The results of one such analysis have been compressed and summarized using the tool of computer graphics (see Figure 6.2). An oil expert could look at one result and instantly interpret it as a promising site for an oil reservoir. A trained eye can almost "see" the oil in the computer graphics representation of the seismic data. Such analysis is an effective collaboration between the computational power of machines and the ability of humans to see and understand.

In the next two sections we will use both numbers and pictures to explore data. We will organize our thinking by moving from simple descriptions of a single variable to complex descriptions of the relations between several variables.

Describing a Single Variable

Baseball fans have long memories for the statistics of the game (see Box 6.1). Let's investigate one of these, the measure that tells us how well major league baseball players hit. First,

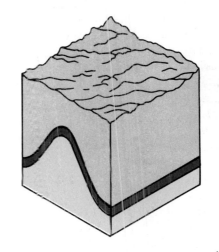

Figure 6.2 Underground structures reconstructed from seismic data.

we must refine the question. The simplest measure of how well a player hits is his batting average, which is simply the proportion of times at bat that the player gets a hit. However, the batting average of a player who has been at bat only a few times may be owed to luck—either good or bad. To be fair, we will consider only players who have been at bat 200 or more times in a season.

BOX 6.1 Baseball: The Great Statistics Game

Baseball fans have a love affair with statistics. There's absolutely no doubt about it: it's the greatest statistics game there is, probably because baseball goes back so far in this country: it's embedded in most of the population. They really understand a home run and an RBI, a batting average and an earned run average—all those basic statistics that have been with baseball throughout its history. The basics have never changed, so people know and love them.

However, in the last 20 years, many new statistics have evolved: hitting with runners in scoring position; the percentages of men driven in with runners on second and third base; a pitchers' saves, as opposed to the percentage of times he has the opportunity to make a save. These are the so-called sophisticated statistics.

DICK BRESCIANI
Red Sox Chief Statistician

George Brett of the Kansas City Royals baseball team hit .390 in 1980, the highest major league batting average since Ted Williams's average of .406 in 1941. [Kansas City Royals.]

There is a whole lore of baseball history involving statistics. The great thing is to compare the players of old with the players of today. Many times on talk shows or panels people will say, "Could Dave Winfield or Jim Rice or George Brett have played with a Ty Cobb or a Mickey Cochran or a Ted Williams or a Joe Dimaggio?" What you have to argue is statistics. You have to go back and examine Dimaggio's years in the big leagues. You look at what he did year by year: he was on average a .300-and some hitter; he drove in so many home runs; he did thus-and-so defensively in the outfield. The statistics are all that remain of the career of that player. So you lay them out against a Brett or a Winfield and try to compare them. That is the fun of the game.

LOU GORMAN
Red Sox General Manager

Figure 6.3 shows the batting averages for all 167 American League players who batted 200 or more times in the 1980 season. As we saw in Chapter 5, a graph showing frequencies of outcome for a single variable is called a *histogram*. In our example, each bar covers a 10-point range of batting averages, such as .295 to .304; the bars are labeled by their midpoints, .300, for example. The height of each bar is the count, or **frequency,** of players with batting averages in the range covered by the bar.

The overall pattern is immediately apparent: almost all major league regulars hit between about .225 and .305. A typical player hit about .270.

Another aspect of the histogram is also immediately apparent. The single observation at .390 is isolated from the rest of the group. An observation that stands apart from the bulk of the data is known as an **outlier** and warrants further investigation.

It is a fact of life that any sizable set of data is likely to contain errors that must be tracked down and corrected. Outliers are often the result of errors and must be carefully investigated. It would have been easy, for instance, to have typed .390 instead of .290 while recording the data. However, this outlier is not a mistake. George Brett of the Kansas City Royals hit .390 in 1980, the highest major league batting average since Ted Williams's average of .406 in 1941. Brett therefore deserves his isolated point on the histogram. He and his more-average colleagues illustrate an important principle in the exploration of data: always look for an *overall pattern,* and then look for *deviations* from that pattern. In short,

$$Data = smooth + rough$$

The *smooth* in any data is an overall pattern. In this histogram, the smooth is the very regular

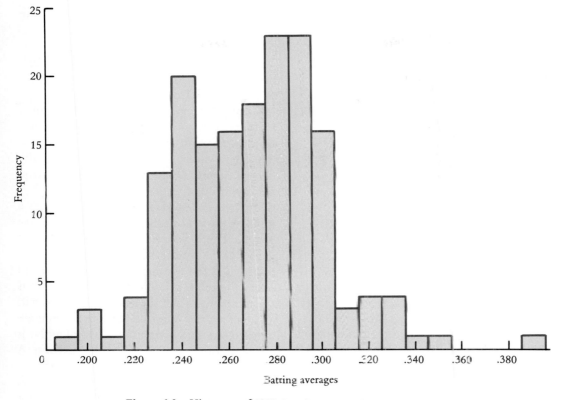

Figure 6.3 Histogram of 1980 American League batting averages.

distribution of batting averages, with a few in the .220s, increasing to a peak at .270 or .280, then falling off to a few in the .330s and above. The *rough* is the outlier recording Brett's performance. Both the smooth and the rough — the pattern and the exception — carry useful information.

In general, the smooth in a histogram is the overall shape of the distribution. A distribution is *symmetric* if the right and left halves are mirror images of each other. A *skewed* distribution has one tail that is much longer than the other. Distributions of real data will of course be only approximately symmetric. We consider Figure 6.3 — without George Brett — to be approximately symmetric. The rough in a

histogram takes the form of *outliers* and *gaps*. George Brett was responsible for an outlier. The next example features a gap.

Spotting patterns in quality-control data can increase a manufacturer's productivity and profitability. Although Japanese products are known today for high reliability, this wasn't always the case. W. Edwards Deming, an American expert in quality control, taught Japanese industrialists about statistical methods some 35 years ago. His efforts were so successful that Japan's annual award for excellence in quality control is named after him.

One of Deming's studies concentrated on the size of steel rods used in one particular manufacturing process. The histogram in Fig-

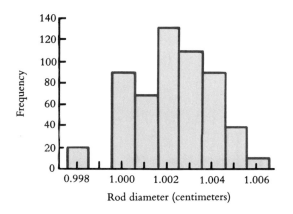

Figure 6.4 Deming's illustration of the effect of improper inspection: a histogram with a gap.

ure 6.4 summarizes this study. It displays the diameters of 500 steel rods, as measured by the manufacturer's inspectors. An overall pattern — the smooth — is apparent in the size of the rods: the distribution is approximately symmetric, centered at 1.002 centimeters and falling off rapidly both above and below. The departure from regularity — the rough — is the empty space at 0.999 centimeter.

One centimeter, or precisely 1.000, is the lower specification limit for these rods. Rods that are any smaller than this will be loose in their bearings and should therefore be rejected. The empty 0.999 cell in the histogram, together with the higher-than-expected spike in the 1.000 cell, provides strong evidence that the inspectors were passing rods that fell just below this limit by recording them in the 1-centimeter class. The histogram alerted Deming to this irregularity. This so-called flinching at the specification level is common occurrence: inspectors don't realize that just $\frac{1}{1000}$ of a centimeter can be crucial.

Once the flinching is corrected, the 0.999-centimeter slot fills in and the distribution is quite regular. In this case the strong visual impact of a graph helped pinpoint an important

problem. Interpreting the smooth portion of the histogram can also be helpful. For instance, it might suggest that the entire process should be adjusted so that more of the distribution (that is, more of the steel rods produced) exceed 1 centimeter in diameter.

Clearly, a histogram is a very useful kind of picture — a way to describe the overall shape of a distribution of values. But we can also describe a distribution with a few carefully chosen numbers, appropriately called *descriptive statistics.*

Perhaps the most obvious way to match a number to a distribution is to look at the center. There you find what might be regarded as the typical value. For example, looking back at the batting averages of American League regulars in 1980, we find that .271 is the midpoint — half the averages fall below it, and half above. When all the observations are arranged in increasing order, this midpoint is the descriptive statistic called the **median.** A similar but different statistic is the **mean,** the ordinary arithmetic average of the data. Both mean and median describe the *location* of a distribution by identifying the center of the data.

But a measure of location alone can be misleading. Consider this example: in 1984, the median income of American families was $26,430. Half of all families earned below $26,430, and half had incomes that were higher. But these figures do not tell the whole story.

Imagine two countries with vastly different economic policies. In Affluentia, everyone earns close to the median income. In Spartany, on the other hand, there are extremes of wealth and poverty. Imagine further that these two countries have the same median income; in both cases, half the incomes are less than $10,000, and half are greater. However, income patterns in the two countries are dramatically different: income in Spartany varies much more widely. The median tells us something about the center of a distribution in the two countries but not much about the spread or *variability* (see Figure 6.5).

In order to capture more of the information displayed in a histogram, it is therefore necessary to measure variability. The simplest way to do this is to calculate the *range,* the difference between the highest and lowest observations.

The highest and lowest 1980 American League batting averages were .390 and .188; therefore, the range is .202. But here we see a weakness in the usefulness of the range: it is determined by the most extreme observations. George Brett alone adds 40 points to the range of 1980 batting averages. For a more useful indication of variability, one that is less influenced by extremes, we can limit the range to the middle 50% of observations.

We find this range by identifying points called **quartiles.** The *lower* quartile is the value below which one-fourth of the observations fall. In Figure 6.3 that point is .244. Similarly, .290 is the *upper* quartile because three-fourths of the observations fall below it. Together, quartiles and the median describe

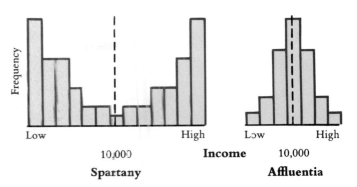

Figure 6.5 Two distributions with the same center but unequal spread.

both the center and the spread of a set of data (see Box 6.2).

A quick description of both the location and the variability of a set of data is provided by the **five-number summary.** This consists of the median, the two quartiles, and the two extremes (the smallest and largest individual observations). The five-number summary of 1980 American League batting averages is: .188, .244, .271, .290, .390. The extremes of a

---

BOX 6.2 Computing Median and Quartiles

Here is a rule for obtaining the numerical value of the median of any set of numbers:

STEP 1. Arrange the numbers in increasing order from smallest to largest.

STEP 2. If there are n numbers in the list, count up $(n + 1)/2$ places from the bottom of the list. This is the location of the median in the ordered list.

For example, the observations

$$6, 1, -2, 0, 3, 4, 1$$

when ordered are

$$-2, 0, 1, \boxed{1,} 3, 4, 6$$

Here $n = 7$, so the location of the median is $(7 + 1)/2 = 4$, or fourth from the bottom. The median is 1. With an odd number of observations, there will always be a middle observation, and this is the median. Here is a second example:

$$-3, 2, 0, -1, 4, 4, 3, 1$$

Ordering the data gives

$$-3, -1, 0, 1, 2, 3, 4, 4$$

Here $n = 8$ and $(n + 1)/2 = \frac{9}{2} = 4.5$. This is location "four-and-a-half," or *halfway between* the fourth and fifth numbers in the ordered list. These numbers are 1 and 2, so the median is 1.5. An even number of observations will always have a middle pair of observations, and the median will be the average of this middle pair.

We can use the rule for the median again to obtain the quartiles. The *first quartile* is the median of all numbers lying to the left of the location of the median in the ordered list. The *third quartile* is the median of all numbers lying to the right of the location of the median.

In our first example, the first quartile Q_1 is the median of

$$-2, 0, 1$$

so that $Q_1 = 0$. The third quartile Q_3 is the median of

$$3, 4, 6$$

so that $Q_3 = 4$. You should check that in the second example $Q_1 = -0.5$ (halfway between -1 and 0) and $Q_3 = 3.5$ (halfway between 3 and 4).

In applying these rules, be sure to write down each individual observation in your set of data, even if many observations repeat the same value. And be sure to arrange the observations in order of size before locating the median because the middle observation in the haphazard order in which the data come has no significance.

---

very large data set are often so far out as to tell us little about the bulk of the data. When this is the case, the lowest and highest **deciles** often replace the extremes in the five-number summary. The lowest decile is the point with 10% of the observations falling below it; the highest decile, the point with 10% above it. The middle 80% of a distribution falls between the lowest and highest deciles.

A graph helps to make the five-number summary more vivid. A **box plot** consists of a central box that spans the quartiles, with a line marking the median and whiskers extending out from the box to the extremes or deciles. Box plots are particularly helpful for comparing several distributions. Figure 6.6, for example, displays the earnings of male and female workers in England. Because there are millions of workers, the deciles are used in place of the extremes in this five-number summary.

Figure 6.6 quickly conveys a lot of information. You can tell at a glance that women in general earn far less than men. Indeed, the median earnings for full-time female workers in England is less than the lower quartile of male earnings. You can see, too, that the distribution of women's earnings is much more compressed. In particular, the upper decile for women does not extend far up the income scale. Figure 6.6 illustrates, once again, the effectiveness of visual displays.

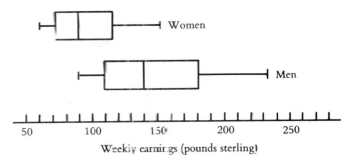

Figure 6.6   Box plot of the earnings of male and female workers in England.

Displaying Relations Between Two Variables

Using pictures to describe data is an important part of statistics. However, for only one variable, such as baseball batting averages, box plots or histograms don't reveal any new patterns. A careful reading of a table or list could provide the same information. The graph simply makes it easier to see the pattern that is already evident in the table.

Even in these simple examples, however, we are doing something that is very important for the statistician — we are exploring the data, seeing what they reveal, and drawing conclusions in an informal way. This technique really begins to unfold when we are interested in relationships between two or more factors.

In the baseball statistics we looked at earlier, there was only one variable, a batter's performance at the plate. And the picture revealed to us by the data was correspondingly simple. With two variables, however, the exploration of data has to be more elaborate. Using only a

table, we would find it difficult to spot relationships between two variables; we therefore require a graph or some descriptive statistics.

Let's say that Bob, a homeowner, is concerned about the amount of energy he uses and has decided to keep a record of the natural gas consumed over a period of nine months. Because the months are not all equally long, he divides by the number of days in the month to get cubic feet of gas used per day. Then from local weather records, he obtains the number of degree days for each month, and converts to degree days per day. Degree days are a measure of demand for heating. The weather service has a special formula for computing degree days: the point midway between the high and low temperature for the day is taken, and one degree day is accumulated for each degree below 65°F. For example, a month with 35 degree days per day had an average temperature of 30°F.

Table 6.1 shows Bob's data: nine measurements of the two variables. Looking at the

TABLE 6.1 Bob's household consumption of natural gas compared to the need for heat

	Oct	Nov	Dec	Jan	Feb	Mar	Apr	May	June
Degree days per day	15.6	26.8	37.8	36.4	35.5	18.6	15.3	7.9	0.0
Gas consumed per day (in cubic feet)	5.2	6.1	8.7	8.5	8.8	4.9	4.5	2.5	1.1

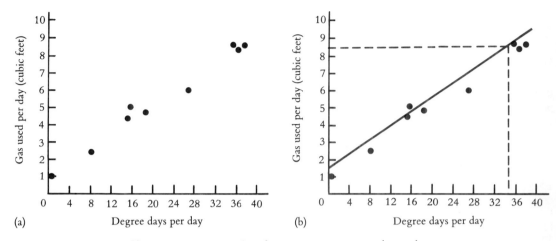

Figure 6.7   (a) Scatterplot of gas consumption versus degree days.
(b) A fitted line and its use for prediction.

numbers in the table, we can see that more degree days go with higher gas consumption; but the shape and strength of the relationship are not fully clear. They can be made more clear with a **scatterplot,** a plot of all data points in two dimensions. Figure 6.7a shows a scatterplot for our example. By convention, we plot the explanatory, or causal, variable on the horizontal scale and the dependent, or response, variable on the vertical scale. Degree days appear on the horizontal and units of gas consumption on the vertical scale because the weather affects gas consumption; gas consumption does not explain the weather.

Notice that the relationship between these variables is very strong and takes the form of a straight line. The straight line in Figure 6.7b captures this relationship. The smooth here is represented by the straight line drawn through the points; the rough is the scatter of points about the line. In this case, the number of degree days explains most of the variation in gas consumption. The scattered points above and below the line reflect the effects of other factors, such as use of gas for cooking or turning down the thermostat when the family is away from home. These effects are relatively small.

When a scatterplot shows a strong straight-line relationship, it is easy to fit a line by simply stretching a black thread over the scatterplot. But we can also fit the line by using a technique called *least-squares regression.* These calculations are usually carried out on a computer using a statistical software package. This technique finds the line that has the smallest sum of the squares of the vertical distances of the data points from the line (see Box 6.3). The least-squares **regression line** is designed to predict the gas consumption that will result from any given number of degree days. That explains why the *vertical* distances (distances in the gas-consumption direction) are minimized in fitting the line.

After drawing a scatterplot and fitting a line, we can use the line for making predictions, using only a ruler and pencil. To predict gas consumption in a month with 35 degree days per day, first locate 35 on the horizontal axis. Draw a vertical line up to the fitted line, then move horizontally to the gas consumption scale. As Figure 6.7b shows, we predict that slightly more than 8 cubic feet per day will be consumed. A computer program for least-squares regression, however, would report the

BOX 6.3 Simple Linear Regression

When a scatterplot of n observations on two variables x, y shows an approximately linear pattern, we often want to fit a straight line to the data in order to predict y from x. The *least-squares regression line* is the line that makes the sum of the squares of the vertical distances from the points to the line as small as possible. Here is the equation of this line:

$$y = b_0 + b_1 x$$

$$b_1 = \frac{\Sigma xy - (1/n)(\Sigma x)(\Sigma y)}{\Sigma x^2 - (1/n)(\Sigma x)^2}$$

$$b_0 = \bar{y} - b_1 \bar{x}$$

Here Σ is just an abbreviation for "sum of," so that Σx means the sum of all the x values. The quantities $\bar{x}$ and $\bar{y}$ are the means (ordinary averages) of the x and y values. These algebraic formulas are shorthand for the operations that we will now carry out in the home-heating example. For clarity, let's arrange our work in the form of a table.

x	y	x^2	y^2	xy
15.6	5.2	243.36	27.04	81.12
26.8	6.1	718.24	37.21	163.48
37.8	8.7	1428.84	75.69	328.86
36.4	8.5	1324.96	72.25	309.40
35.5	8.8	1260.25	77.44	312.40
18.6	4.9	345.96	24.01	91.14
15.3	4.5	234.09	20.25	68.85
7.9	2.5	62.41	6.25	19.75
0	1.1	0	1.21	0
$\Sigma x = 193.9$	$\Sigma y = 50.3$	$\Sigma x^2 = 5618.11$	$\Sigma y^2 = 341.35$	$\Sigma xy = 1375.0$

At the bottom of each column we write its sum. These sums are the building blocks for our formulas

STEP 1. Sum the x and y columns and compute the two means. Here

$$\Sigma x = 193.9 \quad \text{and} \quad \bar{x} = \frac{193.9}{9} = 21.54$$

$$\Sigma y = 50.3 \quad \text{and} \quad \bar{y} = \frac{50.3}{9} = 5.59$$

STEP 2. Compute x^2, y^2, and xy for each (x, y) data point, enter in the proper column, and sum. Here

$$\Sigma x^2 = 5618.11 \quad \Sigma y^2 = 341.35 \quad \Sigma xy = 1375.0$$

STEP 3. Substitute these into the formulas

$$b_1 = \frac{1375.0 - (193.9)(50.3)/9}{5618.11 - (193.9)^2/9} = 0.202$$

$$b = 5.59 - (0.202)(21.54) = 1.232$$

The equation of the least-squares regression line is therefore

$$y = 1.232 + 0.202x$$

The number $b_1 = 0.202$ is the slope of the line, that is, the amount by which y increases when x increases by 1. In this case, each added degree day results in consumption of 0.202 additional cubic foot of gas.

---

line in the form of an equation. If x is the number of degree days per day during a month, and y is the number of cubic feet of gas consumed per day, then the least-squares regression line computed from the data in the table is

$$y = 1.2 + 0.2x$$

How much gas will this house consume in a month that averages 35 degree days per day? Our prediction is

$$y = 1.2 + (0.2)(35)$$
$$= 8.2 \text{ cubic feet per day}$$

Although this prediction may not be exactly correct, the past data points lie so close to the line that we can be confident that in a future month with 35 degree days per day, gas consumption will be quite close to 8.2 cubic feet per day.

Bob had a very practical reason for working through this exercise in statistics. He plans to add insulation to his house during the summer and wants to know how much he will have saved in heating costs at the end of next winter. He cannot simply compare before-and-after gas usage, because the winters before and after will not be equally severe. Nor can he do a comparative experiment to weigh the costs of

insulated and uninsulated houses during the same winter, for he has only one house. However, using the fitted line and next winter's degree-day data, Bob can predict how much gas he would have used before insulating. Comparing this prediction with the actual amount used after insulation will show his savings.

Scatterplots are a good tool for exploring the relationships between two or more variables, but they are not often as simple as the one in our home-heating example. In fact, it is frequently necessary to combine these pictures with some of the numerical descriptions mentioned earlier to achieve success in exploring data.

Consider the case of the 1970 draft lottery. To eliminate distinctions among men eligible for the draft, Congress decided to allow chance to determine who would be selected for military service. Birth dates for all men born between 1943 and 1952 were to be drawn at random and assigned selection numbers in the order drawn. Men whose birth dates matched selection number 1 were drafted first. They were followed by men with numbers 2, 3, 4, and so on.

This procedure sounds fair, but the random drawing was mishandled. Birth dates were placed into identical capsules, and the capsules were placed in a drum for the drawing. But the capsules were not mixed thoroughly enough. December dates, which were added last, remained on top and had a greater chance of being drawn early. Thus, men born in December were drafted and sent to Viet Nam in greater numbers than those born in January.

Let's explore the result of the faulty draft lottery to try to see the inequity. Figure 6.8 is a scatterplot of all 366 birth dates and their corresponding selection numbers. Having a low selection number, nearer the bottom of this graph, means being drafted earlier. But the alleged association between birth dates late in the year and low selection numbers isn't easily seen. Our scatterplot alone is not very helpful because the graph looks much like a random scatter of points. A formal analysis by probability would show that a result this unfair to men born late in the year would happen in less than 1 in 1000 truly random lotteries. The inequity is really there, but to see it, we need to be more imaginative in looking at the data.

In Figure 6.9, we combine graphical methods with some basic numerical descriptions to get a better overall picture of the draft lottery. Because medians and quartiles give a compact summary of both the center and spread of a distribution, it will be useful to compute these for each month's selection numbers. We can now replace each month's points on the scatterplot with a box whose ends are at the upper and lower quartiles. These box plots allow us to instantly compare the selection numbers from month to month. A line within each box indicates the median. To make the picture even clearer, we can draw a line connecting the medians. The misfortune of men born late in the year is now evident.

The media noticed this inequity, and in 1971 a new and genuinely random selection process was designed by statisticians at the National Bureau of Standards. Two drums instead of one were used in the 1971 lottery: one contained birth dates; the other, selection numbers. The capsules in both drums were

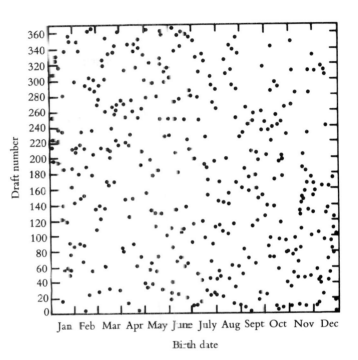

Figure 6.8   Scatterplot of draft number versus birth date for the 1970 draft lottery.

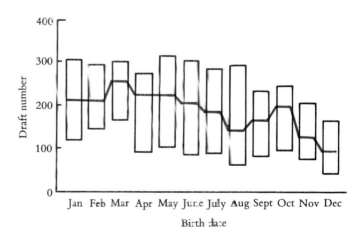

Figure 6.9   Monthly medians and quartiles for the 1970 draft lottery.

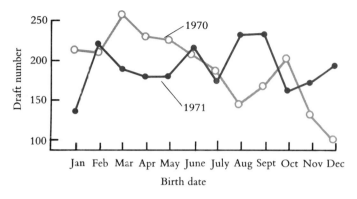

Figure 6.10   Median draft number by month for the 1970 and 1971 lotteries.

Computer graphics makes it possible to see relations and detect outliers in high-dimensional data sets. From most viewing angles, an outlier would appear as part of the main group, blending invisibly with the mass of points. However, if we change the viewing angle by rotating the plot, the outlier would eventually be seen apart — like the Death Star appearing from behind the moon in *Star Wars*. Imagine, for example, a manufacturing process in which three key measurements are plotted in three dimensions. Figure 6.11 shows that from most points of view, the data appear as a cloud,

thoroughly mixed, and then a randomly drawn birth date was paired with a randomly drawn selection number. When we add to our plot the medians of the reformed 1971 lottery, we can see the difference (Figure 6.10). There is a good bit of random variation in the 1971 medians, but no systematic trend, as in 1970.

Graphics in Many Dimensions

Graphic analysis of the draft lottery shows how much insight we can gain using only simple graphs and some basic statistical calculations. Looking at or exploring data in several ways can yield results that we wouldn't see using a single method such as a scatterplot.

But so far, we have looked only at scatterplots for two variables. What if we want to display a third factor, in a more complex situation? Because we have already used the horizontal and vertical directions of the graph, there is really only one space dimension left, moving out of and into the page. Unfortunately, such three-dimensional scatterplots are very hard to see clearly unless color or motion (or both) are used to help us gain perspective. Computer graphics can add color and motion, allowing us to see a scatterplot in three dimensions.

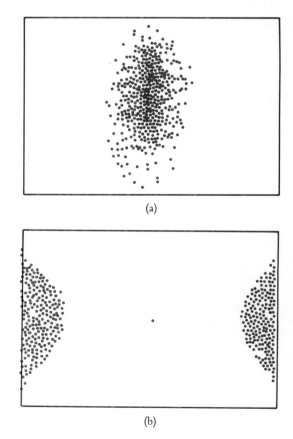

Figure 6.11   The effect of changing the viewing angle in a three-dimensional scatterplot: (a) The outlier is hidden. (b) The outlier is revealed.

almost like gnats swarming in space. But if we rotate the image, we suddenly see an outlier emerge from the pack.

Remember that every point in this picture is positioned according to three measurements. If we tried instead to find an outlier by looking at a long table with three columns of detailed measurements for thousands of points, we would quickly lose our bearings. This is a case where a visualization creates clarity out of the chaos of raw numbers.

Computer graphics allows us to move around the cloud of data points and look at it from virtually any direction. Moreover, it is possible to move around a ten-dimensional cloud of points almost as easily as a three-dimensional one. We do not, of course, see ten dimensions; nor do we actually see three dimensions. In both cases, we view a *projection* of

the cloud of data onto the two-dimensional surface of the video screen. A computer can be instructed to compute and display changing projections that simulate our moving around an imaginary, many-dimensional data plot. Outliers and other important relations among the variables come clearly into view when they are scanned from the correct angle. To reduce the amount of time needed to discover the most meaningful viewing angles, a computer can even be programmed to search for interesting projections.

These multidimensional displays can have very practical applications. At Harvard University, for example, a graphics system designed by Professor Peter Huber aids geologists who are studying the pattern of earthquakes near the Fiji Islands (see Box 6.4). For many years, the only view scientists had of

BOX 6.4 Visual Statistics: Eyeballing the Data

Dr. Peter Huber, Harvard University.

Dr. Peter Huber, a Harvard University statistician, comments on the computer's ability to display complex, multidimensional arrangements of data, giving new meaning to the term *descriptive statistics*:

For most of the twentieth century, statistics focused on mathematical rigor and on small samples. In the past 10 years or so we have seen renewed interest in descriptive statistics, driven by the computer. It would be a mistake, however, to view this new emphasis as just a reaction of statistics to high tech. What we are witnessing now is a return to the descriptive statistics of the nineteenth century, completing unfinished business left over from that era. The computer makes it possible to do things that couldn't be done before.

In my view, descriptive statistics and mathematical statistics are complementary. In descriptive statistics, you see things, but you cannot test them in a formal way. In mathematical statistics, you may test for something preconceived, but you may overlook something you hadn't built into the test. So you simply have to do both so as to complement one with the other.

You also need subject specialists to complement the data analyst. In our experience, doing data analysis usually turns out to be a close conversational collaboration between the scientist (the one with the data) and the data analyst. They sit in front of the screen, discuss what they see, suggest the next action to be taken. Data analysis requires dynamic collaboration between the statistician and the subject-matter specialist.

The first step in any data analysis is data inspection, mainly to get familiar with the data and find extraordinary features. The next step is modification; enhance the picture by lines, maybe color groups, cluster and label selected points, maybe even fit some model to it. This leads to the most important step in data analysis: *comparison.* Without comparing things, you are not able to interact with your data. To help with interpretation, you almost always have to compare things — either several data sets or a data set and a model or different models for the same data set. Interpretation is the next step. And after interpretation, one usually has to begin another round of modeling. Very often the entire cycle starts again.

We are used to seeing three-dimensional structures in natural surroundings. We reconstruct three-dimensional images through motion parallax by moving our head around or by turning the object around. Problems arise when you would like to interact with a picture, say, to identify an outlying point that you see in three dimensions. Then, of course, you have to stop the motion. In certain relatively complex situations, you may lose the special impression that would be necessary for the action you would like to take.

We have found cases where neither stereo pairs nor rotation, when used alone, would suffice to establish the picture. Suppose you have a scatterplot in three dimensions, that is, a lot of points floating around in space and a hole in them — an empty region in space. Such problems can be quite interesting in marketing research, in finding market gaps. A structure such as this is almost impossible to see unless you have the full gamut of stereo cues — stereo pairs and motion. But after you have once seen the three-dimensional scene, you can omit some of the cues — for instance, you can stop rotation — and the three-dimensional impression stays on.

---

earthquake epicenters was a standard map projection showing the two-dimensional location in latitude and longitude (see Figure 6.12). Crucial data about the depths of the epicenters were therefore not reflected in their maps.

A statistical graphics system designed by Dr. Huber allows earthquake locations to be seen in more detail. A fully three-dimensional view of the epicenters can be presented and manipulated by incorporating the depth information. This kind of picture permits geologists to examine the Fiji Island earthquake pattern in light of *plate tectonics,* the geological theory of movements of vast plates that make up the earth's crust. These plate movements give

Figure 6.12 Earthquake epicenters in the Fiji Islands as seen from the earth's surface.

rise to such powerful events as the eruption of volcanoes and the creation of mountain ranges.

Geologists explain that in the region of the Fiji Islands, the plate moving in from the east has bent beneath the plate on the west and that the eastern plate is diving straight down into the earth's mantle. The collision and resulting redirection of the plate accounts for most of the earthquakes in this region. The three-dimensional view that emerges as the computer continuously changes our viewing angle (see Figure 6.13 and Color Plate 12) makes this interpretation clear.

By rotating the viewpoint, geologists can also see the surface features of the plate itself, such as wrinkles in the surface shape. They can locate regions of greater compression or tension, where the stresses are at their maximum. Abundant geological information has become visible.

Statistics makes it possible to look at any number of complicated relationships in an organized, insightful way. All our examples of exploring data have fundamental themes and procedures in common.

First, when looking at a set of data, progress from the simple to the complex. Examine a single variable, perhaps by means of a histogram. Then, if the problem warrants it, look at the relationship between two variables, using a scatterplot or other methods. Then move on to the complexity of picturing and understand-

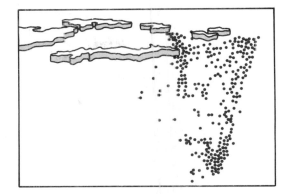

Figure 6.13 Earthquake epicenters after changing the viewing angle to show depth below the surface.

ing the relationship between three or more variables.

Second, remember that data equals smooth plus rough. Organize your thinking to look for an overall pattern and then focus on deviations from that pattern.

Third, remember the power of a picture. The human eye is the best known device for seeing both the smooth and the rough.

Today, the computational power and display features of computers allow us to move our point of view freely; more and more, computers will be doing our arithmetic and drawing our pictures for us. But it remains for the human eye to see and the human brain to understand.

REVIEW VOCABULARY

Box plot A graph of the five-number summary. A box spans the quartiles, with an interior line marking the median. Whiskers extend out from this box to the extremes or deciles.

Exploratory data analysis The practice of examining data for unanticipated patterns or ef-

fects, as opposed to seeking answers to specific questions.

Five-number summary A summary of a distribution of values by the median, the upper and lower quartiles, and either the highest and lowest deciles or the largest and smallest observations.

Frequency The number of times an outcome or group of outcomes occurs in a set of data.

Histogram A graph of the frequencies of all outcomes (often divided into groups) for a single variable. The height of each bar is the frequency of the group of outcomes covered by the base of the bar. All bars should have the same width.

Mean The ordinary arithmetic average of a set of observations; to find the mean, divide the sum of all the observations by the number of observations summed.

Median The midpoint of a set of data; half the observations fall below the median and half fall above.

Outlier A data point that falls well outside the overall pattern of a set of data.

Quartiles The lower quartile is the point with 25% of the observations falling below it; the upper quartile is the point with 75% below it.

Regression line A line drawn on a scatterplot that shows a straight-line pattern. The regression line can be used to predict the response y for a given value x of the explanatory variable.

Scatterplot A graph of the values of two variables as points in the plane; the value of the explanatory variable is plotted on the horizontal axis, the corresponding value of the response variable on the vertical axis.

EXERCISES

1. In 1798 the English scientist Henry Cavendish measured the density of the earth in a careful experiment with a torsion balance. Here are his 23 repeated measurements of the same quantity (the density of the earth relative to that of water) made with the same instrument.

5.36	5.62	5.27	5.46
5.29	5.29	5.39	5.30
5.58	5.44	5.42	5.75
5.65	5.34	5.47	5.68
5.57	5.79	5.63	5.85
5.53	5.10	5.34	

[Source: S. M. Stigler, "Do robust estimators work with real data?" *Annals of Statistics* 5:1055–1078 (1977).]

   a. Make a histogram of these data.

   b. Describe the overall shape of the distribution. Is it approximately symmetric, or distinctly skewed?

   c. Are there gaps or outliers in the data?

2. Table 6.2 lists the amount spent per student by the public schools of each state.

   a. Make a histogram of these data

   b. Are there any outliers? Can you explain them?

TABLE 6.2 Average dollars expended per pupil in public elementary and secondary day schools, 1981

State	Dollars	State	Dollars	State	Dollars
Alabama	1384	Kentucky	1569	North Dakota	2062
Alaska	5010	Louisiana	1972	Ohio	2143
Arizona	1914	Maine	2055	Oklahoma	2007
Arkansas	1571	Maryland	2541	Oregon	3049
California	2594	Massachusetts	3174	Pennsylvania	2798
Colorado	2656	Michigan	2461	Rhode Island	2559
Connecticut	2697	Minnesota	2484	South Carolina	1560
Delaware	2781	Mississippi	1536	South Dakota	1995
District of Columbia	3682	Missouri	2079	Tennessee	1458
Florida	2262	Montana	2948	Texas	1955
Georgia	1652	Nebraska	2105	Utah	1742
Hawaii	2121	Nevada	2179	Vermont	2017
Idaho	1780	New Hampshire	2033	Virginia	2223
Illinois	2441	New Jersey	2791	Washington	2653
Indiana	1793	New Mexico	2219	West Virginia	1816
Iowa	2560	New York	3358	Wisconsin	2769
Kansas	2714	North Carolina	1992	Wyoming	2596

Source: *Statistical Abstract of the United States, 1982–83.*

c. When the outliers are omitted, are the data roughly symmetrical (like the batting averages) or are they skewed (one tail much longer than the other)?

3. Choose a set of interesting data from the *Statistical Abstract* or an almanac (for example, populations of the states, per capita incomes of nations, number of home runs hit by the league leader each year). Make a histogram of the data and describe the pattern and any outliers.

4. Here are the percentages of the popular vote won by the successful candidate in each of the presidential elections from 1960 to 1984. Find the median and the mean of these data.

$$49.7, 61.1, 43.4, 60.7, 50.1, 50.7, 58.8$$

5. Find the mean and the median per student expenditures on public education from Table 6.2.

6. Find the median and quartiles of Cavendish's measurements of the density of the earth in Exercise 1. Then give a five-number summary. How is the symmetry of the distribution reflected in the five-number summary?

7. Find the range, the quartiles, and the deciles of the data in Table 6.2. Which states are in the top 10% in public-school expenditures? The bottom 10%?

8. You should be aware that the mean is heavily influenced by outlying observations, whereas the median is not. Here, for example, are the number of moons circling each planet in our solar system.

Mercury	0	Saturn	22
Venus	0	Uranus	5
Earth	1	Neptune	3
Mars	2	Pluto	1
Jupiter	17		

Find the mean and the median number of moons. How many planets have more than the mean number of moons?

9. A study of the size of jury awards in civil cases (such as injury, product liability, and medical malpractice) showed that the median award in Cook County, Illinois, was about $8000. But the mean award was about $69,000. Explain how this great difference between two measures of location can occur.

10. Give an example of a small set of data whose mean is larger than the upper quartile.

11. Make a box plot of the public-school-expenditures data in Table 6.2. (For small sets of data, it is convenient to extend the whiskers at the end of the box to the smallest and largest observations, rather than stop at the deciles. Do this here and in the following exercises. There are many variations on box plots.)

12. Joe DiMaggio played center field for the Yankees for 13 years. He was succeeded by Mickey Mantle, who played for 18 years. Here are the number of home runs hit each year by DiMaggio:

$$29, 46, 32, 30, 31, 30, 21, 25, 20, 39, 14, 32, 12$$

and by Mantle:

$$13, 23, 21, 27, 37, 52, 34, 42, 31, 40, 54, 30, 15, 35, 19, 23, 22, 18$$

Compute the five-number summary for each player, and make side-by-side box plots of the home run distributions. What does your comparison show about the relative effectiveness of DiMaggio and Mantle as home run hitters?

13. Using Table 6.2, consider the states east of the Mississippi, including Louisiana and Mississippi, through which the river flows. Omit the District of Columbia and divide the other states into two groups, the Northeast and the Southeast. Make box plots of the public-school expenditures for both groups and compare the two distributions. Be sure to list the states that you place in each group.

14. Table 6.3 gives the survival times (in days) of 72 guinea pigs after they were infected by tubercle bacilli in a medical study. Make a histogram of these data. Is the survival-time distribution approximately symmetric or strongly skewed? Compute the five-number summary for this distribution. How is the shape of the distribution reflected in the distances of the two quartiles and the two extremes from the median?

TABLE 6.3 Guinea pig survival times

43	45	53	56	56	57	53	66	67	73
74	79	80	80	81	81	81	82	83	83
84	88	89	91	91	92	92	97	99	99
100	100	101	102	102	102	103	104	107	108
109	113	114	118	121	123	126	128	137	138
139	144	145	147	156	162	174	178	179	184
191	198	211	214	243	249	329	380	403	511
522	598								

Source: T. Bjerkedal, "Aquisition of resistance in guinea pigs infected with different doses of virulent tubercle bacilli," *American Journal of Hygiene* 72:130–148 (1960).

15. We saw that the least-squares regression line for the home heating data of Table 6.1 is $y = 1.2 + 0.2x$. Use this line to predict the daily gas consumption for this home in a month averaging 20 degree days per day and in a month averaging 40 degree days per day.

16. Table 6.4 shows the true number of calories in 10 common foods and the average number of calories estimated for these same foods in a sample of 3368 people.

TABLE 6.4 True and estimated calories in 10 common foods

Product	True calories	Estimated calories
8 oz. whole milk	159	196
5 oz. spaghetti with tomato sauce	163	394
5 oz. macaroni and cheese	269	350
1 slice wheat bread	61	117
1 slice white bread	76	136
2-oz. candy bar	260	364
Saltine cracker	12	74
Medium-size apple	80	107
Medium-size potato	88	160
Cream-filled snack cake	160	419

Source: *USA Today*, October 1983.

a. Make a scatterplot of these data, with true calories on the horizontal axis. Is there a general straight-line pattern? Which foods are outliers from the pattern?

b. Stretch a thread through a pattern (ignoring the outliers). If a food product contains 200 calories, what would you guess the general public's estimated calorie level for that food to be?

c. The least-squares regression line, computed without dropping the outliers, is

$$y = 58.6 + 1.30x$$

(y = estimated calories, x = true calories.) Draw this line on your scatterplot. Use it to predict y when $x = 200$.

The difference between your results in **b** and **c** reflects the influence of the outliers. It is often difficult to decide whether to include outliers in making a prediction.

17. The following scatterplot of the average SAT math score (y) versus the average SAT verbal score (x) for high-school seniors in each of the 50 states in 1983 shows a strong linear pattern. The least-squares regression line for scores is

$$y = 27 + 1.03x$$

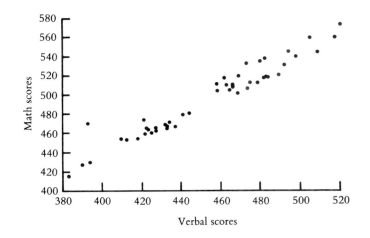

a. Do math SAT scores tend to be higher or lower than verbal SAT scores?

b. The average verbal SAT score in New York was 422. Use the regression line to predict New York's average math score. New York's actual math score was 466. What is the residual (observed minus predicted)?

c. The only outlier among the states is Hawaii, where $x = 393$ and $y = 471$. Is Hawaii's math score higher or lower than would be predicted from its verbal score? Can you suggest why Hawaii might be an outlier?

18. Table 6.5 gives the price and weight of 70 automobiles of the 1986 model year. The models marked with an asterisk are made by foreign companies. Make a scatterplot of weight (y) versus price (x), with domestic and foreign distinguished either by different colors or by different symbols (such as x and o). Discuss carefully what the plot shows about the relationship between the three variables: weight, price, and domestic versus foreign make.

TABLE 6.5 Price (dollars) and weight (pounds) of 1986 cars (four-door sedan-type models only)

Model	Price	Weight	Model	Price	Weight
Chevrolet Nova	7,435	2250	*Volkswagon Quantum	13,595	2640
Chevrolet Spectrum	6,928	1920	*Volvo	14,370	2920
*Honda Civic	7,798	2035	*Audi 5000S	18,065	2915
*Isuzu I-Mark	7,249	1920	Buick Century	10,228	2775
*Mazda 323	7,495	2240	Cadillac Seville	26,756	3425
*Mitsubishi Tredia	7,849	2400	Chevrolet Celebrity	8,931	2800
*Nissan Sentra	7,169	2090	Chrysler Fifth Avenue	14,910	3575
Plymouth Colt	6,862	2185	Chrysler Le Baron	10,127	2585
*Renault Alliance	6,199	2010	Chrysler New Yorker	13,409	2720
*Subaru	7,391	2295	Dodge 600	9,370	2600
*Toyota Corolla	7,348	2270	Dodge Aries	7,301	2535
*Volkswagon Jetta	8,545	2310	Dodge Diplomat	10,036	3575
AMC Eagle	10,719	3480	Ford Taurus	9,645	2870
*Audi 4000s	14,230	2360	Lincoln Continental	24,556	3800
*BMW 325	20,055	2550	Mercury Sable	10,700	2870
Buick Skyhawk	8,073	2500	Olds Cutlass Ciera	10,354	2745
Buick Skylark	9,620	2645	Olds Cutlass Supreme	10,872	3355
Cadillac Cimarron	13,128	2620	Plymouth Caravelle	9,241	2720
Chevrolet Cavalier	6,888	2555	Plymouth Gran Fury	10,086	3575
Ford Tempo	7,508	2600	Plymouth Reliant	7,301	2535
*Honda Accord	9,679	2590	Pontiac 6000	9,729	2535
*Mazda 626	8,995	2600	Pontiac Bonneville	10,249	3350
*Mercedes-Benz 190	25,080	2745	*Saab 9000	22,145	3005
Mercury Topaz	8,235	2600	*Volvo 740	18,240	2980
*Mitsubishi Galant	13,219	2840	*Volvo 760	22,960	3065
*Nissan Maxima	14,459	3150	Buick Electra	15,588	3300
*Nissan Stanza	10,069	2440	Buick Le Sabre	12,511	3195
Oldsmobile Calais	9,478	2645	Cadillac De Ville	19,990	3345
Oldsmobile Firenza	8,035	2535	Chevrolet Caprice	10,243	3630
*Peugeot 505	12,651	3075	Ford LTD Crown Victoria	12,562	3885
Pontiac Grand Am	8,749	2725	Lincoln Town Car	20,764	4060
Pontiac Sunbird	7,495	2435	Mercury Grand Marquis	13,504	3930
*Saab 900	12,985	2845	Olds Delta 88 Royale	12,760	3195
*Toyota Camry	9,678	2690	Olds 98 Regency	15,989	3265
*Toyota Cressida	16,630	3335	Pontiac Parisienne	11,169	3630

* Manufactured by foreign companies.

19. A study of sewage control measures the oxygen demand of decomposing solid wastes. If y is the logarithm of the oxygen demand (milligrams per minute) and x is the total solids (milligrams per liter of waste), measurements on 20 occasions give the following data:

x	7.2	7.8	7.1	6.4	6.4	5.1	5.9	5.3	5.0	5.0
y	1.56	0.9	0.75	0.72	0.31	0.36	0.11	0.11	-0.22	-0.15

x	4.8	4.4	4.3	3.7	3.9	3.6	4.4	3.3	2.9	2.8
y	0.0	0.0	-0.09	-0.22	-0.4	-0.15	-0.22	-0.4	-0.52	-0.05

a. Make a scatterplot of these data. Is the relationship roughly linear? Are there outliers?

b. Compute the least-squares regression line for predicting y from x. Draw your fitted line on the scatterplot.

c. Use the equation you obtained in **b** to predict the log of the oxygen demand y when $x = 4$ milligrams per liter of solids.

Probability: The Mathematics of Chance

Have you ever wondered how gambling, which is a recreation or an addiction for individuals, can be a business for the casino? A business requires predictable revenue from the service it offers, even when the service is a game of chance. Individual gamblers may win or lose; they can never say whether a day at Lake Tahoe or Atlantic City will turn a profit or a loss. But the casino itself does not gamble. Casinos are consistently profitable, and lotteries are now an important source of revenue for many state governments.

It is a remarkable fact that the aggregate result of many thousands of chance outcomes can be known with near certainty. The casino need not load the dice, mark the cards, or alter the roulette wheel. It knows that in the long run, each dollar bet will yield its five cents or so of revenue. It is therefore good business to concentrate on free floor shows or inexpensive bus fares to increase the flow of dollars bet. The flow of profit will follow.

Gambling houses are not alone in profiting from the fact a chance outcome many times repeated is firmly predictable. For example, although a life insurance company does not know *which* of its policyholders will die next

Figure 7.1 The more dollars bet, the more money a casino is guaranteed to take in. [Las Vegas News Bureau]

year, it knows very clearly *how many* will die. It sets its premiums by this knowledge, just as the casino sets its jackpots.

A phenomenon is called *random* if individual outcomes are unpredictable but the long-term pattern of many individual outcomes is predictable. Many phenomena, both natural and of human design, are random. The lifetimes of insurance buyers and the hair color of children are examples of natural randomness. Indeed, quantum mechanics asserts that at the sub-atomic level the natural world is inherently random. Probability theory, the mathematical description of randomness, is therefore essential to much of modern physics.

The odds in a game of dice are examples of carefully planned randomness. We are primarily concerned with randomness deliberately produced by human effort. The casino's dice are carefully machined, and their drilled holes (called pips) are filled with material equal in density to the plastic body. This guarantees that the six-side has the same weight as the opposite side, which has only one pip. Thus, each side is equally likely to land upward. All the odds and payoffs of dice games rest on carefully planned randomness.

Statisticians and casino managers have the same vested interest in planned randomness, although statisticians use tables of random digits rather than dice and cards. All the careful reasoning of statistical inference rests on planned randomness and on the mathematics of probability, the same mathematics that guarantees the profits of casinos and insurance companies. This mathematics of chance is the topic of this chapter.

What Is Probability?

The mathematics of chance, the mathematical description of randomness, is called the *theory of probability*. Probability describes the predictable long-run patterns of random outcomes.

Toss a coin in the air. Will it land heads or tails? It lands sometimes heads and sometimes tails. We cannot say what the next fall will be. Perhaps you would argue that the coin has an equal chance of falling heads or tails on the next toss. You might explain your personal opinion by saying that if forced to bet on the next toss, you would accept even odds on the two outcomes. Probability as the expression of personal opinion is a sensible idea. But we have in mind a different meaning, one based on *observation* of random phenomena.

Suppose that we toss a coin not once but 10,000 times. John Kerrich, an English mathematician, actually did this while interned by the Germans during World War II. Figure 7.2 shows Kerrich's results. Kerrich's first 10 tosses gave 4 heads, a proportion of 0.4. The proportion of heads increased to 0.5 after 20 tosses and to 0.57 after 30 tosses. Figure 7.2 shows how this proportion changes as more and more tosses are made. In a small number of tosses, the proportion of heads fluctuates — it is still essentially unpredictable. But many tosses will produce a smoothing effect. A proportion of 0.507, or 50.7%, heads resulted after Kerrich threw the coin 5000 times. And

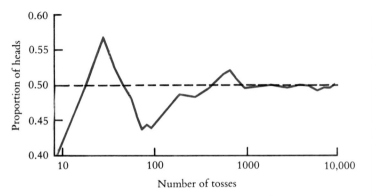

Figure 7.2 Percent of heads versus number of tosses in Kerrich's coin-tossing experiment. [David Freedman et al., *Statistics* Norton, 1978.]

in all 10,000 trials, he scored 5067 heads, again 50.7% of the total. After many trials, the proportion of heads settled down to a fixed number. This number is the **probability** of a head.

The probability of an outcome is the proportion of trials in which the outcome occurs in a very long run of trials. Strictly speaking, Kerrich's 0.507 only estimates the probability that his coin will come up heads. The proportion after 100,000 tosses would be even closer to the true probability. In fact, we can't even say that Kerrich's coin is weighted in favor of heads. There is still enough unpredictability in the results of 10,000 tosses that 70 excess heads could easily occur even if heads and tails were equally probable.

Mathematical Description of Probability

Gamblers have known for a long time that the fall of coins, cards, or dice stabilizes into definite patterns in the long run. France gave birth to the mathematics of probability when gamblers in the seventeenth century turned to mathematicians for advice (see Box 7.1). The idea of probability rests on the observed fact

---

BOX 7.1 The Mathematical Bernoullis

Few families have made more contributions to mathematics than the Bernoullis of Basel, Switzerland. No fewer than eight Bernoullis, over three generations spanning the years between 1680 and 1800, were distinguished mathematicians. Five of them helped build the new mathematics of probability.

James (1654–1705) and John (1667–1748) were sons of a prosperous Swiss merchant and studied mathematics against the will of their practical father. Both were among the finest mathematicians of their times, but it was James who concentrated on probability. Several seventeenth-century mathematicians had started the study of games of chance, concentrating on counting outcomes to find chances. James Bernoulli was the first to see clearly the idea of a long-run proportion as a way of measuring chance. He proved that if in a very large population (say size N), K members have a property A, then the proportion of members of a sample of size n having property A must approach K/N (the probability of A) as the sample size n increases. This *law of large numbers* helped to connect probability to the study of sequences of chance outcomes observed in human affairs.

John's son Daniel (1700–1782) and James and John's nephew Nicholas (1687–1759) also studied probability. Nicholas saw that the pattern of births of male and female children could be described by probability. Despite his own rebellion against his father's strictures, John tried to make his son Daniel a merchant or a doctor. Daniel, undeterred, became yet another Bernoulli mathematician. He studied mainly the mathematics of flowing fluids (later applied to designing ships and aircraft) and elastic bodies. In the field of probability, he worked on the fair price of games and gave evidence for the effectiveness of innoculation against smallpox.

The Bernoulli family in mathematics, like their contemporaries the Bachs in music, are an unusual example of talent in one field appearing in successive generations. Their work helped probability to grow from its birthplace in the gambling hall to a respectable tool with worldwide applications.

that the average result of many thousands of chance outcomes can be known with near certainty. But a definition of probability as "long-run proportion" is not clear enough. Who can say what "the long run" is? Instead, we give a mathematical description of how probabilities behave, based on our understanding of long-run stability.

Our first task in assigning probabilities to outcomes is to list all the possible outcomes. This set of outcomes is called the **sample space,** S. The set S may be very simple or very complex. In the case of a coin, there are only two outcomes, heads and tails. So the sample space is $S = \{H, T\}$. If we draw a random sample of 1500 U.S. residents age 18 and over, as the Gallup poll does, the sample space contains all possible choices of 1500 of the over 180 million adults in the country. This S is extremely large — to count its members would require a number over 8000 digits long! Each member of S is a possible Gallup poll sample, which explains the term *sample space.*

Suppose that we roll two dice and count the pips on the up faces. There are 11 possible outcomes, from snake eyes (the sum is 2) through 3, 4, and 5 and on up to 12. We cannot say what any single roll of two dice will bring, but we are certain to get some member of the sample space

$$S = \{2, 3, 4, 5, 6, 7, 8, 9, 10, 11, 12\}$$

The next step is to assign probabilities to the outcomes that make up the sample space. In the case of throwing two dice, we can see that a 7 is more likely than a 12. After all, a 7 can result from a 5 on one die and a 2 on the other, or from a 4 and 3, or from a 6 and 1. There is only one way to throw a 12. So the probability of a 7 is greater than the probability of a 12. Exactly what these probabilities are is still a problem.

To solve this problem, we first take a more general look at probabilities. There are two

laws that any assignment of probabilities to outcomes must satisfy. Let $P(s)$ stand for the probability of any outcome s in the sample space S. Here are the fundamental laws of probability:

1. Every probability $P(s)$ is a number between 0 and 1.

2. The sum of the probabilities $P(s)$ over all outcomes s in S is exactly 1.

These laws are based on our understanding of probability as long-run proportion. The first law says that if an outcome *never* occurs, it has probability 0; if it *always* occurs, the probability is 1; if it *sometimes* occurs, the proportion of trials producing the outcome is a number between 0 and 1. The second law says that some outcome must always occur, so that the sum of all their probabilities is 1.

Laws 1 and 2 describe any legitimate assignment of probabilities to outcomes. An assignment that does not satisfy these laws does not make sense. **A probability model** consists of a sample space S and an assignment of probabilities that satisfies laws 1 and 2.

Let's look at tossing a single die to see what a probability model looks like. The possible outcomes (see Figure 7.3) are

$$S = \{1, 2, 3, 4, 5, 6\}$$

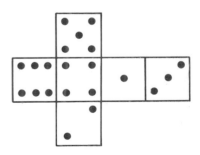

Figure 7.3 The possible outcomes of rolling a die.

If the die is a carefully made casino model, each face should be equally likely to land up. Because the probabilities of the six outcomes must sum to 1 by law 2 (and none can be negative, by law 1), we are forced to assign

$$P(1) = P(2) = P(3) = P(4) = P(5) = P(6)$$
$$= 1/6 = 0.166$$

What is the probability of rolling an odd number? The odd outcomes are 1, 3, and 5. So

$$P(\text{outcome is odd}) = P(1) + P(3) + P(5)$$
$$= 3/6 = 0.5$$

Any set of outcomes in S is called an **event.** The probability of any event is found by summing the probabilities of the outcomes that make up the event.

Laws 1 and 2, with the assumption that each face is equally probable, also make it possible to assign probabilities when *two* dice are thrown (see Box 7.2). This solves our original problem.

BOX 7.2 How to Find Probabilities When Outcomes Are Equally Likely

When a casino die is thrown, we are willing to assume that each of the six faces is equally likely. When *two* such dice are thrown, there are 36 possible combinations of faces. They are:

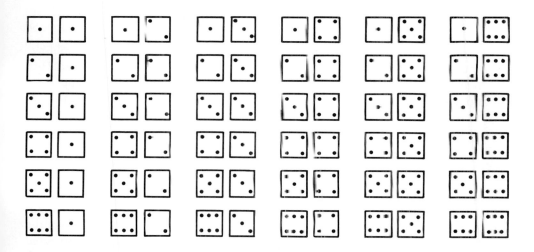

Because each of these 36 outcomes is also equally likely, each must have probability $1/36 = 0.028$. But when we take as our outcome the sum of the pips on the two faces, these outcomes are not equally likely. In fact "roll a 5" is the *event* containing the four outcomes

Each of these outcomes has probability 1/36. Therefore

$$P(\text{roll a 5}) = 1/36 + 1/36 + 1/36 + 1/36 = 4/36 = 0.111$$

Similarly, you can find the probability of rolling a 3, a 7, or any other number, by *counting the outcomes* that produce the desired result.

This example leads to a useful rule: When all outcomes have equal probabilities, the probability of any event A is found by

$$P(A) = \frac{\text{number of outcomes in } A}{\text{number of outcomes in } S}$$

But beware. You must be very careful to choose the right S in order to get equally likely outcomes. The S consisting of the 36 combinations of faces of two balanced dice *does* have equally likely outcomes. But the sample space for tossing two dice *and adding the pips* is

$$S = \{2, 3, 4, 5, 6, 7, 8, 9, 10, 11, 12\}$$

and does *not* contain outcomes having equal probabilities.

It is important to understand what the definition of a probability model does and does not do for us. It does state which assignments of probabilities to outcomes make sense. But it does not say which assignments are correct, in the sense of accurately describing a real die; this can only be determined by actual trial. Professional dice are well described by the equal-probabilities model of the example. But cheap dice with hollowed-out pips fall unequally. The 6 face is lightest and is located opposite the 1 face, which is heaviest. Thus, a cheap die might be described by a model in which lighter faces are more likely, such as

$$P(1) = 0.159 \qquad P(4) = 0.166$$
$$P(2) = 0.163 \qquad P(5) = 0.171$$
$$P(3) = 0.166 \qquad P(6) = 0.175$$

This assignment also satisfies laws 1 and 2. It is legitimate, even if the dice are not.

Sampling Distributions

Sampling, in a way, is a lot like gambling. Both rely on randomness — the deliberate use of chance. To see how probability applies to sampling, let's look again at the population of light and dark beads described in Chapter 5.

We have a large container filled with beads of the same size and shape. How can we estimate the percentage of dark beads in the box? We plunged a scoop with 50 bead-sized hollows into the container and drew out 50 beads. This is a simple random sample of size 50. Of the 50 beads in the sample, 12, or 24% of the sample, were dark. From this we estimated that 24% of the population is made up of dark beads. Our sampling strategy illustrates a basic form of statistical inference: take a simple random sample and use the sample result to estimate the unknown truth about a population. The Gallup poll and the Current Population Survey carry out more elaborate versions of this method.

When we drop the scoop into the box once more, drawing a second sample, perhaps only 8 of the 50 beads, or 16%, will be dark. A third sample of the beads might yield 7 dark ones, or 14%. This chance variation is called **sampling variability.** It is the same sampling variability that we saw earlier when we were looking at surveys. In this chapter, however, we will use the laws of probability to describe it more precisely.

The sample space in the bead-sampling experiment is made up of all the possible samples of 50 beads drawn from the beads in the container. Random sampling guarantees that each possible sample is equally likely to be drawn. So we could find the probability of getting 16% dark beads by seeing how many of the possible samples have 16% dark beads. This procedure would be hopelessly tedious; we therefore need a shortcut.

Let's return to actually watching the outcomes of many samples to get a clue about what the probabilities will look like. Figure 7.4a records the results of 200 samples. For example, 21 of the 200 samples had exactly 16% (8 of 50) dark beads. So we estimate the probability of getting 16% dark beads to be about 21/200 or 0.105.

Look at the overall pattern of the histogram of estimated probabilities in Figure 7.4a. The percentages of dark beads range from 6% to 38%. The histogram is fairly symmetric. The center is close to the 20% bar (representing 10 dark beads out of 50). This is also the outcome that occurred most often. There are no extreme outliers.

Statisticians call a number that is computed from a sample a **statistic.** The percentage of dark beads in the sample is a statistic. The histogram displays the sampling variability of this statistic. In fact, the histogram assigns probabilities to each value of the statistic. These probabilities make up the **sampling distribution** of the statistic. Of course, the probabilities are only approximate because they are

based on only 200 trials. But the overall pattern of the sampling distribution is already clear.

Let's try a second experiment in the hope of confirming the pattern. Remember that public opinion polls often rely on a sample of about 1500 people. We will take 200 simple random samples of 1500 beads each. To do this, we must imagine a very large container of beads. In fact, we instruct a computer to simulate the experiment. Figure 7.4b displays the outcomes.

The histogram in Figure 7.4b differs from that in Figure 7.4a. You can see that the one in Figure 7.4b is taller and narrower. It is nar-

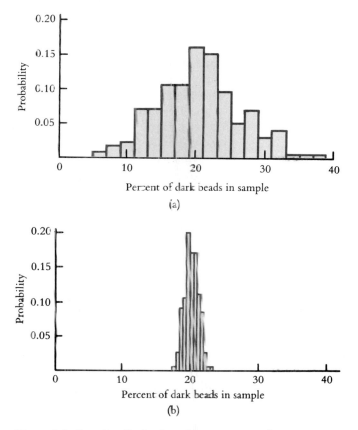

Figure 7.4   Sampling distribution of the percentage of dark beads in samples of (a) size 50 and (b) size 1500.

rower because the sampling variability is greatly reduced. All 200 outcomes fall between 17% and 23% dark beads. And if we add up the estimated probabilities near the middle, we find there is a probability 0.93 that a sample will fall between 18.25% and 21.75%. We can rely much more confidently on a single sample of size 1500 than on a single sample of size 50. In fact, all these samples were drawn from a population with 20% dark beads. Samples of size 1500 almost always give estimates quite close to this true value.

Normal Distributions

Although they differ in variability, the histograms in Figures 7.4a and 7.4b have similar shapes in other respects. Both are symmetric, with centers close to 20%. The tails fall off smoothly on either side, with no outliers. Suppose that we describe the shape of each histogram by drawing a smooth curve through the tops of the bars. If we do this carefully — using the actual probabilities of each outcome rather than estimates from only 200 samples — the curve we obtain will be quite close to two members of the family of *normal curves.* All

normal curves contain total area 1, corresponding to the fact that all outcomes together have probability 1. These appear in Figure 7.5.

The normal curves are symmetric and bell-shaped, with tails that fall off rapidly. That the distribution of the sample proportion from a simple random sample is closely described by a normal curve is not just a matter of artistic judgment. It is a mathematical fact, first proved by Abraham DeMoivre in 1718.

A histogram makes probability visible. The height of any bar represents the probability of the outcomes spanned by the base of that bar. Because all bars are of equal width, their area also represents probability. It is area that our eyes respond to, and it is as area under the curve that the normal curves represent probability. Such an assignment of probabilities to outcomes is a **normal probability distribution.** The sampling distributions of such common statistics as means and proportions are approximately normal, and many sets of data are also approximately described by normal distributions. These distributions therefore deserve more detailed study.

Let's look first at the center of a normal distribution. As we saw earlier, the *median* is the point half the outcomes fall above and half below; it lies at the center of the curve. The *mean* — the arithmetic average outcome — lies at the same point. Both these facts are due to the symmetry of the curve. Thus, the mean and median of any normal distribution are the same.

This is not true of all distributions. Figure 7.6, for example, shows a skewed distribution.

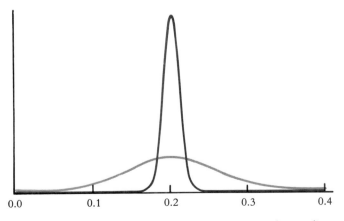

0.0   0.1   0.2   0.3   0.4

Figure 7.5 Normal curves representing the sampling distribution of the percentage of dark beads in samples of size 50 (light) and 1500 (dark).

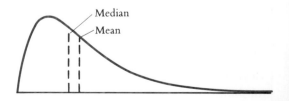

Figure 7.6 The mean and median of a skewed distribution.

The right tail of the curve is much longer than the left. The prices of new houses are an example of a skewed distribution — there are many moderately priced houses and a few extravagantly priced mansions out in the right tail. Those mansions pull the mean — or average price — up, so that it is greater than the median. Sure enough, the mean price of a new house in 1985 was $100,600, but the median price was only $84,100.

What about the spread of a normal curve? Figure 7.7 shows the distribution of heights of American women ages 18 to 24. The shape of the curve is normal, with the mean (and median) height at 64 inches. In Chapter 6, we learned that the quartiles indicate the spread of a distribution. The quartiles marked in Figure 7.7 contain between them the middle 50% of women's heights; they fall at 62.3 and 65.7 inches. Because the curve is symmetric, the quartiles are each 1.7 inches from the mean of 64 inches.

The location and spread of normal curves have a special property: the shape of a normal distribution is completely specified once the mean and the spread are given. This information is not sufficient to determine the exact shape of most distributions of data, but it is enough when the distribution is normal. What is more, there is a single number that measures the spread of a normal curve. This measure of spread is called the **standard deviation.** The mean and the standard deviation, taken together, determine at once all the percentiles of a normal curve.

You can find the standard deviation by running a pencil along a normal curve from the center outward. Look for the point where the curve changes from coming down ever more steeply to sloping down ever more gently. There is one such point on either side of the curve. The distance of each of these points from the mean is the standard deviation of the curve. Try finding the standard deviation on the normal curve in Figure 7.8. The two

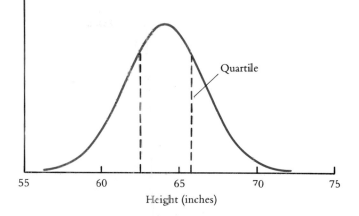

Figure 7.7   The normal distribution of women's height with the quartiles marked

change-of-curvature points are at 61.5 inches and 66.5 inches. The standard deviation of women's heights — the difference between each of these points and the mean — is 2.5 inches.

Changing the mean of a normal curve does not change its shape; it only slides the curve to a new location. Changing the standard devia-

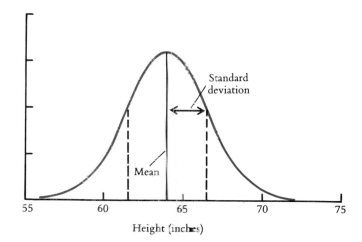

Figure 7.8   Locating the mean and standard deviation on a normal curve.

tion does change its shape. A normal curve with a smaller standard deviation is taller and narrower (has less spread) than one with a larger standard deviation. You can see this by comparing the two normal curves for our bead-sampling experiments in Figure 7.5. Both normal curves have the same mean, but the curve for samples of size 1500 has the smaller standard deviation.

Let's look more closely at the way the mean and standard deviation completely specify a normal distribution. The probability of an outcome falling within one standard deviation on either side of the mean is 0.68. If we go out two standard deviations from the mean, we find a probability of 0.95. Finally, the probability of falling within three standard deviations of the mean is almost 1, 0.997 to be exact. These facts hold for any normal distribution but not for distributions with other shapes. That is, in terms of percentages,

• 68% of the observations fall within one standard deviation of the mean.

• 95% of the observations fall within two standard deviations of the mean.

• 99.7% of the observations fall within three standard deviations of the mean.

Figure 7.9 illustrates this *68-95-99.7 rule* for normal distributions.

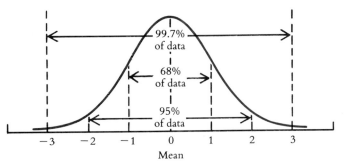

Figure 7.9 The 68-95-99.7 rule for normal distributions.

This is a useful set of numbers to remember. By using these three numbers, we can quickly derive helpful information about any normal distribution. For example, we learned that the heights of American women between the ages of 18 and 24 are roughly normally distributed, with a mean of 64 inches and standard deviation 2.5 inches. From this we can immediately tell that about 68% of women are between 61.5 and 66.5 inches tall; 95% are between 59 and 69 inches tall; and almost all women have heights of between 56.5 and 71.5 inches. Very few women are 6 feet tall or over.

Many other sets of data fit into a normal distribution. The distribution of scores on tests such as the Scholastic Aptitude Test (SAT), for example, is close to normal. SAT scores are adjusted so that for a standard reference population, the mean score is 500 and the standard deviation is 100. Thus, only 5% of the scores are above 700 or below 300; the other 95% fall within two standard deviations of the mean. Because of the symmetry of the normal curve, this 5% is divided evenly above and below the mean, so that a score above 700 places a student in the top 2.5% of the reference population.

The Central Limit Theorem

The significance of normal distributions is explained by a key fact in probability theory: the **central limit theorem.** This theorem says that the distribution of any random phenomenon tends to be normal if we average it over a large number of repetitions. This notion is strikingly powerful. It means that we can analyze and predict the results of chance phenomena if we average over many observations.

We have already seen the central limit theorem at work in our bead-sampling experiment. A single bead drawn at random is either dark or light. Only two outcomes are possible, and there is no normal curve in sight; however, the percent of dark beads when 50 beads are drawn at random roughly follows a normal distribution. When 1500 beads are drawn, the percent of dark beads represents an average of "dark or light" over a larger number of beads and is even closer to a normal curve.

Our sampling experiment showed that samples of 1500 beads have much less spread than samples of 50. Spread can be described by the standard deviation of the normal distribution of outcomes. The central limit theorem makes this explicit. It says not only that an average over n observations is approximately normal but that the standard deviation decreases with the square root of the number of observations, $\sqrt{n}$. So an average over 100 observations ($\sqrt{100} = 10$) has a standard deviation half as large as an average over 25 observations ($\sqrt{25} = 5$) from the same population. A sample of 1500 beads is 30 times as large as a sample of 50 beads. So the standard deviation of the percentage of dark beads is $\sqrt{30}$, or 5.5, times smaller for samples of 1500 than for samples of 50. The two normal curves in Figure 7.5 display exactly this difference in standard deviations.

The central limit theorem can help to answer our opening question: How can gambling be a business for a casino? Let's look at just one of the many bets that a casino offers.

One of the casino's simplest wagers is choosing between red and black in roulette. An American roulette wheel has 38 slots, of which 18 are black, 18 are red, and 2 are green (see Figure 7.10). When the wheel is spun, the ball is equally likely to come to rest in any of the slots. A bet of one dollar on red will pay off an additional dollar if the ball lands in a red slot; otherwise, the player loses.

In fact, a number of different bets can be placed in roulette. We could bet on any one number, on groups of numbers, or on odd and even numbers, as well as on red or black. But when we bet on red or black, the two green slots, zero and double-zero, belong to the house. Gamblers collect when red or black comes up, depending on their wagers, but the house takes everything bet on one spin when the ball lands in a green slot.

If we decide to bet on red, there are only two possible outcomes: win or lose. We win if the ball stops in one of the 18 red slots; we lose otherwise. Because casino roulette wheels are carefully balanced so that all slots are equally likely, the probabilities are

$$P(\text{win } \$1) = 18/38 = 0.47$$

$$P(\text{lose } \$1) = 20/38 = 0.53$$

Notice that laws 1 and 2 are satisfied by this assignment of probabilities to the outcomes.

Just as when only one bead is selected from the container of beads, there is no normal curve in sight when only one bet is made. But let's figure out the average winnings per bet when we play all evening, placing 50 bets. This is simply our overall gain (or loss) divided by 50. If we win 30 and lose 20 times, the overall gain is $10, an average winnings of $.20 per bet.

Figure 7.10 In the game of roulette, each gambler may win or lose, but the casino always makes a profit. [Las Vegas News Bureau.]

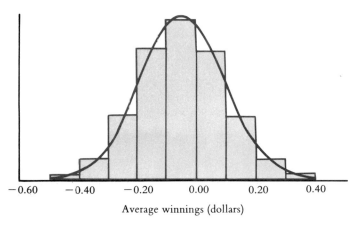

Figure 7.11   The distribution of winnings in repeated bets on red or black in roulette.

The central limit theorem applies to the average winnings of many bets. If we continue to gamble night after night, placing 50 bets each night, our average winnings per bet will vary from night to night. But a histogram of these values will follow a normal distribution (see Figure 7.11).

From Figure 7.11, we see that our income varied between $.47 lost and $.37 won. The mean of the curve is at a loss of $.05 (that is, at −0.05). This mean represents the expected result of many bets: on the average, we will lose 5 cents per bet. It is possible to calculate such expected values directly from the probabilities without actually making any bets (see Box 7.3). This more accurate calculation gives a mean of −0.053, an expected loss of 5.3 cents per dollar bet.

We can also see the standard deviation in Figure 7.11. The full spread of outcomes observed was −0.47 to 0.37, or 0.84 in all. By the 99.7 part of the 68-95-99.7 rule, the outcomes should span about three standard deviations on each side of the mean. The standard deviation is therefore about one-sixth of 0.84, or 0.14 (14 cents). Try checking this by locating the change-of-curvature points of the normal curve.

Using −0.053 as the mean, we expect almost all average winnings after 50 bets to fall between

$$-0.053 + (3)(0.14) = 0.367$$

and

$$-0.053 - (3)(0.14) = -0.473$$

So after 50 bets we will have overall winnings between

$$(0.367)(50) = 18.35$$

and

$$(-0.473)(50) = -23.65$$

We may win as much as $18.35 or lose as much as $23.65. Gambling is exciting because the outcome, even after an evening of bets, is uncertain. It is possible to walk away a winner. It's all a matter of luck.

The casino, however, is in a different position. It doesn't want excitement, just a steady income. The house bets with all its customers — perhaps 100,000 individual bets on black or red in a week. The distribution of average customer winnings on 100,000 bets is very close to normal, and the mean is still −0.053, or a loss of 5.3 cents per dollar bet.

The central limit theorem says not only that the average winnings closely follow a normal distribution, but that the standard deviation of this distribution decreases with the square root of the number of bets over which we are averaging. Now, 100,000 is 2000 times as much as 50. So the standard deviation of the casino's distribution (average winnings over 100,000 bets) is

$$\sqrt{2000} = 44.72$$

BOX 7.3 Expected Value and the Law of Large Numbers

When the possible outcomes of a random phenomenon are numbers, we might ask what the average outcome will be after many repetitions. For example, the outcomes of a bet on red or black in roulette are -1 (lose \$1) and 1 (win \$1). Now, -1 occurs in $20/38 = 0.53$ of all bets in the long run, and 1 occurs in $18/38 = 0.47$ of all bets. That's what probability means. So the long-run average outcome should be

$$(-1)(20/38) + (1)(18/38) = -0.053$$

Here is the general definition:

If the possible outcomes $s_1, s_2, \ldots, s_m$ in a sample space S are numbers, and the probability of outcome s_j is $P(s_j)$, then the *expected value* of the random outcome is

$$E = s_1 P(s_1) + s_2 P(s_2) + \cdots + s_m P(s_m)$$

To find the expected value, multiply each possible outcome by its probability and sum over all possible outcomes. The *law of large numbers* says that the average of the outcomes actually obtained in many trials gets ever closer to the expected value as the number of trials increases. So in many trials, the proportion of each outcome will be close to its probability, and the average outcome will be close to the expected value computed from these probabilities.

Here is another example. Suppose that you will roll a fair die and win as many dollars as the number of pips on the up face. Your expected winnings are

$$(1)(1/6) + (2)(1/6) + (3)(1/6) + (4)(1/6) + (5)(1/6) + (6)(1/6) = 21/6 = \$3.50$$

---

times as small as the standard deviation of the gambler's distribution (average over 50 bets). The gambler has standard deviation 0.14; the casino therefore has standard deviation

$$\frac{0.14}{44.72} = 0.003$$

There you have it. The individual gambler will experience wide variation in winnings; he or she gets excitement. The casino experiences very little variation; it has a business. Here is what the spread in the casino's average winnings per bet looks like after 100,000 bets:

Spread $=$ mean $\pm$ 3 standard deviations
$= -0.053 \pm (3)(0.003)$
$= -0.053 \pm 0.009$
$= -0.044$ to -0.062

Because the casino covers so many bets, the standard deviation of the average winnings per bet becomes very narrow. And because the mean is negative, almost all outcomes will be negative. Thus, the gamblers' losses and the casino's winnings are almost certain to average between 4.4 and 6.2 cents for every dollar bet.

The gamblers who collectively placed those 100,000 bets will lose money. We are now in a position to estimate the probable range of their losses:

$$(-0.044)(100{,}000) = -4400$$

$$(-0.062)(100{,}000) = -6200$$

The gamblers are almost certain to lose—and the casino is almost certain to take in—between $4400 and $6200 on those 100,000 bets. What's more, we have seen from the central limit theorem that the more bets that are made, the narrower is the range of possible outcomes. That is how a casino can make a business out of gambling. The more money that is bet, the more accurately the casino can predict its profits.

We have seen how probability describes the long-run pattern of outcomes of a random phenomenon. According to the central limit theorem, that pattern takes the shape of a normal distribution when each outcome is itself the average of a number of trials.

Normal distributions are completely determined if we know the mean and the standard deviation. Using the mathematics of normal distributions, we can make remarkably accurate predictions about the pattern of many events.

All this knowledge can be applied to gambling, genetics, and any other situation where the mathematics of chance comes into play. The theory of probability has given us tools to describe patterns and interpret data. These rules can help us to make reliable predictions and inferences about the world in which we live.

REVIEW VOCABULARY

Central limit theorem   The average of many independent random outcomes is approximately normally distributed. When we average n independent repetitions of the same random phenomenon, the resulting distribution of outcomes has standard deviation proportional to $1/\sqrt{n}$.

Event   Any collection of possible outcomes of a random phenomenon. An event is a subset of the sample space.

Normal distributions   A family of probability models that assign probabilities to events as areas under a curve. The normal curves are symmetric and bell-shaped. A particular curve is completely described by giving its center (the mean) and a measure of spread (the *standard deviation*).

Probability   A number between 0 and 1 that gives the long-run proportion of repetitions of a random phenomenon on which an event will occur.

Probability model   A sample space S together with an assignment of probabilities $P(s)$ to all outcomes s in S. The probabilities $P(s)$ must satisfy two laws:

1. For every outcome s, $0 \le P(s) \le 1$.

2. The sum of the probabilities $P(s)$ over all outcomes s is exactly 1.

Sample space   A list of all possible outcomes of a random phenomenon.

Sampling distribution   An assignment of probabilities to the possible values of a statistic. This distribution describes the sampling variability of the statistic.

Statistic   A number computed from a sample. In random sampling, the value of a statistic will vary in repeated sampling. This variation is called *sampling variability*.

EXERCISES

1. Some probabilities might be guessed in advance. For example, you would suspect that the probability of heads in tossing a coin is close to 0.5. Here are some random phenomena for which the distribution of probability is not so obvious. Repeat each at least 100 times to estimate probabilities.

   a. Toss a thumbtack on a hard surface. What is the probability of landing point up?

   b. Hold a penny upright on its edge under your forefinger on a hard surface, then snap it with your other forefinger so that it spins for some time before falling. What is the probability of heads?

2. Describe a reasonable sample space for each of the following random phenomena.

   a. You toss a coin 10 times and count the number of heads observed.

   b. You toss a coin 10 times and compute the percentage of heads among the outcomes.

   c. A female lab rat is about to give birth. You count the number of offspring in the litter.

   d. Subjects in a clinical trial are assigned at random to either treatment A or treatment B. For the next subject, we record treatment or control, male or female, and smoker or nonsmoker.

3. Which of the following are legitimate probability models for tossing three (possibly unfair) coins? Explain your answer in each case.

Outcome	Model A	Model B	Model C	Model D
H, H, H	0.125	0	0.125	0.250
H, H, T	0.125	0.375	0.250	0.125
H, T, H	0.125	0	−0.125	0.250
H, T, T	0.125	0.125	0.125	0.125
T, H, H	0.125	0	0.125	0.250
T, H, T	0.125	0.375	0.125	0.125
T, T, H	0.125	0	0.250	0.250
T, T, T	0.125	0.125	−0.125	0.125

4. You deal a single card from a bridge deck and record the face value of the card dealt (ace, king, queen, jack, ten, nine, . . . , two). Give an assignment of probabilities to these outcomes that should be correct if the deck is thoroughly shuffled. Give a second assignment of probabilities that is legitimate (that is, obeys laws 1 and 2) but differs from your first choice. Then give a third assignment of probabilities that is *not* legitimate, and explain what is wrong with this choice.

5. Exactly one of Brown, Chavez, and Williams will be promoted to partner in the law firm that employs them all. Brown thinks that she has probability 0.25 of winning the promotion and that Williams has probability 0.2. What probability does Brown assign to the outcome that Chavez is the one promoted?

6. Suppose that A and B are events that have no outcomes in common and thus cannot occur simultaneously. For example, in tossing three coins we could have $A = \{$First coin gives $H\}$ and $B = \{$First coin gives $T\}$. Starting from the definition of a probability model and the fact that the probability of any event is the sum of the probabilities of the outcomes making up the event, explain why

$$P(A \text{ or } B \text{ occurs}) = P(A) + P(B)$$

must always be true for two such events.

7. If a fair die is rolled once, it is reasonable to assign probability $1/6$ to each of the six faces. If we accept this probability model, what is the probability of rolling a number less than 3?

8. Box 7.2 gives a probability model for rolling two fair dice and recording the two up faces. Use this model to answer the following questions. What is the probability that the sum of the faces is 7? What is the probability that the sum of the faces is either 7 or 11?

9. Starting with the model in Box 7.2 for rolling two fair dice and recording the two up faces, give a probability model for rolling two fair dice and recording the sum of the faces showing.

10. Return to Exercises 6 and 7 of Chapter 5. Working as a team with other students, draw 100 simple random samples of size 5 from this population. Compute the sample proportion $\hat{p}$ of females in each sample. What probability model, based on your experiment, describes the sampling distribution of $\hat{p}$? Make a histogram of this distribution. (Note: Because both population and samples are small, and because the 100 samples will not be independent of each other if overlapping parts of the table of random numbers are used, your results will be only approximately correct.)

11. The distribution of heights of adult American men is approximately normal, with mean $\mu = 69$ inches and standard deviation $\sigma = 2.5$ inches. Draw a normal curve on which this μ and σ are correctly located. (Hint: Draw the curve first, then mark the horizontal axis.)

12. The figure that follows is a probability distribution that is not symmetric. The mean and median do not coincide. Which of the points marked is the mean of the distribution, and which is the median?

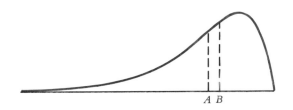

A B

13. Using the normal distribution described in Exercise 11, answer the following questions about the heights of adult American men.

    a. What percent of men are taller than 74 inches?

    b. Between what heights do the middle 95% of American men fall?

    c. What percent of men are shorter than 66.5 inches?

14. The upper and lower quartiles for any normal distribution are located 0.67 standard deviations above and below the mean, respectively. What are the quartiles of the distribution of heights of American men in Exercise 11?

15. The deciles of any normal distribution are located 1.28 standard deviations on either side of the mean. Answer the following questions about the distribution of SAT scores in the reference population($\mu = 500$, $\sigma = 100$).

    a. What score is needed to place you in the top 10%?

    b. Draw a median-quartile-decile box plot for the distribution of scores.

16. The concentration of the active ingredient in capsules of a prescription painkiller varies according to a normal distribution with mean 10% and standard deviation 0.2%. Use the 68-95-99.7 rule and the additional information about normal distributions given in Exercises 14 and 15 to answer the following questions.

    a. What range of concentrations covers the middle 95% of all the capsules? What range covers the middle half of all capsules?

    b. What is the median concentration? Explain your answer.

    c. What percent of all capsules have a concentration higher than 10.4%? Higher than 10.6%?

17. In Exercise 9, you found a probability model for rolling two fair dice and counting the pips on the two up faces. What is the expected number of pips obtained?

18. Teachers in the Lost Valley Central School District are allowed up to 7 days of paid sick leave each year. Here is the distribution of the number of days of sick leave taken by the teachers last year. What is the expected number of days of sick leave that a teacher will take in a year?

Days taken	0	1	2	3	4	5	6	7
Percent of teachers	15	15	10	10	10	12	8	20

19. Consider probability model A of Exercise 3 for tossing three coins. From this information, find the probability distribution of the number of heads obtained. Then compute the expected number of heads.

20. Another common casino game is Keno, in which 20 numbers between 1 and 80 are chosen at random and gamblers attempt to guess some of the numbers in advance. As in roulette, a bewildering variety of Keno bets are available. Here are some of the simpler Keno bets. Give the expected winnings for each.

a. A $1 bet on "Mark 1 number" pays $3 if the single number you mark is one of the 20 chosen; otherwise you lose your dollar.

b. A $1 bet on "Mark 2 numbers" pays $12 if both your numbers are among the 20 chosen. The probability of this is about 0.06. Is Mark 2 a more or a less favorable bet than Mark 1?

Statistical Inference

Inference is the process of reaching conclusions from evidence. Evidence can come in many forms. In a murder trial, evidence might be presented in an alibi, by a witness, or with a weapon. Evidence can also be more subtle. For example, if we walk into an office filled with papers, journals, library books, class notes, and a computer terminal, we might infer that the office belongs to a college professor. But in the case of statistical inference, the evidence is numerical.

Informal statistical evidence is based on a visual presentation of data. Formal evidence requires a statement of probabilities. Often informal evidence is enough to move us to take action. The histogram of steel-rod diameters in Chapter 6 plainly called for an investigation of the inspection process. In many other cases, we are looking for an overview or pattern rather than for specific answers. For example, the histogram of American League batting averages (p. 117), and the box plot comparing men's and women's earnings (p. 121) both provide informative overviews of data.

It often is difficult to reach a firm conclusion with informal evidence. We saw from Figure 6.9 in Chapter 6 that the 1970 draft lottery appeared to favor men born early in the year. However, that inequity was not large enough to be visible in a scatterplot, and we might well ask whether the 1970 outcome was simply due to chance rather than to systematic bias in the lottery. After all, any lottery will show some deviation from perfect uniformity due to the play of chance.

In many situations, we want to verify appearance by calculation. Formal statistical inference can be compared to a structural engineer's calculations of the load on a beam — we are more confident after the mathematics is done than we are if the engineer merely assumes that a beam will be large enough.

In the case of the 1970 draft lottery, calculation shows that a trend as strong as the one actually observed would occur less than 1 in 1000 times in a truly random lottery. One of the most intriguing aspects of statistical inference is the fact that *chance*—which usually causes more chaos than stability—is the ally rather than the enemy of confident conclusions.

At first, however, the opposite seems to be true. Suppose that the Gallup poll decided to take its weekly public opinion survey twice, separately and simultaneously selecting two random samples, sending out interviewers, and asking the same questions.

Two random samples, each selecting 1500 of 165 million U.S. residents aged 18 and over, will, of course, contain different people. And these different people will hold somewhat different opinions. Therefore, Gallup's announcement that 45% of Americans are afraid to go out at night for fear of crime really only refers to the 1500 people in one particular sample. And if the Gallup poll did take a simultaneous second survey, no doubt the results would be different. Random sampling may eliminate *bias,* but it can't eliminate *variability.*

How can we trust the results of a random sample, knowing that a second sample may yield a different result? For that matter, how can we trust the results of an experiment? As we saw in Chapter 5, the Physicians' Health Study tests the effects of aspirin and beta-carotene on reducing heart disease and cancer. We know, however, that different people react differently to drugs. What is more, the people in the four treatment groups were assigned by chance. A second trial with other participants would distribute drug reactions and persons with a high risk of cancer and heart disease differently among the treatments. We must ask whether the conclusions drawn from this particular study are really sound.

This chapter will address the issue of confidence in statistical conclusions. Formal statistical inference, in fact, enables us to quantify our confidence in the results of random surveys and randomized experiments and to verify our impressions by calculation.

Sampling Distributions and Confidence Intervals

We can use a simplified version of the Gallup survey to make some probability calculations. Suppose the Gallup poll drew a simple random sample of 1500 Americans and discovered that 45% were afraid to go out at night because of crime. The *sample proportion* $\hat{p} = 45\%$ is an estimate of the unknown proportion (call it p) of adult Americans in the entire population who stay home at night for fear of crime. A number such as p that describes a population is called a **parameter;** a number such as $\hat{p}$ that describes a sample is called a **statistic.**

In order to find out how close to the p the estimate $\hat{p}$ is, we must calculate the *sampling distribution.* This is the distribution of values taken by the sample proportion as it varies from sample to sample in a large number of repeated samples. If the sample size is relatively large, such as the one in this hypothetical poll, the sampling distribution will be very close to a normal curve (see Figure 8.1). The mean of the curve will be the true proportion p of people afraid to go out at night, and the standard deviation can be calculated by a simple formula based on p and the sample size n:

$$\text{Standard deviation of } \hat{p} = \sqrt{\frac{p(100 - p)}{n}}$$

(Throughout this chapter, p and $\hat{p}$ are measured in percent.)

Thus, if we repeated the sampling many times and sent out waves of interviewers across the nation, each time we would get a value of the sample proportion somewhere along this curve, and the values would vary according to

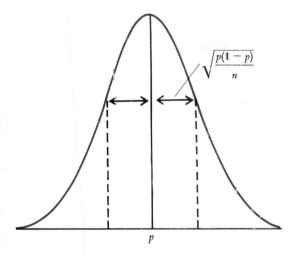

$$\sqrt{\frac{p(1-p)}{n}}$$

Figure 8.1   The sampling distribution of p.

this distribution. This is why it's called the sampling distribution. The value of the standard deviation will determine how spread out our normal curve is, or how often our sample proportion $\hat{p}$ will be close to the actual proportion p in the total population.

Our next step is to calculate the standard deviation in our example. Unfortunately, the formula uses the parameter p, and we don't know the value of p. If we did know p, we would not need to take a sample to estimate it! Fortunately, the standard deviation changes very slowly with p when the sample size n is large; therefore, knowing the exact value of p isn't essential.

For example, if $p = 40\%$ and our sample size is $n = 1500$, then the standard deviation of $\hat{p}$ is

$$\sqrt{\frac{(40)(60)}{1500}} = \sqrt{1.6} \approx 1.26$$

If $p = 50\%$ instead, then the standard deviation of p is

$$\sqrt{\frac{(50)(50)}{1500}} = \sqrt{1.67} \approx 1.29$$

You can see that these numbers are close together. Because $\hat{p}$ is close to p for large n, we can simply substitute $\hat{p}$ for p in the formula without changing the result very much.

Using this information, we can now calculate the standard deviation of the Gallup poll result. We know that $n = 1500$, and for p we can use the estimate $\hat{p} = 45\%$, based on our survey. Here the standard deviation is

$$\sqrt{\frac{45(100-45)}{1500}} \approx 1.28$$

The result is 1.28. This is called the *standard error of the estimate* $\hat{p}$. If we plot the mean and the standard error on the normal curve, then by the 68-95-99.7 rule, we know that with probability 95% the sample $\hat{p}$ will be within two standard deviations of the true p, that is, within about 2×1.28, or 2.6%.

In mathematical terms, the probability is 0.95 that the sample proportion will fall within $\pm 2.6\%$ of the unknown true fraction of people in the total population afraid to go out at night because of crime. In statistical shorthand, we can now say, "We are 95% confident that the true proportion lies in the interval $45\% \pm 2.6\%$." We now know the probability that our method gives a correct result. This *confidence statement* asserts how confident we can be that a specific application of our method is correct.

We can combine all these calculations into a single formula. If a simple random sample of size n is drawn where n is large, then a 95% **confidence interval** for the population proportion p is:

$$\hat{p} \pm 2\sqrt{\frac{\hat{p}(100-\hat{p})}{n}}$$

Remember that both p and $\hat{p}$ are measured in percent.

We can see how confidence intervals work by returning to the picture of the sampling

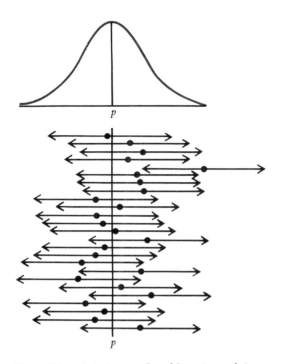

Figure 8.2 The behavior of confidence intervals in repeated sampling.

distribution of $\hat{p}$. In repeated samples, $\hat{p}$ varies according to this distribution. Each value of $\hat{p}$ produces a confidence interval. As Figure 8.2 shows, most of these intervals cover the true p. But one sample produced an interval that misses p. Of all possible samples, 95% give intervals covering p, and only 5% miss.

The confidence interval has two essential pieces: the confidence level (95%) and the margin of error. Knowing this, we can read our newspapers or listen to TV news broadcasts more carefully. Although reputable sample surveys such as Gallup, Lou Harris, or the Current Population Survey announce confidence intervals to help us interpret their results, editors often cut this information from their stories. Often a news report gives the margin of error without the confidence level; we need to know both figures because a lower confidence level allows a smaller margin of error. However, almost all polls of public opinion announce 95% confidence intervals. So if the story gives the margin of error without the confidence level, you can usually assume 95%. Box 8.1 summarizes the procedure for finding a 95% confidence interval.

The Bureau of Labor Statistics, on the other hand, chooses to announce its unemployment-survey results at a 90% level of confidence. Basing its conclusions on a monthly survey of 60,000 people, the bureau states that the proportion unemployed is within ±0.2% (two-tenths of 1 percent) of the figure they would get if they counted everyone. So when the headlines announce a 7.9% unemployment rate, the Bureau is saying—with 90% confidence—that between 7.7% and 8.1% of the labor force is out of work.

Statistics allow us to make inferences about a population on the basis of a sample, and as we have seen, the larger the sample size, the more confident we can be that our inference is close to the truth. Remember that a confidence interval has two essential parts: a confidence level and a margin of error. If 95% confidence is not sufficient, we can make a statement with 99.7% confidence if we adjust the margin of error, allowing three, rather than two, standard errors.

There is an important trade-off between confidence and margin of error. Higher confidence demands a larger margin of error as long as the sample size n remains fixed. Larger samples give smaller margins of error at the same confidence level. However, the square root of n that appears in the calculations shows that in order to reduce our margin of error by half, we need a sample size four times bigger. To obtain a very small margin of error, the Bureau of Labor Statistics goes to the trouble to interview a sample of 60,000 people, compared with the Gallup poll's 1500. The Gallup poll can afford to be 3% off. The unemployment rate must be more exact because so many economic and political decisions depend upon it.

BOX 8.1 Inference for a Population Proportion

We wish to estimate the proportion p in a population having some characteristic. Take a simple random sample of size n from this population and compute the sample proportion $\hat{p}$. The statistic $\hat{p}$ estimates the unknown parameter p, but we need an indication of how accurate the estimate is.

When the sample size n is large, the sampling distribution of $\hat{p}$ is approximately normal, with mean equal to p and standard deviation $\sqrt{\dfrac{p(100-p)}{n}}$. From this, a 95% confidence interval for p is given by

$$\hat{p} \pm 2 \sqrt{\frac{\hat{p}\,(100-\hat{p})}{n}}$$

Both p and $\hat{p}$ are expressed in percent in these formulas.

Example. A simple random sample of 100 seeds from a new lot are tested for germination; 87 germinated. Here $\hat{p} = 87/100 = 87\%$. A 95% confidence interval for the proportion p of all such seeds that will germinate is

$$87 \pm \sqrt{\frac{(87)(13)}{100}} = 87\% \pm 3.4\%$$

Statistical Process Control

Statistical inference has many practical uses other than polling the public and testing the effects of new medications. The basic tools of statistical inference can also be used to decide whether an industrial process is working properly. For this reason, statistics is in constant use in thousands of factories around the world.

At its Oklahoma City plant, AT&T Technologies manufactures the computerized electronic switches that interconnect our telephones. These switches are largely composed of complex electronic elements called circuit packs. AT&T needs efficient, accurate methods to check these newly manufactured circuit packs for errors. The best method is to monitor the manufacturing process to catch problems early rather than wait to inspect the product when it is done (see Box 8.2).

In one important step in this manufacturing process, a wave-soldering machine completes the electrical connections between the components and the printed wiring boards, using a wave of molten solder. In high-density circuit-pack manufacturing, there are about 2000 connections per circuit pack. Therefore, it is more advantageous to solder all the leads at once than to form each joint separately as in the traditional method. As the boards move along a conveyor, they pass through a wave of solder formed by a special nozzle that forces the solder to flow at such a velocity and angle that the boards are almost perfectly soldered. The process is delicate. Bad soldering will produce bad connections, both in the circuit pack

BOX 8.2 Check the Process Before the Product

Process control engineer
Connie Moore of AT&T.

At its Oklahoma City plant, AT&T Technologies uses statistical
sampling techniques to ensure smooth production of circuit packs.

Connie Moore, a process control engineer at AT&T, gives her views on ensuring
product quality.

> If we looked at 100% of the product it would take more time, more people, and
> would not give us any better infomation about the process. There was a time when
> industry thought that a quality control department's function was to inspect quality
> at the end of the line. Now we know that the only reasonable philosophy is to build
> it right the first time.
>
> I've been told that unless you measure how you're doing as you go along, you'll
> never know if you're done or if you succeeded. That's why I think that statistics and
> people are such an important combination. Statistics is the tool that tells us how
> we're doing as we go along and people are the force that drives us until we've
> succeeded.

and in our telephone conversations. AT&T
therefore monitors the performance of the
wave-soldering machine constantly and takes
immediate action if something goes wrong.

To accomplish this goal, workers called
process checkers take a sample of five circuit
boards every hour and inspect them carefully
for defects. Once a board is checked, a process

checker will calculate a number that expresses
the quality of soldering for that pack. The
sample mean of these quality scores is plotted
on a **process control chart.** This point is a
sample estimate of the quality of that hour's
production.

After plotting the points on this chart, it is
easy to determine from the positioning of

points near the center line if the manufacturing process is going smoothly. A point far from the center indicates an unnatural variation in the process and alerts the checker to investigate the irregularity. It may have been caused by a new operator who hasn't been properly trained or by a malfunction in the machine.

The purpose of process control sampling is not to check the function of the circuit boards; they will be rigorously tested when completed. The goal of sampling at this point is to monitor the soldering process and correct any malfunctions quickly. It is not practical to check every circuit board at every stage of manufacture. Instead, statistical sampling techniques give a quick and economical way to keep the process running smoothly.

Any industrial process will produce some variability. When the wave-soldering machine is performing its task properly, the quality index will vary according to a normal distribution. Let's imagine that we know from long experience that the mean of this distribution should be 100 and the standard deviation 4. In this case, almost all quality index values will fall within three standard deviations, or

12 points, on either side of the mean — between 88 and 112. AT&T knows that this range represents excellent soldering quality. But suppose something goes wrong. Perhaps the conveyor moves at the wrong speed or the solder temperature is too high or the solder doesn't flow properly. The mean of the distribution may drop below 100, and the quality will deteriorate.

AT&T collects five circuit packs each hour and uses the data on these to determine if the mean of the process is maintaining an acceptable value. If not, the process must be adjusted. Suppose that one sample has a mean quality index of 98.17 for the five packs and that another sample has a mean of 94.2. Should AT&T adjust the process in either case?

Once again, we are using a statistic (the mean of the sample) to estimate a parameter (the mean quality index of all circuit packs produced). And once again, a knowledge of sampling distributions helps us to estimate with confidence. The sample mean (call it $\bar{x}$) has a standard deviation smaller than that of individual measurements by a factor of the square root of the number of packs in the sample (see Box 8.3). Because the distribution has

---

BOX 8.3 Inference for a Population Mean

We wish to estimate the mean, μ, of a population (μ is the Greek letter mu and is the usual symbol for a population mean). Take a random sample of size n and compute the sample mean $\bar{x}$. This is our estimate of μ.

When n is large, the sampling distribution of $\bar{x}$ is approximately normal with mean μ. The standard deviation of this distribution is $\sigma/\sqrt{n}$, where σ is the standard deviation of the population (σ is the Greek letter sigma, the usual symbol for a population standard deviation).

If σ is known, a 95% confidence interval for μ is given by

$$\bar{x} \pm 2\frac{\sigma}{\sqrt{n}}$$

If σ is not known, it can be estimated from the sample. We will not study this case.

Example. Experience shows that the Scholastic Aptitude Test (SAT) scores for seniors in a large city have a mean that varies from year to year but a standard deviation that remains close to $\sigma = 87$. A simple random sample of 225 SAT scores from this year's senior class has sample mean $\bar{x} = 416$. A 95% confidence interval for the mean SAT score of all the city's seniors is

$$416 \pm (2)\left(\frac{87}{\sqrt{225}}\right) = 416 \pm 11.6$$

standard deviation 4, the sample mean will have standard deviation

$$\frac{4}{\sqrt{5}} = 1.79$$

If the mean for all the circuit packs stays at 100, then the sample means will have a normal distribution, also with a mean of 100, and with a standard deviation of 1.79. Therefore, 97.5% of all samples will have a mean higher than 96.42, two standard deviations below the mean. Another way to look at this relationship is to note that if one hour's sample has a smaller mean, there is only a 2.5% chance that it

would result from normal variation. So a sample mean below 96.42 is a clear signal that something in the wave-soldering process is amiss.

The calculation we made is very similar to the one used in Box 8.3 to establish a confidence interval, but instead of estimating the process mean, our purpose was to make a decision about its value. We can combine this calculation with the process control chart that plots the behavior of the process over time. This control chart tracks the sample means against time, at a rate of one per hour.

The center line at 100 in Figure 8.3 shows the target value for the process, and the line at 96.42 is called the *control limit*. It indicates when action needs to be taken. Most control charts have both upper and lower control limits lying at equal distances above and below the center line. The upper control limit for the quality-index chart would be at

$$100 + (2)(1.79) = 103.58$$

Because only decreases in the quality index concern us here, the upper control limit is not plotted. We have used 2σ limits for our control charts, that is, control limits located two standard deviations above and below the target mean μ. It is more common in industry to use 3σ limits to minimize the number of false alarms.

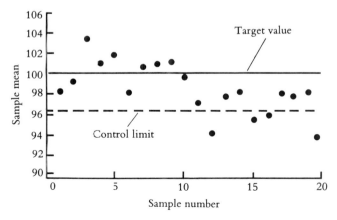

Figure 8.3   A control chart.

Control charts such as these bring together several aspects of statistics: First, they combine formal statistical inference with a graphic display of data in a form that can be used by someone with little statistical training. Second, control charts not only help to keep the quality of the final product at a high level but also keep down costs by catching malfunctions quickly, allowing a faulty process to be corrected immediately. This eliminates the need to repair or scrap parts at the end of the assembly line.

We can simulate sample means for 20 consecutive samples of five circuit packs, and we can plot the results on our hypothetical control chart. Let's concentrate on the later samples of the chart in Figure 8.3. We can see a downward shift in these means; they are systematically lower than those in the earlier samples. Not only are the means for samples 12, 15, 16, and 20 outside the control limits, but the last 11 means are all below the center line. In the long run, only half these means should be below the center line if the process mean is really 100.

It appears that the process quality shifted downward at about sample 9 or 10. The out-of-control point at sample 12 would trigger an investigation to find and correct the cause of this trend. Once again, a graph or picture can show vividly that something is wrong, and statistical analysis can help us make informed judgments about what we are seeing.

The Perils of Data Analysis

Statistical designs for collecting data may, like the Physicians' Health Study, involve experimentation. Or they may use sampling procedures such as those used in the unemployment rate and process control studies. In both cases, we rely on randomization and the mathematics of probability to compute sampling distributions. From sampling distributions we can obtain results that have known levels of confidence.

However, formal statistical inference, as reflected in levels of confidence, is secondary to well-designed data collection and to insight into the behavior of data. The effects of *hidden variables* can make even an apparently clear inference misleading. We saw in Chapter 5 that a well-designed experiment can control for the effects of hidden variables. However, if an experiment is not possible, we need to do statistical detective work. Let's look at a hypothetical example.

We will base our example on a study of admissions to graduate programs at the University of California at Berkeley but call our imaginary campus Metro University. A recent demonstration at Metro against sexual discrimination seemed to be based on some dramatic evidence. Eighty men applied to limited-enrollment courses at Metro, and 35 were admitted—almost half. Of the 60 women who applied, however only 20 were accepted, or one-third. These variables can be shown in a **two-way table:**

	Male		Female	
	No.	Percent	No.	Percent
Admit	35	44	20	33
Deny	45	55	40	66
	80	100	60	100

Because the two variables are simply categories (male and female) rather than measurements, using a scatterplot is not possible. The two-way table, however, can serve the same purpose, namely, to display the relationship between the variables. The table shows that men are much more likely than women to be admitted. About 44% of the men who applied were admitted, whereas only 33% of the women were selected. Probability tells us that the difference between the sexes is too large to be reasonably ascribed to chance. Is Metro University discriminating against its female applicants? Metro is concerned and wants to protect its good reputation as a school that supports sexual equality. Therefore, they decide to take another look at the data.

Men and women tend to apply to different departments. The organic chemistry course is very rigorous. Forty women applied to it, but only 10 got in, or one-fourth. Twenty men also applied, and 5 were admitted — again, one-fourth. These figures do not seem to point to discrimination.

The only other limited course offered at Metro is the history and sociology of the TV sitcom. Compared with organic chemistry, this course is something of a soft option. Sixty men applied, with 30 admitted, whereas 10 out of 20 female applicants were accepted. Once again, we find no apparent discrimination because 50% of both groups were admitted. We can look at the results in a pair of two-way tables:

Organic chemistry

	Male		Female	
	No.	Percent	No.	Percent
Admit	5	25	10	25
Deny	15	75	30	75

TV sitcom

	Male		Female	
	No.	Percent	No.	Percent
Admit	30	50	10	50
Deny	30	50	10	50

When added together, these tables give exactly the single two-way table for total admissions that we examined earlier. The numbers are remarkable. Each program appears to make no distinction between men and women. Yet when we consider the totals only, we see that 44% of the men who applied were admitted, compared with 33% of the women. This happened because two-thirds of the women applied to the more difficult course, whereas three-quarters of the men signed up for the easier sitcom class. The hidden variable we discovered explains the apparent inequity.

This discovery provides a lesson about statistical methods that we can apply to our understanding of real-world examples. Without walking through our hypothetical example, how could we know that data showing equality in each of several cases can appear as evidence of inequality when they are lumped together in a two-way table? At a time when statistical evidence of all kinds is increasingly being used to formulate social policy and resolve legal disputes, it is crucial that we select, analyze, and interpret our data with great care.

Relationships discovered by exploratory analysis can be misleading. Moreover, formal inferences, such as confidence statements, cannot be valid if the data were poorly collected or if the presence of hidden variables was not detected.

Even if carefully collected data are properly analyzed, there is no guarantee of success. There is always some chance, however small, that random selection will lead to a false conclusion. The strength of statistical inference is that the chance of a false conclusion is known and can be controlled by setting the confidence level as high as we think necessary.

Statistics does not produce proof. But in a world where proof is always wanting and most evidence is uncertain, statistical evidence is often the best evidence available.

REVIEW VOCABULARY

Confidence interval   An interval computed from a sample by a method that has a known probability of producing an interval containing the unknown parameter. This probability is called the *confidence level.*

Parameter   A number that describes the population. In statistical inference, the goal is often to estimate an unknown parameter or make a decision about its value.

Process control chart   A graph showing the value of a statistic for successive samples (for example, one sample each hour or one sample each shift). The graph also contains a *center line*

at the target value for the process parameter and *control limits* that the statistic will rarely fall outside of unless the process drifts away from the target. The purpose of a control chart is to monitor a process over time and signal when some unusual source of variation interferes with the process.

Two-way table   A table showing the frequencies (counts) or percentages of outcomes that are classified according to two variables (such as applicants classified by both sex and admission decision.)

EXERCISES

1. Identify each of the boldface numbers in the following examples as either a parameter or a statistic.

a. A random sample of male college students has a mean height of **64.5** inches, greater than the **63**-inch mean height of all adult American women.

b. A sample of students of high academic ability under 13 years of age was given the SAT mathematics examination, which is usually taken by high-school seniors. The mean score for the females in the sample was **386,** whereas the mean score of the males was **416.**

c. About 4% of all U.S. households are without telephones, but another **20%** have unlisted numbers.

2. Exercise 1 of Chapter 5 describes three samples. Find the sample proportion $\hat{p}$ in each case.

3. Each of the following statistics has a normal sampling distribution (at least approximately). Give the mean and standard deviation of the distribution in each case.

a. About 35% of residential telephones in the San Francisco area have unlisted numbers. A telephone sales organization uses random-digit dialing to dial a random sample of 200 residential telephone numbers. The percent of these that are unlisted is the statistic of interest.

b. A shipment of machined parts has a critical dimension that is normally distributed with mean 12 centimeters and standard deviation 0.01 centimeter. The acceptance sampling team measures a random sample of 25 of these parts; the sample mean $\bar{x}$ of the critical dimension for these parts is the statistic of interest.

4. A Gallup poll of a random sample of 1540 adults asked, "Do you happen to jog?" Fifteen percent answered "yes." The news item stated that these results have a 3% margin of error. Explain carefully, to someone who knows no statistics, what is meant by a "3% margin of error."

5. Suppose that the poll in Exercise 4 had used a simple random sample of size 1540, of whom 15% answered "yes." Give a 95% confidence interval for the percent of all adults who would have answered "yes" if asked.

6. A Gallup poll of 1514 adults taken between July 30 and August 2, 1983, asked, "Do you approve of the way Ronald Reagan is handling his job as President?" Of these, 41% said "yes."

> **a.** If the poll had used a simple random sample, what would have been the margin of error in a 95% confidence interval?
>
> **b.** The actual mar ;in of error for a Gallup poll of this same size is ±3%. Why does this not agree witl your result in **a**?

7. The standard error of $\hat{p}$ varies with p. Fortunately, it does not vary greatly unless p is near 0% or 100%. Suppose that $n = 1500$. Evaluate the standard error of $\hat{p}$ for $p = 30\%$, 40%, 50%, 60%, and 70%. (This is a secondary reason why the Gallup poll announces a larger margin of error than our formula gives. Rather than change the margin for every change in $\hat{p}$, the poll uses a conservative value good for all p from 30% to 70%.)

8. The following quotation from *Organic Gardening,* August 1983, describes a test of jars for home canning. Verify the statistical statement by giving a 95% confidence interval.

"The mayonnaise jars didn't do badly—only 3 out of 100 broke. Statistically this means you'd expect between 0% and 6.4% to break."

9. Electrical pin connectors for use in computers are gold plated for better conductivity. The specified plating thickness is 0.001 inch. Due to variations in the plating process, the actual plating thickness on different pins has a normal distribution with mean 0.001 inch and standard deviation 0.0001 inch. What range of plating thickness contains 95% of all pins?

Quality control samples of four pins are taken regularly during production. The plating thickness is measured and the sample mean of the four measurements is recorded on a control chart. What range of plating thickness contains 95% of the recorded sample means?

10. Explain carefully how the confidence interval for μ is obtained from the sampling distribution of $\bar{x}$ in Box 8.3.

11. A laboratory scale is known to have a standard deviation of $\sigma = 0.001$ gram in repeated weighings. Suppose that scale readings in repeated weighings are normally distributed, with mean equal to the true weight of the specimen. Three weighings of a specimen give (in grams)

$$3.412, 3.414, 3.415$$

Give a 95% confidence interval for the true weight of the specimen.

12. An automatic lathe machines shafts to specified diameters as part of a manufacturing operation. Due to small variations in the operation of the lathe, the actual diameters produced follow a normal distribution with standard deviation 0.0005 inch. The shafts now being produced are supposed to have a diameter of 0.75 inch. You measure 10 such shafts and find that they have a sample diameter of $\bar{x} = 0.7505$. Give a 95% confidence interval for the true mean diameter μ of the shafts being produced.

13. The deciles of any normal distribution are located 1.28 standard deviations above and below the mean.

    a. Use this information to give a recipe for an 80% confidence interval for p (for large n) and for μ (for normal populations).

    b. Give an 80% confidence interval for the situation described in part **a** of Exercise 6.

    c. Give an 80% confidence interval for the situation described in Exercise 11.

14. Give the center line and control limits for a control chart for means $\bar{x}$ of samples of size 4 in the gold plating process described in Exercise 9. Use 2σ limits, as in the text example.

15. It is common for laboratories to keep a control chart for a measurement process based on regular measurements of a standard specimen. Suppose that you are maintaining a control chart for the scale in Exercise 11 by weighing a 5-gram standard weight three times at regular intervals. What should be the center line of your chart? What are the control limits if you decide to use 3σ limits?

Use 3σ limits in the control charts of Exercises 16 through 18. That is, use control limits that are three standard deviations on either side of the mean μ. Notice that the center line and control limits are the same for all three charts.

16. In the data set below are $\bar{x}$'s from samples of size 4 with $\mu = 101.5$ and $\sigma = 0.2$. Only random variation is present. Make an $\bar{x}$ chart of these data using the given μ and σ. Are any points out of control?

Sample	$\bar{x}$	Sample	$\bar{x}$	Sample	$\bar{x}$
1	101.627	8	101.458	15	101.429
2	101.613	9	101.552	16	101.477
3	101.493	10	101.463	17	101.570
4	101.602	11	101.383	18	101.623
5	101.360	12	101.715	19	101.472
6	101.374	13	101.485	20	101.531
7	101.592	14	101.509		

17. The following set of $\bar{x}$'s for samples of size 4 illustrates the effect of a shift in the standard deviation. The first 10 samples have $\mu = 101.5$ and $\sigma = 0.2$, whereas the last 10 have $\mu = 101.5$ and $\sigma = 0.3$. Make a control chart for these data using the given μ and the original σ. Are any points out of control? Is the increase in σ visible in any way on the chart?

Sample	$\bar{x}$	Sample	$\bar{x}$	Sample	$\bar{x}$
1	101.602	8	101.453	15	101.756
2	101.547	9	101.446	16	101.707
3	101.312	10	101.522	17	101.612
4	101.449	11	101.664	18	101.628
5	101.401	12	101.823	19	101.603
6	101.608	13	101.629	20	101.816
7	101.471	14	101.602		

18. The following set of $\bar{x}$'s for samples of size 4 illustrates the effect of a steady drift in the mean of the population. The first 10 samples have $\mu = 101.5$ and $\sigma = 0.2$, whereas the last 10 have $\sigma = 0.2$ and μ increasing by 0.04 in each successive sample, reaching 101.7 at sample 15 and 101.9 at sample 20. Make an $\bar{x}$ chart for these data. Are any points out of control? Is the upward drift in μ visible in any way on the chart?

Sample	$\bar{x}$	Sample	$\bar{x}$	Sample	$\bar{x}$
1	101.458	8	101.695	15	101.408
2	101.618	9	101.351	16	101.616
3	101.507	10	101.555	17	101.350
4	101.494	11	101.398	18	101.508
5	101.533	12	101.522	19	101.432
6	101.334	13	101.448	20	101.674
7	101.547	14	101.612		

19. The U.S. government publication *Science Indicators 1980* shows that the average salary of women in all science and engineering fields is only 77% of the average salary for all male engineers and scientists. But the same source shows that in every individual field of science and engineering, the average female salary is at least 92% of the average male salary. Explain how this apparent discrepancy can come about.

20. Here is a two-way table:

Degrees earned in 1980, by level and sex (thousands)

	Bachelor's	Master's	Doctorate
Male	473.6	150.7	22.9
Female	455.8	147.3	9.7

Source: *Statistical Abstract of the United States, 1982–83.*

Use this table to answer the following questions.

    a. What percent of all bachelor's degrees were earned by women?

    b. How many master's degrees were awarded in 1980?

    c. What percent of all degrees earned by women were doctorates?

    d. The table shows that women earned almost as many master's but fewer bachelor's degrees and less than half as many doctorates as men earned. A breakdown by field of study shows that 103,453 master's degrees in education were awarded, of which women earned 72,578. What do you conclude about master's degrees in other fields of study?

SOCIAL CHOICE

A polyhedral torus as part of a central projection of a four-dimensional cube (a hypercube; see page 202). [Computer graphics image by Thomas F. Banchoff and his associates at Brown University.]

A revolution currently taking place in the field of mathematics is the successful use of mathematics as a fundamental tool to study human beings — their behavior, values, interactions, conflicts, organizations, fair allocations, and decision making, as well as their interface with modern technology and complex organizations. This latter revolution could eventually prove to be as far reaching as the turning of mathematics to study physical objects and their motion some three centuries ago. As mathematics and computers play an increasingly important role in understanding our social institutions, a new profession is emerging devoted to thinking mathematically about human affairs.

In particular, human decision making is being influenced profoundly by modern mathematics, and several particular mathematical subjects have been created primarily to assist in arriving at good decisions. Many aspects involved in arriving at a decision are, of course, nonquantitative in nature. These may relate to history, past experience, instinct, judgment, morality, and so forth. As a consequence one often refers to decision making as an art rather than as a science. On the other hand, many ingredients in contemporary decision making are mathematical in nature, and one can also view this activity as a scientific subject.

A decision maker will begin by listing the options over which he or she has some control, and the likely outcomes resulting from these choices. The person may attempt to identify all relevant variables and the relationships among them, and may associate quantitative measures when possible. Moreover, the decision maker must clarify his or her own values, identify desired goals, and spell out explicitly any limiting resources or social constraints. One then seeks the best possible result obtainable.

The situation is typically confounded by a variety of different uncertainties involving data and forecasting that typically cannot be completely resolved in advance. The effect that other decision makers may have on the outcome, and the best contingent responses to their moves, should be predetermined. Various ethical concerns such as fairness may well need consideration, and ways to ascertain group opinions may be necessary. Finally, decision makers must study the social and political context in which the decision will be implemented.

Several different mathematical subjects have been introduced since World War II primarily for the purpose of assisting individuals or groups in arriving at good or equitable decisions. As an illustration, a dozen major aspects of making decisions with the corresponding mathematical specialties are listed in the table on this page. All of these fields, except for continuous optimization, statistics, and probability theory, have developed mostly in recent decades and in the context of mathematics applied to human actions and organization.

In this part of the text we will illustrate a few of the mathematical techniques available to assist decision makers. In Chapter 9 we discuss the important problem of social choice. How does a group of individuals, each with his or her own set of values, select one outcome from a list of possibilities? This problem arises frequently in any democratic society, and even in more authoritarian institutions where deci-

A decision maker's concerns	Related mathematical subjects
1. Identify and measure strategic variables	Theory of measurement
2. Understand a complex system	System analysis, graph theory
3. Quantify one's preferences	Utility theory
4. Formulate objectives and constraints	Mathematical programming
5. Acquire data and forecast results	Statistics
6. Determine the most efficient outcomes	Optimization theories
7. Deal with uncertainty	Probability theory
8. Resolve conflicts	Game theory
9. Group decision mechanisms	Social choice theory
10. Equity considerations	Fair division theory
11. Make decisions using a multidisciplinary approach	Operations research, management science
12. Make decisions in an institutional setting	Policy science

sions are made by more than one person. We will learn that all voting systems have inherent flaws and that agenda designed for ascertaining the collective group will are often subject to manipulation. Group decision making is inherently a strategic encounter, and every citizen should be aware of the difficulties and pitfalls that can arise in this arena.

In Chapter 10 we consider decision-making bodies in which the individual voters or parties do not have equal power. In particular we will look at weighted voting systems such as stockholders in a corporation or political parties in a national assembly in which the voters cast different numbers of votes. We first observe that power in such systems is not necessarily proportional to the voters' weights. The notion of power is of fundamental importance in political science, although it is typically difficult to quantify. We will nevertheless describe one popular index for measuring power for weighted voting systems, an index that has proved useful in rulings by courts regarding local governments whose elected officials represent districts having different populations. It allows one to assign weighted votes to the legislators from constituencies of varying sizes so that the individuals in the districts are represented in an equitable manner.

Chapter 11 introduces the mathematical field called game theory, which describes situations involving two or more decision makers seeking different goals. Game theory provides a collection of models to assist in the analysis of conflict and cooperation. It prescribes optimal strategies for games of total conflict in which one's gain is another's loss. It also provides insights into purely cooperative situations as well as encounters of partial conflict that involve aspects of both competition and cooperation. Some particular games such as those known as "prisoners' dilemma" and "chicken" provide us with insights into certain social paradoxes that we routinely meet in our daily lives.

A general theme throughout this part concerns the idea of fairness in decision making. In Chapter 12 this becomes most explicit. Here we describe some fair division schemes in which a group of individuals with different tastes can be assured of each receiving what he or she views as a fair share when dividing up "smooth" objects like cakes or "chunky" goods such as estates. We then discuss the apportionment problem that is concerned with rounding fractions in an equitable manner. It arises in many fair allocation problems in which the things to be allocated must be multiples of some basic unit. This occurs, for example, in political representation, personnel assignments, and when adding capacity to a transportation system. The listing of reasonable assumptions for fair division schemes leads naturally to discussion of axioms needed for mathematical models to always achieve a fair distribution.

Social Choice: The Impossible Dream

The basic question of **social choice,** of how groups can best arrive at decisions, has occupied social philosophers and political scientists for centuries. Social-choice theory arose to help explain voting and other decision-making processes. Voting is a subject that lies at the very heart of representative government and participatory democracy. Surprisingly, voting poses difficult problems in both theory and practice.

The fundamental problem is to turn individual preferences for different outcomes into a single choice by the group as a whole. This situation arises whenever government representatives pass a bill, stockholders decide on a course of action, a political party nominates a presidential candidate, or a community elects members to serve on the school board. Few people realize that the voting method they use can significantly affect the outcome of an election.

The notion of **majority rule** still holds in some cases, but it is truly effective only in elections in which only two candidates are competing for a single office. The process is then quite simple: the candidate with the larger share of votes wins. But where there are more than two candidates, it is possible that the person with the highest tally will not actually hold a majority of all votes cast.

Most civilized societies have developed a variety of voting procedures to single out particular options from a longer list of feasible alternatives. How do we decide which of these voting schemes should be used?

All voting methods have some inherent faults and any method of voting we use will occasionally give rise to rather paradoxical results. So no single voting method is either universally applicable or the best overall: we have to select for each situation one of the many available voting schemes, knowing that the method we choose may greatly influence the result of the election (see Box 9.1).

The surprisingly large number of different voting procedures we can choose include majority rule, plurality wins, elimination and runoffs, sequential pairwise comparisons, various weighted or scoring schemes, approval voting, and a host of various other partitioning schemes that choose successively between subsets of potential outcomes. All these methods can produce at times some disconcerting results (see Box 9.2).

BOX 9.1 Historical Highlights of Voting Methods

Historically, popular voting methods have often lead to counterintuitive results. In the eighteenth century, certain voting paradoxes were brought to light by Jean-Charles de Borda (1733–1799) and Marie Jean Antoine Nicholas Caritat (1743–1794), the Marquis de Condorcet. Many scholars in the nineteenth and twentieth centuries also expressed concern over the lack of adequate voting methods to support the emerging representative forms of government and group decision making. For example, Lewis Carroll, whose real identity was the Oxford mathematician Reverend Charles Lutwidge Dodgson (1832–1898), appeared to deplore the tendency of voters to adapt a

principle of voting which makes an election more of a game of skill than a real tool of the wishes of the electors, and . . . my own opinion is that it is better for elections to be decided according to the wish of the majority than of those who happen to have the most skill at the game.

A review of historical developments in this area is given in *The Theory of Committees and Elections* (1958) by the English economist Duncan Black.

Kenneth Arrow's "impossibility" result in 1951 demonstrated that all attempts to arrive at a perfectly suitable technique of voting for all occasions were doomed to failure (see Box 9.3). His theorem inspired three different approaches to analyzing group decision procedures.

First, *social-choice* theory emphasizes normative models and the axiomatic approach. It is concerned with the conditions under which particular voting schemes will guarantee desirable outcomes. *Collective Choice and Social Welfare* (1970) by Oxford political economist A. K. Sen describes the social-choice perspective.

Second, the more descriptive approach called the theory of *public choice* explains the way individuals actually behave in existing group-decision forums to best obtain their aims. Among other things, public-choice theorists study the frequency of incidents involving undesirable outcomes or insincere voting. William H. Riker of the University of Rochester and James M. Buchanan of George Mason University (winner of the 1986 Nobel Memorial Prize in Economic Science) have been leading figures in these advances, and the journal *Public Choice* publishes research in the field.

A third direction is shown in the technical volume *Game Theoretical Analysis of Voting in Committees* (1984) by Bezalel Peleg. This approach accepts the strategic nature of group decision making as a given and uses existing subjects such as game theory to learn how to better compete when such instances arise.

BOX 9.2 Unattainable Ideals

The discoveries of mathematicians typically come in two forms: First, *theorems* provide statements of fact that are deduced from more basic premises. Second, *counterexamples* give particular cases showing that some conjectured statements are indeed not true in general. Further, some mathematical discoveries demonstrate that a certain presumed situation does not exist or that some highly desirable outcome cannot always be attained.

The Platonists of ancient Greece knew that the number $\sqrt{2}$ was *irrational*, in the sense that it could not be written as a ratio p/q for any two integers p and q. Such insights often have the effect of terminating the search for some ideal state or for certainty within a subject. They may place theoretical limits on what can be achieved by means of the scientific method.

In 1927, the German physicist Werner K. Heisenberg (1901–1976) announced the *uncertainty principle,* which states that it is impossible to determine precisely both the position and momentum of a particle at a given time. The product of the uncertainties in these two variables always exceeds a particular constant. This result played havoc with the popular deterministic philosophy of the nineteenth century.

In 1931, the Austrian-American mathematician Kurt Gödel (1906–1978) published his paper on "formally undecidable propositions." It proved that given any set of axioms, there would always be statements within the system ruled by these axioms that could be neither proved nor disproved on the basis of these axioms. This showed that the paradoxes that had been disturbing mathematical logicians for the previous half-century were unavoidable. Gödel showed, against the hopes of some, that the totality of mathematics could not be deduced from any single system of axioms.

In 1951, Kenneth J. Arrow (1921–) listed five highly desirable properties that one would expect any reasonable voting system to possess. He then went on to prove that no possible voting method could satisfy these properties in all situations. Every voting scheme will, at times, exhibit shortcomings. This *impossibility theorem* destroyed the dream of social philosophers who had sought fair and nonmanipulative social-choice mechanisms for more than a century. It also forced social scientists to use more rigorous methodologies in their analyses.

In 1980, Michel L. Balinski (1933–) and H. Peyton Young (1945–) showed that there is no general method for rounding a set of fractions to integers with a given sum that will always satisfy three very natural conditions. Therefore, allocating seats to states in the U.S. House of Representatives or seats to parties in parliament has no completely satisfactory solution. The two-century search by the U.S. Congress and other representative bodies was doomed from the start (see Chapter 12). More generally, attempts to apportion discrete objects in an equitable manner can result in undesirable allocations.

The outcome of an election can be affected by the type of voting method employed or by the order or formatting of the questions. The agenda itself can be a signficant factor in determining which motion survives, and agenda are usually subject to manipulation. For example, insincere amendments can be introduced for diversionary purposes or to kill off popular motions at an early stage. Voters may sometimes benefit by falsifying their preferences and misleading others. The final outcome of an election can be altered by any of these **strategic-voting** maneuvers.

As we focus on decision procedures in this chapter, we will see that it is no easy matter to ascertain the "true will of the people." The problem of unifying individual preferences into one particular choice for the whole group is essentially unsolvable. We will now examine a few actual cases to illustrate these difficulties in voting.

Plurality Vote

Almost 2000 years ago, the Roman historian Pliny the Younger was grappling with a voting dilemma:

A debate arose in the Senate concerning the freedmen of the consul Afranius Dexter; it being uncertain whether he killed himself, or whether he died by the hands of his freedmen; and again, whether they killed him from a spirit of malice, or of obedience.

It appears that there were questionable circumstances surrounding the death of Consul Dexter. If his freedmen (former slaves and servants) did in fact execute him, it is not clear whether it was an act of "mercy killing," according to Dexter's own wishes, or murder. His death may also have been the result of suicide. The three possible circumstances are shown in Figure 9.1, with appropriate verdicts for the freedmen indicated in parentheses. Pliny continues in his letter:

One of the senators (it is of little purpose I tell you I was the person) declared that he thought these freedmen ought to be put to the question, and afterwards released. The sentiments of another were, that the freedmen should be banished, and of another, that they should suffer death. It was impossible to reconcile such a diversity of opinions.

Pliny describes three groups in the Roman Senate:

- Group A believed the freedmen were innocent and thus favored their *acquittal*.

- Group B considered them guilty to some extent and thought the appropriate punishment was *banishment*.

- Group C believed the freedmen guilty of the crime of murder and felt that they should be *condemned* to death.

We can use A, B, and C to represent these three groups and a, b, and c to denote the three actions: acquittal, banishment, and condemnation. To be more specific, let's suppose that these three groups consist of 40%, 35%, and 25% of the senators, respectively. We can arrange this information in a table:

	Group		
	A	B	C
Preferred action	a	b	c
Percent	40	35	25

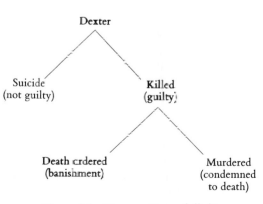

Figure 9.1   How was Dexter killed?

The outcome of this trial could depend as much on the voting procedure the senators use as on the kind of information they have, as well as the strategies or arguments they employ. We'll consider four agenda in this chapter.

If the prisoners' fate were determined by a **plurality** (agenda 1), then at first glance it looks as if they would go free: in a plurality vote, the position with the most votes is declared the winner. Our table shows that outcome *a* is favored by the largest number of the senators, 40%, including Pliny himself.

However, the senators may choose to vote differently. What if they know the distribution of potential votes in advance? Suppose they know that the numbers 40%, 35%, and 25% would be in favor of *a*, *b*, and *c*, respectively. Perhaps those in group *C* would then compromise on their hard-line position (the death penalty) and vote instead for banishment. The result would be 40% for *a* and 35% + 25% = 60% for *b*. Banishment would then carry the day.

In many cases, we can reasonably expect group *C* to vote in this way, that is, in favor of outcome *b* rather than their first choice, *c*. Such **strategic,** or **insincere, voting** is not at all uncommon.

How does strategic voting occur in this case? Suppose that we have additional information about the groups *A*, *B*, and *C*. Assume, for example, that we know their second and third choices as well. It seems reasonable that group *A* would prefer *a* over *b* and *b* over *c* and that group *C* would prefer the reverse order: *c* over *b* and *b* over *a*. It may well be that those in group *B*, who favor banishment *b*, would be divided in their second choice. Some may prefer acquittal *a* to the harsh penalty *c*, whereas others who are against the death penalty may nevertheless feel that the guilty must be punished in some manner, even if it means execution. For the sake of simplicity, however, let's assume that *all* members of group *B*

favor *b* over *a* and *a* over *c*. This **schedule of preferences** is summarized as follows:

	Group		
	A	*B*	*C*
First choice	*a*	*b*	*c*
Second choice	*b*	*a*	*b*
Third choice	*c*	*c*	*a*
Percent	40	35	25

It is fairly clear that the senators in group *A* have no real choice other than to vote for acquittal. So we'll assume that *A* votes for *a*. Group *B* is free to choose either *a*, *b*, or *c*, as is group *C*. We can represent these strategies and the resulting outcomes as follows:

A votes	*a*	C votes		
		a	*b*	*c*
B votes	*a*	*a*	*a*	*a*
	b	*a*	*b*	*a*
	c	*a*	*a*	*c*

In the preceding table, the three rows correspond to the three choices for group *B*, the three columns indicate the three options for group *C*, and the letters in the table itself are the outcomes when the corresponding strategies are chosen by these groups. For example, if *A* voted for *a*, *B* voted for *b*, and *C* voted for *c*, the result would be the boldfaced *a*, found in the second row and third column in this table. This result, in which each senator has voted for his most preferred outcome, is called **sincere** voting.

A closer examination of the table shows that both groups *B* and *C* will benefit if *C* votes for

b instead of c. If group C switches its vote, C will achieve its second choice b instead of its third choice a, and group B will achieve its first choice b.

		C votes		
A votes	a	a	b	c
		a	a	a
B votes	b	a	b	a
	c	a	a	c

This analysis suggests that group C should vote *insincerely* and select b. Groups B and C will in effect have formed a *coalition* against A. However, neither collusion nor communication need take place to bring this about. Not only do both C and B benefit from voting for b, but once they have done so there is no possible further switch that can do better for either. Thus this action is self-reinforcing since no voting group can now deviate unilaterally and expect to gain from its action.

Sequential Voting

One major objection to plurality voting is that when there are more than two outcomes, the final outcome may be favored by less than half of those voting — only 40% in the case of A. In order to make sure that the ultimate decision receives a majority vote, it may be necessary to resort to a **runoff** election or to some other type of **sequential voting** — a procedure that requires a majority vote at each step. We can now consider two such agenda that the Roman Senate could have used.

In agenda 2, the Senate will vote first between innocent and guilty; only when a guilty verdict results will they decide the punishment, b or c. This agenda is depicted in Figure 9.2.

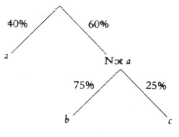

Figure 9 2   A runoff election.

For this scheme, a loses to "not a" 40% to 60% (35% + 25%) on the first round, and b beats c by 75% (40% + 35%) to 25% on the second ballot. You can see that banishment will win if the voters vote sincerely at each decision point. There is no maneuvering or collusion that A or C can undertake in this case that will produce a better outcome for them. Whereas insincere voting was the optimal strategy in agenda 1, the best strategy in this case is to vote in a sincere manner.

In agenda 3, we assume that the Senate moves to decide *first* upon the appropriate punishment, b or c, before it addresses the question of guilt. This agenda is pictured in Figure 9.3. Sincere voting would result in b winning over c by 75% to 25% on the first round and then b winning over a by 60% to 40%. It appears as though b should win.

However, it is not clear that A will vote sincerely, Group A may well vote for outcome c on the first ballot, which could result in the

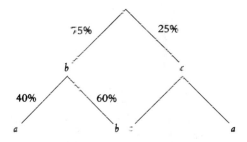

Figure 9.3   The outcome under sincere voting.

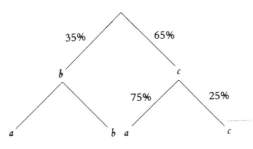

Figure 9.4   An outcome under strategic voting.

middle position *b* being eliminated on the first vote by 65% (40% + 25%) to 35%. After eliminating *b*, group *A* would change from *c* to *a* on the second ballot; thus, preference *a* would ultimately prevail over *c* by 75% (40% + 35%) to 25% (see Figure 9.4). When forced to choose between *a* and *c*, group *B* should go with *a*, their second choice, rather than *c*, their least-preferred outcome.

On the other hand, the groups may not vote this way after all. *C* is well aware that group *A*, in an effort to eliminate *b* at an early stage, may vote insincerely on the first round. In effect, *A* and *C*, representing the extreme positions *a* and *c*, have temporarily united to eliminate the middle position *b*. However, *C* need not go along with this ploy by *A*. To avoid eliminating *b*, *C* may actually vote for *b* instead of *c* at the first tally.

In short, in the first round *A* may vote for *c* rather than *b* and *C* may actually vote for *b* rather than *c*. This is the type of "deplorable" behavior that Pliny and Lewis Carroll must have had in mind! Strategically speaking, however, this type of thinking could be viewed as simple common sense.

Condorcet Winners

Let's make one more attempt to resolve Pliny's problem. We'll call this agenda 4. Consider what would happen if we held an election between each pair of outcomes: *a* versus *b*, *b* versus *c*, and *c* versus *a*. Under sincere voting,

- *b* beats *a* 60% to 40%
- *b* beats *c* 75% to 25%
- *a* beats *c* 75% to 25%

It seems as though *b* should be the winner because *b* beats either of the other positions when they meet head to head. Such a winner, if it exists, is called a **Condorcet winner.**

Once again, however, those in group *A* could frustrate the victory for outcome *b* by pretending to alter their "preference" schedule from

$\underline{A}$		$\underline{A}$
a	to	*a*
b		*c*
c		*b*

That is, they will vote for *c* over *b* whenever this pair comes up for a vote. As a result, we wind up with

- *b* beats *a* 60% to 40%
- *a* beats *c* 75% to 25%
- *c* beats *b* 65% to 35%

So we end up in a tie, indicated by the cycle

- *b* beats *a* beats *c* beats *b*

This last result is said to violate the **law of transitivity,** which says that if *x* is preferred to *y*, and if *y* is preferred to *z*, then *x* is preferred to *z*. Our example shows the **paradox of voting,** or the **Condorcet paradox:** even if individuals hold to the law of transitivity, the voters as a group may not satisfy it. It is, however, only one of a variety of different paradoxes that occur in voting situations (see Box 9.3).

Having discussed only a few of the many possible agenda that could have been used by the Roman Senate in the trial of Dexter's freedmen, we can appreciate Pliny's dilemma:

BOX 9.3 Kenneth J. Arrow

For centuries, mathematicians have searched for a perfect voting system. Finally, in 1952, economist Kenneth Arrow proved that finding an absolutely fair and decisive voting system is impossible. Kenneth Arrow is the Joan Kenney Professor of Economics, as well as a professor of operations research at Stanford University. In 1972, Arrow received the Nobel Prize in Economic Science for his outstanding work in the theory of general economic equilibrium. His numerous other honors include the 1986 von Neumann Theory Prize for his fundamental contributions to the decision sciences. He has served as president of the American Economic Association, the Institute of Management Sciences, and other organizations.

Dr. Arrow talks about the process by which he developed his famous impossibility theorem and his ideas on the laws that govern voting systems:

My first interest was in the theory of corporations. In a firm with many owners, how do the owners agree when they have different opinions, for example, about the prospects of the company? I was thinking of stockholders In the course of this, I realized that there was a paradox involved — that majority voting can lead to cycles. I then dropped that discussion because I was frustrated by it.

I happened to be working with the RAND Corporation one summer about a year or two later. They were very interested in applying concepts of rationality, particularly of game theory, to military and diplomatic affairs. That summer, I felt not like an economist but instead like a general social scientist or a mathematically oriented social scientist. There was tremendous interest in game theory, which was then new.

Someone there asked me, "What does it mean in terms of national interest?" I said, "Oh, that's a very simple matter," and he said, "Well why don't you write us a little memorandum on the subject." Trying to write that memorandum led to a sharper formulation of the social-choice question, and I realized that I had been thinking of it earlier in that other context.

I think that society must choose among a number of alternative policies. These policies may be thought of as quite comprehensive, covering a number of aspects: foreign policy, budgetary policy, or whatever. Now, each individual member of the society has a preference, or a set of preferences, over these alternatives. I guess that you can say one alternative is better than another. And these individual preferences have a property I call *rationality* or *consistency,* or more specifically, what is technically known as *transitivity*: if I prefer a to b, and b to c, then I prefer a to c.

Imagine that society has to make these choices among a set. Each individual has a preference ordering, a ranking of these alternatives. But we really want society, in some sense, to give a ranking of these alternatives. Well, you can always produce a ranking, but you would like it to have some properties. One is that, of course, it be responsive in some sense to the individual rankings. Another is that when you finish, you end up with a real ranking, that is, something that satisfies these consistency, or transitivity, properties. And a third condition is that when choosing between a number of alternatives, all I should take into account are the preferences of the individuals among those alternatives. If certain things are possible and some are impossible, I shouldn't ask individuals whether they care about the impossible alternatives, only the possible ones.

It turns out that if you impose the conditions I just stated, there is no method of putting together the individual preferences that satisfies all of them.

The whole idea of the axiomatic method was very much in the air among anybody who studied mathematics, particularly among those who studied the foundations of mathematics. The idea is that if you want to find out something, to find the properties, you say, "What would I like it to be?" [You do this] instead of trying to investigate special cases. And I was really accustomed to this approach. Of course, the actual process did involve trial and error.

But I went in with the idea that there was some method of handling this problem. I started out with some examples. I had already discovered that these led to some problems. The next thing that was reasonable was to write down a condition that I could outlaw. Then I constructed another example, another method that seemed to meet that problem, and something else didn't seem very right about it. Then I had to postulate that we have some other property. I found I was having difficulty satisfying all of these properties that I thought were desirable, and it occurred to me that they couldn't be satisfied.

After having formulated three or four conditions of this kind, I kept on experimenting. And lo and behold, no matter what I did, there was nothing that would satisfy these axioms. So after a few days of this, I began to get the idea that maybe there was another kind of theorem here, namely, that there was no voting method that would satisfy all the conditions that I regarded as rational and reasonable. It was at this point that I set out to prove it. And it actually turned out to be a matter of only a few days' work.

It should be made clear that my impossibility theorem is really a theorem [showing that] the contradictions are possible, not that they are necessary. What I claim is that given any voting procedure, there will be some possible set of preference orders for individuals that will lead to a contradiction of one of these axioms.

But you say, "Well, okay, since we can't get perfection, let's at least try to find a method that works well most of the time." Then when you do have a problem, you don't notice it as much. So my theorem is not a completely destructive or negative feature any more than the second law of thermodynamics means that people don't work on improving the efficiency of engines. We're told you'll never get 100% efficient engines. That's a fact—and a law. It doesn't mean you wouldn't like to go from 40% to 50%.

In the same vein, one way of developing research on voting is to seek systems that do break down occasionally. Just try to do it in such a way that it doesn't happen too often.

---

the fate of the freedmen depends to a large extent on game playing!

Bogus Amendments

We can use still another simple election situation to show how diversionary amendments, when strategically introduced, can mislead some voters into acting against their own interest.

Assume that three representatives, *A*, *B*, and *C* each have the choice of voting in favor of or against a new bill *N*. Voting against this new law means that the old law *O* will prevail. Assume that two of the three voters do prefer the proposed bill N over the existing law O, as indicated in this table of preferences:

	Voter		
	A	B	C
First choice	N	N	O
Second choice	O	O	N

In a direct comparison between the outcomes N and O, N will win by a vote of 2 to 1. Nevertheless, voter C may attempt to defeat N by the following maneuver. He proposes to modify the new bill N with an amended version called M. C selects the amendment so that A prefers M most of all, whereas B prefers M least of all. This may be done, perhaps, by merely shifting some of the proposed reward in bill N from B to A. Meanwhile C pretends to prefer O over M and M over N. The schedule of preferences now becomes:

	Voter		
	A	B	C
First choice	M	N	O
Second choice	N	O	M
Third choice	O	M	N

The new agenda and voting appear in Figure 9.5. When voting between N and M is sincere at the first decision point, M wins over

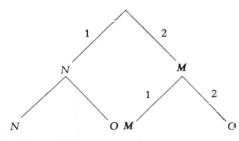

Figure 9.5 Voting on a bogus amendment.

N by 2 to 1. At the second step, O beats M by 2 to 1. Voter C has tricked the others into defeating N and maintaining the status quo O. Voter A should have noticed this tactic and resisted the temptation to initially vote for the fleeting amendment M.

In the course of a long, complex agenda and heated debate, we must be continuously on guard to avoid being manipulated into voting against our own long-range interests. Those designing the agenda can often rig it in their own favor. For example, one contingent might stack up a larger number of popular outcomes and pit them against a *single* highly desired one in an attempt to eliminate this single outcome at an early stage of the agenda. As a general rule of thumb, it's best to enter the more preferred outcomes at a later stage of the agenda. The chances of survival may increase when there are fewer competing alternatives and fewer remaining votes to be taken.

Everyone Wins

We have seen how the outcome of an election may very well depend on the voting procedure or agenda as well as on strategic choices by the voters. To show an extreme case of how the method chosen might affect the results in a realistic situation, we will consider an example of a political party convention at which five different voting schemes are adopted. Assume that there are 55 delegates to this national convention, at which five of the party members, denoted by A, B, C, D, and E, have been nominated as the party's presidential candidate. Each delegate must rank all five candidates according to his or her choice. Although there are 5! = 120 possible rankings, many fewer will appear in practice because electors typically split into blocks with similar rankings. Let's assume that our 55 delegates submit only six different preference schedules, as indicated in the table on the next page.

Delegates preference schedules

	Number of delegates					
	18	12	10	9	4	2
First choice	A	B	C	D	E	E
Second choice	D	E	B	C	B	C
Third choice	E	D	E	E	D	D
Fourth choice	C	C	D	B	C	B
Fifth choice	B	A	A	A	A	A

We see from the preceding table that the 18 delegates who most favor nominee A rank D second, E third, C fourth, and B fifth. Although A has the most first-place votes, he is actually ranked last by the other 37 delegates. Note that the 6 electors who most favor nominee E split into two subgroups of 4 and 2 because they differ between B and C on their second and fourth rankings. We will assume that our delegates must stick to these preference schedules throughout the following five voting methods. That is, we will not allow any delegate to switch preference ordering in order to vote in a more strategic manner.

1. *Plurality.* If the party were to elect its candidate by a simple plurality, nominee A would win with 18 first-place votes, in spite of the fact that A was favored by less than one-third of the electorate and was ranked dead last by the other 37 delegates.

2. *Runoff.* On the other hand, if the party decided that a runoff election should be held between the top two contenders, who together received a majority of the first-place votes in the initial plurality ballot, then candidate B outranks A on 37 of the 55 preference schedules and is declared the winner in the runoff.

3. *Eliminate the loser.* Another approach that could be used is holding a sequence of ballots and eliminating at each stage the nominee with the fewest first-place votes. The last to

survive this process becomes the winning candidate. We see in our example that E, with only 6 first-place votes, is eliminated in the first round. E can then be deleted from our table of preferences, and *all* 55 delegates will vote again on successive votes. On the second ballot, the number of first-place votes for the four remaining nominees is

$$\begin{array}{cccc} A & B & C & D \\ 18 & 16 & 12 & 9 \end{array}$$

Thus, D is eliminated. Note that the 6 delegates who most favored E earlier now vote for their second choices, that is, 4 for B and 2 for C. On the third ballot the nine first-place votes for D are reassigned to C, their second choice, giving

$$\begin{array}{ccc} A & B & C \\ 18 & 16 & 21 \end{array}$$

Thus, B is eliminated. On the final round, 37 of the 55 delegates favor C over A, and therefore C wins by this method.

4. *Borda count.* Given that they now have the complete preference schedule for each delegate, the party might instead choose to use a straight Borda count to pick the winner. This could be done, for example, by assigning 5 points to each first-place vote, 4 points for each second, 3 points for a third, 2 points for a fourth, and 1 point for a fifth. The highest total score of

$$191 = (5)(9) + (4)(18) + (3)(12 + 4 + 2) \\ + (2)(10) + (1)(0)$$

is achieved by D, who then wins. Note that A has the lowest score (127) and B the second worst (152).

5. *Condorcet.* In the Condorcet method, each nominee is matched head-to-head with every other. There are 10 such competitions, and each candidate appears in 4 of them. Assuming sincere voting, we can easily see that E wins out over

* A by a vote of 37 to 18
* B by a vote of 33 to 22
* C by a vote of 36 to 19
* D by a vote of 28 to 27

In this case, the Condorcet method does produce a winner, namely E.

In summary, our political party has employed five different common voting procedures and has come up with five different winning candidates. We see from this illustration that those with the power to select the voting method may well determine the outcome.

Approval Voting

We know from Arrow's theorem (Box 9.3) that there can never be a perfect voting system. To select a voting system is to compromise between the different shortcomings inherent in each. Nonetheless, social-choice theorists continually strive to create (or perhaps rediscover) better voting schemes in an attempt to minimize such flaws.

One voting method that shows great promise for electoral reform is called **approval voting.** It is particularly suitable for elections in which several candidates typically compete, such as the party primary elections for the President of the United States. Many existing multicandidate-election procedures should be reviewed with a mind toward adopting this simple and practical method.

In approval voting, each voter is allowed to give *one* vote to each of the candidates on the multicandidate slate. No limit is set on the number of candidates an individual can vote for: Voters can approve of as many choices as they like and show disapproval by withhold-

ing a vote on that candidate. This system replaces the traditional "one person, one vote" by "one candidate, one vote."

The winner in approval voting is the candidate who receives the largest number of approval votes. This approach is also appropriate in situations where more than one candidate or outcome may win, for example, in electing new members to an exclusive society such as the National Academy of Sciences or the Baseball Hall of Fame.

In recent years, several important officials in New York State have been elected with much less than a majority of the vote. Clearly, some of these contests would have been reversed if approval voting had been used. The 1970 U.S. Senate race in New York State gave James Buckley 39%, Richard Ottinger 37%, and Charles Goodell 24% of the vote. Buckley may well have been last in approval voting or been the Condorcet loser in a head-to-head battle with either of the other two candidates. The result in 1980 was Senator Alfonse D'Amato 45%, Elizabeth Holtzman 44%, and incumbent Jacob Javits 11%. Polls indicated that more of Javits' supporters preferred Holtzman to D'Amato and that she probably would have won in a runoff or under approval voting. John Lindsay was reelected mayor of New York City in 1969 with only 42% of the vote. In 1977, Edward Koch beat Mario Cuomo for mayor in a runoff after they initially won only 19.8% and 18.6% of the vote, respectively. (Four others also got over 10% of the vote in that initial election.)

Approval voting may well prove to be particularly effective in presidential primaries when several contestants are entered. The result of the 1980 New Hampshire elections for the Republican Party were

Ronald Reagan	50%
George Bush	23%
Howard Baker	13%
Others	14%

An ABC News exit poll indicated that approval voting may have given the following tally:

Ronald Reagan	58%
George Bush	39%
Howard Baker	41%

It is very possible that such results would have delayed the withdrawal of Senator Baker from the race. Perhaps he would have become Vice President.

Approval voting, like any other voting method, is not entirely free of faults. It, too, is subject to strategic manipulation and can give rise to counterintuitive outcomes. Analyses to date indicate that it is no more vulnerable than other known methods to insincere voting. On the other hand, it is practical, simple, and easy to implement. It gives the voters greater freedom in expressing themselves without requiring more-complicated ranking schemes, which have their own inherent problems. Approval voting has a lot in its favor and seems ripe for widespread implementation.

REVIEW VOCABULARY

Agenda An ordered list of the ballots to be taken in order to arrive at a decision.

Amendment A proposed change, deletion, or addition to a motion under consideration.

Approval voting Each voter indicates approval or disapproval for each candidate or issue on a ballot, as opposed to voting for only one candidate or issue.

Arrow's impossibility theorem The discovery by Kenneth J. Arrow that any voting system can give undesirable outcomes.

Borda count Assigning points to voters' preferences and summing the points for each candidate to determine the winner.

Condorcet (or voting) paradox Candidate A beats B, B beats C, and C beats A.

Condorcet winner A candidate who beats every other candidate in a one-on-one ballot.

Insincere voting Voting contrary to one's true preferences in order to obtain a better outcome in the long run.

Law of transitivity If A wins over B and B wins over C, then A must win over C.

Majority More than half of the votes cast.

Plurality The case where a candidate with the most votes in a multicandidate race is declared the winner. The number of votes for the winner could be less than half.

Preference schedule A list of possible outcomes in the order a voter most prefers them.

Sequential voting A voting procedure in which successive ballots are taken for the purpose of eliminating some candidates or issues before the final vote.

Sincere voting Voting in a manner consistent with one's preference schedule. One always votes for the most-preferred candidates or outcomes on each ballot.

Strategic voting Voting insincerely on a ballot in an attempt to achieve a more preferable final outcome than could have resulted by voting sincerely.

EXERCISES

1. How many different ways can a voter:

   a. rank 3 choices (when ties are not allowed)?

b. rank 4 alternatives (without ties)?

c. rank n potential outcomes (without ties)?

d. rank these possibilities when ties are allowed?

2. Consider the example in the text with 55 delegates at a national convention voting on five candidates for the party's presidential nominee. Determine the winning candidate if they used a voting method that eliminates the loser at each step: at each ballot, eliminate the candidate with the most last place votes, and then continue with successive ballots with all 55 delegates voting each time.

3. The 10 members of a party's platform committee must pick one issue to receive the highest priority in the upcoming campaign. The three contenders are defense D, education E, and health H, and their preference schedules are as follows:

	Number of members		
	4	3	3
First choice	D	E	H
Second choice	E	H	D
Third choice	H	D	E

a. Which issue wins if they first vote between E and H, and then vote between this initial winner and D?

b. Which issue wins if they first vote between D and E, and then vote between this initial winner and H?

c. Could those who most prefer E vote insincerely in some way to change the outcomes in **a** or **b** in a way that benefits them?

4. One hundred voters who are to elect one of the three candidates A, B, or C have the following preference schedules:

	Number of voters			
	38	30	25	7
First choice	A	C	B	B
Second choice	B	A	C	A
Third choice	C	B	A	C

a. Which candidate wins an election using the plurality method?

b. Who wins if there is a runoff election between the top two finishers in the initial plurality ballot in **a**?

c. Who would win in **a** and **b**, respectively, if the 7 voters who prefer B over A and A over C were to switch their preference ranking to A over B over C?

d. If the 45 voters who now prefer *A* over *B* over *C* (after the switch made in **c**) knew everyone's preference schedule, could they vote more strategically to ensure a victory for *A* when the voting method in **b** is used?

5. Thirteen students decide to vote on whether to play baseball *B*, soccer *S*, or volleyball *V* at their picnic. Their schedule of preference is as follows:

	Number of students			
	5	2	4	2
First choice	B	S	V	V
Second choice	S	V	B	S
Third choice	V	B	S	B

a. Which sport wins if they use the plurality method?

b. Which one wins if they use a Borda count that assigns, 3, 2, and 1, points to each first, second, and third choice, respectively?

c. Which one wins if they first eliminate the one with the fewest first-place votes and hold a runoff between the other two?

d. Which one wins if they first eliminate the one with the most last-place votes and have a runoff between the other two? Is the method decisive in this case?

e. Which one wins in **d** if the last two students misrepresent their preference ranking and pretend it is *V* over *B* over *S* rather than *V* over *S* over *B* as listed in the table?

f. Would there be a Condorcet winner if the students did vote sincerely?

6. One hundred sports writers with the following preference schedules are to pick the best college football team among Alabama *A*, Michigan *M*, and Washington *W*.

	Number of writers		
	52	38	10
First choice	W	M	A
Second choice	M	W	M
Third choice	A	A	W

a. Which team wins if the election is by a Borda count that assigns 3, 2, and 1 points to each first, second, and third choice, respectively?

b. If those who most favor Michigan suspected that Washington would win and thus voted insincerely for Alabama as their second choice, what would the outcome be?

c. If the supporters of Washington believed that the insincere voting in b might take place, could they still vote so as to guarantee that Washington wins?

7. Eleven students must decide whether to dine together at a Chinese, Italian, or Mexican restaurant. Their preference schedules are as follows:

	Number of students		
	5	2	4
First choice	Chinese	Mexican	Italian
Second choice	Mexican	Italian	Mexican
Third choice	Italian	Chinese	Chinese

a. What choice will the group make if they vote sincerely according to the following methods:

(i) The plurality method.

(ii) Eliminating the restaurant with the fewest first-place votes and having a runoff between the other two.

(iii) Eliminating the restaurant with the most last-place votes and having a runoff between the other two.

b. Is any restaurant a Condorcet winner?

c. What choice will be made if they use a Borda count that assigns x points to each first choice, y points to each second choice, and z points to each third choice when

(i) $x = 3$, $y = 2$, and $z = 1$?

(ii) $x = 4$, $y = 2$, and $z = 1$?

(iii) $x = 5$, $y = 2$, and $z = 1$?

d. Is there any way to pick the points x, y, and z in c with $x > y > z$ so that the Italian restaurant wins the Borda count?

8. Given that three members of a four-person committee prefer a newly proposed bill N over the old existing law O, can you suggest the type of amendment M to N that the advocate of O should propose in an attempt to defeat N (and M). Note that M or N must receive three or four of the four votes cast in order to pass, whereas the existing law wins on a tie vote of two to two.

9. Assume that the members A, B, and C of a three-person committee have the following preference schedules over the three possible outcomes a, b, and c:

	Member		
	A	*B*	*C*
First choice	*a*	*b*	*c*
Second choice	*b*	*c*	*a*
Third choice	*c*	*a*	*b*

Each member can vote secretly for one outcome, and the majority rules. Furthermore, *A* is the chairman and has the power to break tie votes.

   a. What would the result be if each member voted sincerely for his or her most-preferred outcome?

   b. What do you expect to actually happen in this situation?

   c. Can you explain why this example is often referred to as *the chairman's paradox?*

10. Can you prove that $\sqrt{2}$ is an irrational number? Hint: Assume to the contrary that $\sqrt{2} = p/q$, where p and q are integers and p/q is a fraction expressed in *lowest terms.* Then note that $p^2 = 2q^2$ and so $p = 2r$ is an even number, and thus $p^2 = (2r)^2 = 4r^2 = 2q^2$ and so q is also an even number.

11. Consider the following class project: Pick some upcoming election involving more than two alternatives. For example, select a few of the leading candidates for a major party's presidential nominee. Compare the class results for the following different voting methods:

   a. Vote for only one candidate and select the winner by the plurality method.

   b. If the winner in **a** does not have a majority, then hold a runoff ballot.

   c. Have each voter provide his or her preference schedule (that is, each ranks the candidates) and then select the winner by a Borda count.

   d. Use the method of approval voting where the winner is the one with the largest number of approval votes.

CHAPTER 10

Weighted Voting Systems: How to Measure Power

In many voting situations, each citizen has an equally weighted vote, called "one man, one vote." Each person clearly asserts equal influence on the outcome. Each is equally powerful. On the other hand, many voting situations occur in which the participants have different numbers of votes. Such systems are referred to as **weighted voting systems.** In these instances, some individuals may have greater influence than others on the results. Those individuals who cast more heavily weighted votes may or may not turn out to be more powerful, and blocs of voters, called coalitions, can typically increase their power (see Box 10.1).

For example, the shareholders' votes in a public corporation are weighted by the number of shares of stock they own. In the U.S. Electoral College, each state typically casts its total bloc of votes to one presidential candidate. The number of votes per state varies from 3 for states with small populations to 47 for

California. Legislative bodies with strong party discipline can be viewed as weighted voting systems in which the parties, rather than the individual representatives, are the voting units.

In this chapter, we will be primarily interested in the notion of *power*. We will attempt to arrive at a mathematical measure for the power of an individual voter or of a bloc of voters in a weighted voting system. We will see that a person's influence is not always proportional to the fraction of the votes he or she casts. In a formal sense, power is a much more irregular function of the distribution of weights among the voters as well as of the fraction, or **quota,** of votes necessary to pass legislation and the total number of voters. We will first develop an index for measuring power, which is commonly referred to as the **Banzhaf index,** and then return to several familiar institutions that actually make use of

BOX 10.1 Sets and Subsets

The most primitive notion in contemporary mathematics is the idea of a set. A **set** is a collection of distinct objects of our imagination. The objects in the set are called **elements.** The great German mathematician Georg Cantor (1845–1918), who advocated this approach, described a set as

a bringing together into a whole of definite well-distinguished objects of our perception or thought — which are to be called the elements of the set.

Our concern in this chapter is with sets whose elements are voters and whose subsets are the various coalitions or blocs of voters.

It should be emphasized that one does not give definitions of the words "set" or "element." A theory must begin with some undefined terms, plus some rules of logical inference, before it proceeds to introduce new, defined terms and initial axioms and finally goes on to deduce new theorems.

We can talk about a set of physical objects in the external world, such as the set of students in the classroom, the set of cars in the parking lot, or the set of planets in our solar system. In mathematics we usually speak about mental objects, and we represent them with abstract symbols. We represent a set of particular elements by listing these elements between braces. Here are some examples of sets that arise in mathematics:

1. The set of integers from 1 to 3, which is written $\{1, 2, 3\}$.

2. The set of vertices, or corner points, of a unit square. Using the typical Cartesian coordinates in a plane, we can express these as $\{(0, 0), (0, 1), (1, 0), (1, 1)\}$.

3. The set of edges of a cube.

4. The set of roots of the equation $x^2 - 5x + 4 = 0$, which is $\{1, 4\}$.

5. The set of points in the plane that are inside the circle described by the equation $x^2 + y^2 = 4$. This can be expressed $\{(x, y): x^2 + y^2 < 4\}$.

We also consider the set that contains no elements whatsoever. This is called the **empty set** and will be denoted by the symbol $\varnothing$.

The number of distinct elements in a set is called its **cardinality.** For example, the cardinalities of the set examples 1 through 5 are 3, 4, 12, 2, infinity, respectively.

A set T is called a **subset** of a set S if every element in T is also an element of S. For example, the set $\{1\}$ is a subset of the set $\{1, 2, 3\}$ in example 1 and of the set $\{1, 4\}$ in example 4. Any set is considered a subset of itself, whereas the empty set $\varnothing$ is a subset of every set. A set with only one element, that is, a set of cardinality 1, has 2 subsets: itself and the empty set. The set of all subsets of $\{1, 4\}$ is $\{\{1, 4\}, \{1\}, \{4\}, \varnothing\}$. The set $\{1, 2, 3\}$ has 8 subsets: $\{1, 2, 3\}, \{1, 2\}, \{1, 3\}, \{2, 3\}, \{1\}, \{2\}, \{3\}$, and $\varnothing$. A set with four elements has 16 subsets. In general, a set of cardinality n has 2^n subsets.

One can picture the subsets of a set with cardinality 1, 2, 3, and 4 with the help of geometrical figures. The relationship "is a subset of" is indicated by the edges in the diagrams that follow. For example, the eight subsets of {1, 2, 3} can be represented by a cube. The edge arising from {1, 3} indicates that it is a subset of {1, 2, 3} and the two edges connecting {1, 3} to {1} and {3} show that the latter are subsets of the former.

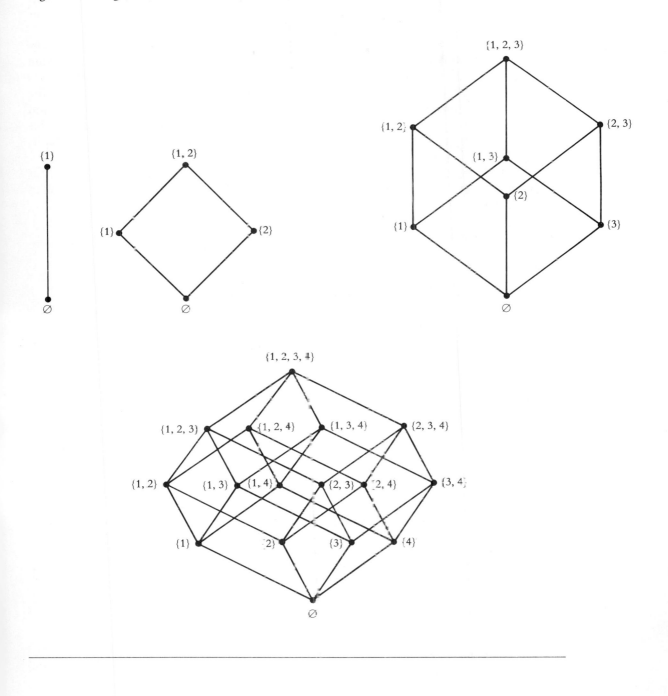

weighted voting. The same mathematical concepts have many other applications besides measuring power in voting systems (see Box 10.2).

How Weighted Voting Works

To see how weighted voting works, let's consider an actual example. In the late 1950s, New York's Nassau County had a County Board of Supervisors consisting of six representatives from five separate districts. Two of the representatives were elected at large from one district, the city of Hempstead. Because the districts had unequal populations, the representatives' votes were weighted accordingly. The names of the five districts and the number of votes for each supervisor are listed in Table 10.1.

A close look at these numbers reveals a surprising fact. The two representatives from Hempstead share a total of 18 votes. This is greater than 15, or over half of the total for the entire board. The two representatives from Hempstead could therefore pass any issue requiring only a simple majority quota, that is, 16 of the 30 votes.

(Actually, the charter for Nassau County at the time had a secondary condition that in

---

BOX 10.2 Binary Systems

A great number of different types of systems have inputs of a binary nature, which in turn determine an output of a binary type. In this chapter we are concerned with voting systems in which individuals vote "yes" or "no" and the resulting combination of votes determines whether an issue passes or fails. This system is illustrated by the following diagram:

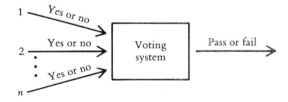

There are many other instances in which the inputs are either one of two alternatives and the resulting output or consequence is also twofold. Some examples of this common phenomena follow:

1. An examination is passed or failed depending on which questions are answered correctly or incorrectly. The questions may in turn be of a true or false category.

2. A machine runs or stalls out, depending on which component parts function properly or break down.

3. A message gets through or is cut off, depending on which channels are functioning or impeded.

4. Whether a vehicle successfully reaches its destination or is halted depends on which routes are passable or blocked. This may depend on which drawbridges are up or down along the river that divides a city.

5. A battle is won or lost, depending on which supply links are usable or interdicted. In turn, "victory" or "defeat" in a war depends on which battles are won or lost.

6. A facility is open for business or shut down, depending on which employees are present or absent.

7. A project is feasible or impossible, depending on what talent or equipment is available or tied up.

8. An electrical signal is present or absent, depending on whether various contacts are open or closed.

9. A security system is safe or breachable, depending on which detectors are sensing or inactivated — for example, which guards are alert or sleeping.

The techniques developed in this chapter for measuring power in voting systems can be applied to other types of binary systems. We can use them to arrive at some measure of the system's reliability, the importance of a particular component in it, or the probability that the system will or will not function.

Advances in engineering and materials have lead in turn to major progress in the ability of modern digital computers to perform complex calculations rapidly. The design, electronics, inner logic, networking, and programming activities related to these computing machines rely heavily upon *discrete mathematics*. In particular, the mathematics of binary systems arises frequently in the construction and use of computers. These discrete forms of mathematics contrast with the more continuous mathematics of calculus, advanced by Sir Isaac Newton (1642–1727) and Gottfried W. Leibnitz (1646–1716), which has proved so useful to the physical sciences and traditional engineering.

If modern technology had instead taken us in the direction of some sort of analog computing device, then the more continuous approaches of classical mathematics might have played a more fundamental role in arriving at today's computer science and computer engineering as well.

---

fact disallowed any one district from passing a bill whenever it is opposed by the other four districts. Because the first two representatives both come from the same municipality, Hempstead, this provision actually prevented this particular pair from forming a winning coalition. Although this secondary rule alters the analysis presented here, including the existence of dummy voters described below, for purposes of illustration we will ignore this secondary condition in our discussion. Other weighted voting systems used by counties in

TABLE 10.1 The Nassau county board in the 1950s

Municipality	1954 population	1958 number of votes
Hempstead ⎫	618,055	⎧ 9
Hempstead ⎭		⎩ 9
North Hempstead	184,060	7
Oyster Bay	164,716	3
Glen Cove	19,296	1
Long Beach	17,999	1
Totals	1,004,135	30

New York State and elsewhere have had powerless voters with no such secondary provision to save them from this fate.)

The supervisor from North Hempstead, who has 7 votes, can join with *either one* of the two from Hempstead, and the resulting sum of 16 is also a simple majority (unaffected by the secondary provision). Thus, any two of those first three representatives have sufficient weight to pass a bill. Apart from the special condition, which we are ignoring, these three legislators do indeed share equal power.

Ironically, a close look at the figures for Oyster Bay, Glen Cove, and Long Beach reveals that these last three supervisors possess no power whatsoever. Because they share a total of only 5 votes, they cannot team up with *any* of the other representatives to turn a losing coalition into a winning one. If they joined one of the representatives from Hempstead, their votes would total 14. If they joined the representative from North Hempstead, their votes would total 12. In either case they fall short of the simple majority quota of 16 necessary to pass most issues in this county.

We can see that in the case of a simple majority, these three supervisors are left without any actual voting power, even though they may have other kinds of influence on the decision-making process; for example, they may serve on committees or draw up and defend potential legislation. But as it stands, they are essentially disenfranchised by the initial constitutional structure of the board's voting rules. A voter who cannot turn a losing coalition into a winning one is called a **dummy.** A dummy can likewise never cause a winning coalition to lose by singly defecting from it.

Some time later, the weights for the Nassau County Board were changed (see Table 10.2).

In this case, the two representatives from Hempstead and the one from Oyster Bay now share equal power in the typical case of a simple majority vote of 58. The remaining three supervisors are dummies. Note that between 1958 and 1964, the third representative, the one from North Hempstead, went from having an equal share of the power to dummy status, whereas the one from Oyster Bay had the reverse experience and became one of the three sharing full power.

By 1970 they changed the simple-majority quota necessary to win from 58 to 63 but maintained the same weights they used in 1964. In 1976 they changed the weights to 35, 35, 23, 32, 2, and 2, respectively, and used a winning quota of 71. In these last two cases, the quota is somewhat more than a bare simple majority of the total weights. Nevertheless, it results in at least some power for each representative.

Mathematical Notation for Weighted Voting

We can represent a *weighted voting system* involving n voters by listing $n + 1$ numbers:

$$[q: w_1, w_2, \ldots, w_n]$$

The n numbers $w_1, w_2, \ldots, w_n$ are the respective *weights* for the n voters who are indicated by $1, 2, \ldots, n$. The initial number q is the number of votes necessary to pass an issue and is called the *quota*. For example, the Nas-

TABLE 10.2 The Nassau County board in the 1960s

Municipality	1960 population	1964 number of votes
Hempstead ⎱ Hempstead ⎰	728,625	⎧ 31 ⎨ ⎩ 31
North Hempstead	213,225	21
Oyster Bay	285,545	28
Glen Cove	22,752	2
Long Beach	25,654	2
Totals	1,275,801	115

sau County Board of Supervisors for 1958 can be expressed as

$$[16: 9, 9, 7, 3, 1, 1]$$

and in 1976 as

$$[71: 35, 35, 23, 32, 2, 2]$$

A subset of voters is called a **winning coalition** whenever the sum of its weights equals or exceeds this value q. It is usually assumed that the quota q necessary to win exceeds $\frac{1}{2}w$, where w is the sum of all the weights:

$$w = w_1 + w_2 + \cdots + w_n$$

Consider a small corporation owned by two people who possess 60% and 40% of the stock, respectively. If measures are allowed to pass by a simple majority, we express it as

$$[51: 60, 40]$$

Clearly, the first shareholder maintains full power and is in effect a **dictator.**

Let's examine a different company, where the stock is split among three individuals holding 49%, 48%, and 3%: [51: 49, 48, 3]. There is no dictator in this case. Any coalition of two or more has a simple majority, so that power is equally divided among the three stockholders. Although the third person holds only 3% of the stock, this stockholder has equal influence. His or her 3%, added to either of the other two stockholders' votes, is enough to sway the majority.

In contrast, look at the case where there are four owners with 26%, 26%, 26%, and 22% of the stock: [51: 26, 26, 26, 22]. The last person, with 22% of the stock, is a dummy. This stockholder is not capable of providing a losing coalition with enough power to win. The power in this case is equally divided among the first three shareholders.

Finally, let's assume that the representatives to a national assembly are split along party lines according to the numbers 45, 43, 8, and 4: [51: 45, 43, 8, 4]. Any two of the first three parties can form a winning coalition, whereas the smallest party has dummy status. (It is said that the Liberal Party in Great Britain "lives for the day" when the numbers divide in such a manner.)

These examples clearly illustrate that power need not be even approximately proportional to one's share of the vote. Power has a more complicated relationship to the weights, and we will proceed to examine that relationship.

The Banzhaf Power Index

The notion of power is fundamental to understanding and explaining political events. Power is also a basic ingredient of group decision making. Because power is an illusive concept, many of its aspects are difficult to identify and measure. It is unlikely that any simple mathematical formula can capture much of the essence of power.

Nonetheless, we will attempt to arrive at a numerical index for measuring power in the abstract, at least for such highly structured voting bodies as weighted voting systems. The resulting index should provide some useful insights for the design of equitable voting structures that arise in representative democracies (see Box 10.3).

We have already seen from our examples that individual voters or blocs of voters cannot simply take their fraction of the total vote as a meaningful indication of their share of power. The importance of an individual or of a coalition is directly related to the ability to enforce its own will. Power relates to achieving the desired end — in short, power is winning.

An individual can frequently appear on the winning side, however, without being powerful. For example, not all professional athletes

BOX 10.3 Power Indices

Lloyd S. Shapely (above), along with Martin Shubik, proposed the first widely accepted power index.

James A. Coleman was one of the proposers of the Banzhaf power index.

The first widely accepted numerical index for assessing power in voting structures called *simple games* was proposed by mathematician Lloyd S. Shapley (1923–) of UCLA and the RAND Corporation and economist Martin Shubik (1926–) of Yale University. Their index is a special case of the well-known Shapley value, which plays a fundamental role as a fairness concept in game theory and mathematical economics. A particular voter's power in this case is proportional to the number of different *permutations* (or orderings) of all the voters in which he or she casts the pivotal vote — the vote that first turns losing into winning.

The Banzhaf power index described in this chapter was introduced independently by John F. Banzhaf III (1940–), a Georgetown University lawyer and well-known consumer advocate, and social scientist James A. Coleman (1926–) of the University of Chicago. Their index is the one most often used in actual court rulings, perhaps because early cases were brought to court by Banzhaf. A voter's power is proportional to the number of different possible voting *combinations* (or arrangements) in which he or she plays a pivotal role.

on a team that regularly wins can demand high salaries. The truly powerful, high-salaried players are those who are crucial or decisive in bringing about a win. Similarly, the real significance of a vote is whether it is essential to victory.

So one reasonable measure of formal power is the frequency with which a voter or vote is **marginal,** or **pivotal,** to a decision. Power has to do with the number of different ways one person alone can turn defeat into victory, or vice versa. In other words, it rides on whether the switch in one vote can reverse the whole outcome.

We will therefore take as our relative measure of power the number of *different* ways an individual can join a losing coalition and thereby turn it into a winning coalition. (This is identical to the number of *distinct* winning coalitions that would lose if this person defected.) The number of ways one can swing losing to winning, or equivalently, winning to losing, determines the **Banzhaf power index.**

For example, consider the weighted voting system [3: 2, 1, 1], with quota 3 and weights 2, 1, and 1. This system might be a three-member committee with simple majority rule, except that the first person, the chairman, has veto power over any outcome. The winning coalitions are those with weights summing to 3 or 4: {1, 2}, {1, 3}, and {1, 2, 3}. The coalitions {1, 2} and {1, 3} are **minimal winning,** in the sense that each of the members is essential to remaining winning. The nonwinning *coalitions* are losing: {2, 3}, {1}, {2}, {3}, and $\varnothing$, the empty set.

Voter 1 can join a losing coalition and turn it into a winning coalition in three distinct ways:

1 joins {2} to obtain {1, 2}

1 joins {3} to obtain {1, 3}

1 joins {2, 3} to obtain {1, 2, 3}

Thus, the power index of voter 1 is 3. Voter 2 can turn a loss into a win in only one way:

2 joins {1} to obtain {1, 2}

Similarly, voter 3 can also change losing to winning in just one way:

3 joins {1} to obtain {1, 3}

Thus, voters 2 and 3 each have a power index of 1.

These five pivotal cases can be pictured as follows:

$$\{1, 2\} \qquad \{1, 3\} \qquad \{1, 2, 3\}$$
$$\uparrow 1 \qquad\quad \uparrow 1 \qquad\quad\; \uparrow 1$$
$$\{2\} \qquad\quad \{3\} \qquad\quad \{2, 3\}$$

$$\{1, 2\}$$
$$\uparrow 2$$
$$\{1\}$$

$$\{1, 3\}$$
$$\uparrow 3$$
$$\{1\}$$

We summarize the Banzhaf index for this game as (3, 1, 1). Note that voter 1, the chairman, has an index three times that of the other committee members. It is not so much the absolute magnitude of these numbers that matters, but rather their relative values.

A Pictorial View of Weighted Voting

We have seen that power depends more on which coalitions are winning than on the particular weight the voter holds. A voting scheme is merely a rule for dividing or partitioning the collection of all coalitions into two classes: those coalitions that win and those that lose. This viewpoint leads us to a highly geometrical way to visualize weighted voting systems.

If there are three voters, designated 1, 2, and 3, then there are 2^3, or 8, subsets, as illustrated in Figure 10.1. Note that the subsets of

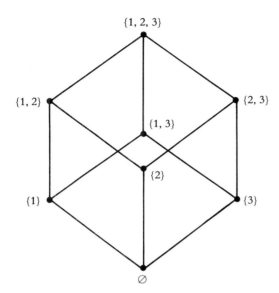

Figure 10.1 The vertices of a cube labeled with the possible subsets of voters 1, 2, and 3.

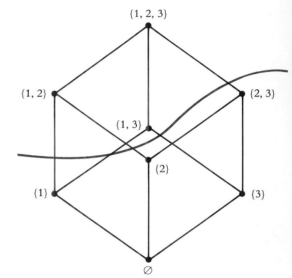

Figure 10.2 The colored line separates winning coalitions from losing coalitions.

{1, 2, 3} appear in Figure 10.1 as the vertices of a cube. A three-person voting rule divides these 8 subsets into the winning coalitions (typically the larger ones), which appear toward the top of the diagram, and the losing coalitions, which appear near the bottom. We assume that the grand coalition {1, 2, 3} is always winning. Such is the case of **unanimity,** whereas the empty set ∅ always loses.

Let's return to the three-person example [3: 2, 1, 1] with quota 3 and weights 2, 1, and 1. Recall that it was characterized by listing the winning coalitions

$$\{1, 2\}, \{1, 3\}, \text{ and } \{1, 2, 3\}$$

and losing coalitions

$$\{2, 3\}, \{1\}, \{2\}, \{3\}, \text{ and } \emptyset$$

This division of the subsets into two classes is indicated by the colored curve in Figure 10.2, which we can call the **quota curve,** or **cut.** In

other words, a voting system corresponds to removing just enough edges of the cube so that the vertices fall into two parts, as indicated in Figure 10.3. That is, the losing coalitions are disconnected from the winning ones.

The power index can be determined directly from Figure 10.4. Each edge in the cube corresponds to exactly one voter. For example, the right-hand edge, running from vertex {2, 3} down to vertex {3}, corresponds to voter 2. We can label each edge in our cube accordingly.

Furthermore, the edges that were removed in Figure 10.3 are precisely the ones that are critical. They are the edges connecting losing to winning coalitions in Figure 10.5. We need only count the number of times a given voter appears on such crucial edges to arrive at his or her power index.

For example, voter 1 appears on the critical edges joining {2} to {1, 2}, {3} to {1, 3}, and {2, 3} to {1, 2, 3}, for an index of 3. Voter 2 appears only once, on the left-hand edge connecting {2} to {1, 2}. Similarly, voter 3 appears

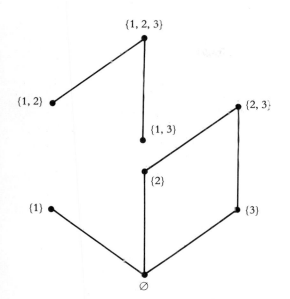

Figure 10.3 The cube after removing the edges connecting the winning coalitions from the losing coalitions.

only on the edge joining {1} to {1, 3}. Again, we arrive at the Banzhaf power index (3, 1, 1).

These lattice-like diagrams, depicting the collection of all possible coalitions of voters, provide a highly pictorial way to describe voting systems, as well as a way to compute power indices. However, such representation becomes rather difficult when we consider more than three voters. In this case we must visualize the vertices and edges of higher-dimensional cubes, the so-called **hypercubes,** which are described in Box 10.4.

Computing the Power Index

It is not difficult to calculate the power index for a weighted voting system when there are only a few voters. We merely list all the theoretically possible ways that the individuals can vote, that is, all the different combinations of "yes" and "no" votes. In the case of n voters there will be 2^n such combinations. We would then examine each combination in turn to see which votes are critical.

Let's return to our three-person example [3: 2, 1, 1]. The following table lists the eight

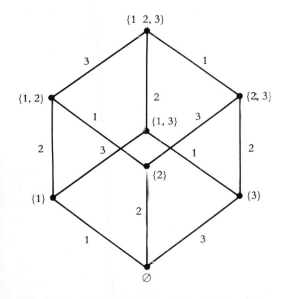

Figure 10.4 Each edge in the cube is labeled with its corresponding voter.

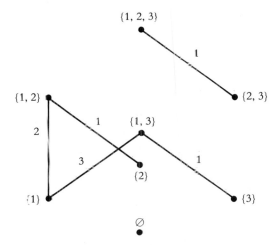

Figure 10.5 The edges connecting winning coalitions to losing coalitions.

BOX 10.4 Hypercubes, Coordinates, and Combinations

Multidimensional Cubes

The cube is one of the most commonly occurring and regular of geometric objects. There is a natural way to extend the idea of a cube to dimensions besides three. We can define a point as a *zero-dimensional cube*. A line segment can be thought of as a *one-dimensional cube;* it is built up by moving a point a unit distance in some direction. Similarly, a square in the plane can be viewed as a *two-dimensional cube*. A square can be obtained from a line segment of length 1, by moving it a unit distance perpendicular to its length. The regular cube in three-dimensional space can be called a *three-dimensional cube,* or a *3-cube.* The cube can be obtained from a square, or *2-cube,* by moving it into the third dimension.

Similarly, we can move a 3-cube a unit distance into a fourth dimension, generating a *four-dimensional cube,* or *4-cube.* (Cubes of dimensions 1, 2, 3, and 4 are illustrated in Box 10.1.) We can continue in this fashion to generate cubes of still higher dimensions, the so-called multidimensional cubes, or *n-cubes;* these are also called *hypercubes.* However, our ability to visualize *n*-cubes geometrically fails us as the dimension *n* continues to increase.

Coordinates for *n*-cubes

In analytic geometry we use an ordered pair of real numbers (x, y) to represent points in the plane. Similarly, we can use triples of numbers, (x, y, z) or (x_1, x_2, x_3), to indicate points in three-dimensional space. We can continue in this manner, thinking of four-dimensional space as the set of all four-tuples of numbers (x_1, x_2, x_2, x_4). In fact, one can generalize in this way to consider *n*-dimensional space as merely the *n-tuples* $(x_1, x_2, \ldots, x_n)$ of real numbers $x_1, x_2, \ldots, x_n$, where *n* can be any integer from 1 onward.

Using this technique, we can represent the vertices of a unit cube of any dimension in terms of coordinates that make use only of the numbers 0 and 1. We can represent the zero-dimension cube (a point) by the number, or coordinate, (0); the vertices of a 1-cube (the endpoints of a line segment) by the coordinates (0) and (1); the vertices of a 2-cube (a square) by the coordinates (0, 0), (1, 0), (0, 1) and (1, 1); and the vertices of the 3-cube by the eight 3-tuples (0, 0, 0), (1, 0, 0), (0, 1, 0), (0, 0, 1), (1, 1, 0), (0, 1, 1), (1, 0, 1), and (1, 1, 1). We can continue in this fashion, representing the vertices of *n*-cubes in higher dimensions by means of 2^n *n*-tuples of coordinates using only 0 or 1.

Voting Combinations

If abstentions are not allowed, there are eight different ways that three voters (1, 2, and 3) can each record either a "yes" or "no" vote. Each voting arrangement, called a *combination,* is shown in the following table, where *Y* indicates a "yes" vote and *N* a "no":

Voter	Combination of votes							
1	Y	Y	Y	N	N	N	Y	N
2	Y	Y	N	Y	N	Y	N	N
3	Y	N	Y	Y	Y	N	N	N

In each such combination, the set of voters partitions into the subset of "yes" voters and the complementary subset of "no" voters. In the case of a secret ballot, however, we can observe only four different outcomes:

| Voting outcome | | Number of |
Y	N	combinations
3	0	1
2	1	3
1	2	3
0	3	1

In the case of n voters, there are 2^n different voting combinations that can result in $n + 1$ possible outcomes. Using the shorthand factorial notation $n! = n(n - 1)(n - 2) \ldots (2)(1)$, these combinations can be enumerated as follows:

| Voting outcome | | Number of |
Y	N	combinations
n	0	$\dfrac{n!}{n!0!} = 1$
$(n - 1)$	1	$\dfrac{n!}{(n - 1)!1!} = n$
$(n - 2)$	2	$\dfrac{n!}{(n - 2)!2!} = \dfrac{n(n - 1)}{2}$
$(n - 3)$	3	$\dfrac{n!}{(n - 3)!3!} = \dfrac{n(n - 1)(n - 2)}{6}$
$\vdots$	$\vdots$	$\vdots$
1	$(n - 1)$	$\dfrac{n!}{1!(n - 1)!} = n$
0	n	$\dfrac{n!}{0!n!} = 1$

The number of different combinations that give rise to each particular outcome corresponds to a binomial coefficient in the expansion

$$2^n = (1 + 1)^n = \sum_{k=0}^{n} \frac{n!}{(n - k)!k!}$$

where $\displaystyle\sum_{k=0}^{n}$ indicates the sum of the $n + 1$ terms when $k = 0, 1, 2, 3, \ldots, n - 1, n$.

This analysis shows that there is a very natural one-to-one *correspondence* between the following four mathematical objects:

1. The vertices of an *n*-cube

2. The *n*-tuples of 0s or 1s

3. The combinations of *Y* or *N* votes

4. The subsets of a set of *n* elements

This correspondence is illustrated for the case when $n = 2$, as follows:

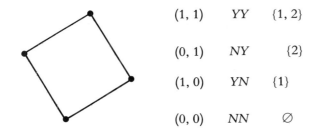

(1, 1)	*YY*	{1, 2}
(0, 1)	*NY*	{2}
(1, 0)	*YN*	{1}
(0, 0)	*NN*	∅

We can interpret a coordinate 1 to mean that the corresponding element is in the subset and interpret a 0 to mean that it is not in the subset. Likewise, a 1 corresponds to a *Y* and a 0 to an *N*. This relationship of the binary coordinates 0 and 1 to subsets and combinations is useful in undertaking large-scale computer calculations when *n* is large. These computations arise in such situations as calculating power indices for voting systems and calculating various probabilities for statistical events.

---

distinct combinations of voters 1, 2, and 3, according to whether they vote "yes," denoted by *Y*, or "no," indicated by *N*. Whether the issue will pass (*P*) or be defeated (*X*) is indicated in the outcomes below each of the combinations.

Voters	Combinations							
1	*Y*	*Y*	*Y*	*N*	*Y*	*N*	*N*	*N*
2	*Y*	*Y*	*N*	*Y*	*N*	*Y*	*N*	*N*
3	*Y*	*N*	*Y*	*Y*	*N*	*N*	*Y*	*N*
Outcome	*P*	*P*	*P*	*X*	*X*	*X*	*X*	*X*

We must examine each column in the table to see which voters are critical to the outcome. This means checking each "yes" or "no" vote in each combination to determine whether a switch of the one vote will cause a change in the result.

For example, the first combination

results in the issues passing by a unanimous vote. But if the first voter changes his or her vote from "yes" to "no"

$$
\begin{array}{ccc}
Y & \longrightarrow & N \\
Y & & Y \\
\underline{Y} & & \underline{Y} \\
P & & X
\end{array}
$$

then the outcome changes to "defeat." This will be indicated by circling the Y in the first row of the first column of the table:

$$
\begin{array}{c}
Y \\
\underline{Y} \\
P
\end{array}
$$

On the other hand, if only voter 2 switches his vote from "yes" to "no" in this first combination, the end result remains the same — the issue still passes by a weight of 3 "for" to 1 "against":

$$
\begin{array}{ccc}
Y & & Y \\
Y & \longrightarrow & N \\
\underline{Y} & & \underline{Y} \\
P & & P
\end{array}
$$

Similarly, voter 3 is not critical in this first combination:

$$
\begin{array}{ccc}
Y & & Y \\
Y & & Y \\
\underline{Y} & \longrightarrow & \underline{N} \\
P & & P
\end{array}
$$

Let's now consider the second column in our table:

$$
\begin{array}{c}
Y \\
Y \\
\underline{N} \\
P
\end{array}
$$

If voter 1 changes from "yes" to "no," the outcome goes from "pass" to "defeat":

$$
\begin{array}{ccc}
\textcircled{Y} & \longrightarrow & N \\
Y & & Y \\
\underline{N} & & \underline{N} \\
P & & X
\end{array}
$$

Likewise, if only voter 2 switches his or her vote, the result again goes from "pass" to "defeat":

$$
\begin{array}{ccc}
Y & & Y \\
\textcircled{Y} & \longrightarrow & N \\
\underline{N} & & \underline{N} \\
P & & X
\end{array}
$$

So voters 1 and 2 are crucial to the second combination, but voter 3 is not critical; the outcome remains the same even if voter 3 decides to switch:

$$
\begin{array}{ccc}
Y & & Y \\
Y & & Y \\
\underline{N} & \longrightarrow & \underline{Y} \\
P & & P
\end{array}
$$

Accordingly, we can proceed to each column in the table, checking whether each voter is critical and circling the corresponding Y and N. For example, in the third combination, 1 and 3 are critical:

$$
\begin{array}{cccc}
\textcircled{Y} \longrightarrow N & Y & Y \\
N & N & N & N \\
\underline{Y} & \underline{Y} & \textcircled{Y} \longrightarrow N \\
P & X & P & X
\end{array}
$$

In the fourth column, only voter 1 is critical, because his or her vote changes "defeat" to "pass":

$$
\begin{array}{ccc}
\textcircled{N} & \longrightarrow & Y \\
Y & & Y \\
\underline{Y} & & \underline{Y} \\
X & & P
\end{array}
$$

In the fifth column, 2 and 3 are critical:

In the sixth and seventh columns, only 1 is critical:

$$\begin{array}{cc}
\text{Ⓝ} \longrightarrow Y & \text{Ⓝ} \longrightarrow Y \\
Y & N \\
\underline{N} & \underline{Y} \\
X \quad P & X \quad P
\end{array}$$

In the last column, no one is critical.

In summary, our original table now takes the following form:

Voters	Combinations							
1	Ⓨ	Ⓨ	Ⓨ	Ⓝ	Y	Ⓝ	Ⓝ	N
2	Y	Ⓨ	N	Y	Ⓝ	Y	N	N
3	Y	N	Ⓨ	Y	Ⓝ	N	Y	N
Outcome	P	P	P	X	X	X	X	X

Finally, if we total up the number of circles in each voter's row, we arrive at the power index of (6, 2, 2). This result of 6, 2, and 2 corresponds to the previously obtained indices of 3, 1, and 1, because our second approach counts each critical vote twice: it counts the number of times a voter can turn winning into losing (P into X) as well as the times this voter can turn losing into winning (X into P). This double counting, however, leaves the *relative* magnitude, or ratios, of the power indices the same.

This procedure for computing the power index is straightforward when the number of voters n is rather small. But as n increases, the number of combinations, 2^n grows rapidly. We would need to enlist the help of a computer to calculate the power index. For still larger values of n, this method of computation may become prohibitive, even with the help of a computer.

Applying the Banzhaf Index

The Banzhaf index for measuring power is useful both in analyzing current voting systems and in designing new ones. Where today's voting systems are concerned, this index can provide insights into existing inequities.

The U.S. Electoral College is used to elect the President of the United States. The power indices in this case show a slight bias in favor of the large states, relative to their populations, when we view the states themselves as 51 voters casting weighted votes, according to the number of Congressmen in each state, with three additional votes allowed for the District of Columbia.

A more in-depth analysis of the power of the *individual voter* in such presidential elections shows a much greater inequity. A voter from California is more than three times as likely to be crucial in electing the President as a voter from the District of Columbia. On the other hand, several of the schemes proposed for reforming the Electoral College give undue influence to small states, according to Banzhaf's index. These results seem counterintuitive in view of the way most states have acted regarding Electoral College reform.

In order to create a new federal law in the United States, a bill must be passed by a simple majority in both the House of Representatives and the Senate, and then signed by the President. Or, in the case of a presidential veto, it must be approved by a two-thirds majority in both houses. Using a different index, Lloyd Shapley and Martin Shubik calculate that each legislative body as a whole is about 2.5 times as powerful as the President. On the other hand, the President has almost 40 times the power of an individual senator and 175 times that of a sole representative in the House.

Power indices also prove useful when new voting systems are designed. Consider the case of a local representative body, such as a county board or a board of education for a unified school district. Two desired objectives are often in conflict. The region may partition in a natural way into submunicipalities with common interests. The first desired objective is that these municipalities be represented by their own elected official. However, this natural division into existing communities may not give rise to districts with similar populations, so that the second desired objective, the popular principle of "one man one vote," cannot be met. One way to avoid redistricting into equal-size districts that cut across current

submunicipality boundaries is to resort to weighted voting.

Weighted voting has been used in many local governments. One cannot, however, merely weight a representative's vote in direct proportion to the number of constituents he or she represents, because such weighting can sometimes lead to strange results, as we saw in the case of Nassau County. Instead, the weights for the representatives should be determined so that their resulting Banzhaf power indices are nearly proportional to their respective populations. Several court rulings in New York State have approved this approach. For example, Table 10.3 gives the weighted voting system introduced in 1982 by Tomp-

TABLE 10.3 Board of representatives, Tompkins County, New York (1982)

District number	Name of municipality	1980 Census population	Relative population	Assigned weights	Relative weights	No. of crucial combinations	Relative power	Discrepancy*
6	Town of Lansing	8317	1.433	404	1.515	4747	1.415	0.0122
14	Town of Dryden-East	7604	1.310	333	1.249	4402	1.316	0.0047
8	Towns of Enfield/Newfield	6776	1.167	306	1.148	3934	1.176	0.0076
3	City of Ithaca, Ward 3	6550	1.128	298	1.118	3806	1.138	0.0085
4	City of Ithaca, Ward 4	6002	1 034	274	1.028	3474	1.039	0.0045
11	Town of Ithaca, Southeast	5932	1 022	270	1.013	3418	1.022	0.0000
1	City of Ithaca, Ward 1	5630	0.970	251	0.979	3218	0.962	0.0080
2	City of Ithaca, Ward 2	5378	0.926	246	0.923	3094	0.925	0.0015
10	Town of Ithaca, Northeast	5235	0.902	241	0.904	3022	0.903	0.0019
9	Town of Groton	5213	0.898	240	0.900	3006	0.899	0.0007
7	Towns of Caroline/Danby	5203	0.896	240	0.900	3006	0.899	0.0027
5	City of Ithaca, Ward 5	5172	0.891	238	0.893	2978	0.890	0.0007
12	Town of Ithaca, West	4855	0.836	224	0.840	2798	0.836	0.0002
15	Town of Ulysses	4666	0.804	214	0.803	2656	0.797	0.0084
13	Town of Dryden, West	4552	0.734	210	0.789	2622	0.784	0.0003
	Totals:	87,085	15.000	3999	15.000	50,191	15.000	0.0000
	Quota necessary to win: 2000							

*Discrepancy = (relative power − relative population) ÷ relative population

kins County, New York, for their 15-member Board of Representatives for the case of a simple majority vote.

More than half of the 63 counties in New York State have actually used weighted voting or seriously considered this possibility over the past 20 years. Several school boards have implemented such systems, especially in the state of Maine. A few governing bodies in New York City have employed weighted voting, and it is currently being proposed for some others. This movement towards weighted voting systems at the local level owes much to lawyer Banzhaf, who initiated court cases in the 1960s in an attempt to arrive at governing bodies that represented their constituents in a more equitable manner than the ones existing at the time. The concepts of weighted voting and power are analogous to similar notions that arise in several other mathematical models, and these ideas have thus found many other applications in these other contexts (see Box 10.2).

REVIEW VOCABULARY

Banzhaf power index   A numerical measure of power for individuals in certain voting situations. It is given by the number of different combinations of yes and no votes in which a person's vote is pivotal.

Binary system   A rule for assigning one of two possible outcomes given the input variables, each of which is also one of two types (for example, the numbers 0 or 1, or a vote in favor of or against an issue).

Cardinality   The number of elements in a set.

Coalition   A subset of a set of voters.

Combination   A partitioning of a set into a subset and its complementary subset, for example, a list of those voters who voted in favor of and against an issue.

Cut (quota curve)   A partitioning of all the subsets of a set into two disjoint classes, for example, those coalitions of voters that win and those coalitions that lose.

Dictator   A person who can pass any issue without the aid of any other voters.

Dummy   A person with no power, in the sense that his or her vote can never be pivotal to any outcome.

Hypercube (n-cube)   The natural extension of the concept of a square in the plane and a cube in space to the analogous configurations in higher-dimensional space. Its vertices can be represented by the coordinate n-tuples $(x_1, x_2, \ldots, x_i, \ldots, x_n)$, where each x_i is either 0 or 1.

Losing coalition   A coalition of voters that does not have enough votes to pass an issue, that is, a nonwinning coalition.

Minimal winning coalition   A winning coalition that is as small as possible in the sense that if any member were to leave the coalition it would become a losing coalition.

Pivotal   A person is pivotal in a particular voting combination if a change in just his or her vote will alter the outcome of the ballot.

Quota   The smallest number of votes necessary to pass an issue.

Veto power   A coalition has veto power if it is a losing coalition and its complementary coalition is also a losing coalition.

Weighted voting system   A voting situation in which various individuals can have different numbers of votes. It can be represented by the symbol $[q: w_1, w_2, \ldots, w_n]$, where the numbers $w_1, w_2, \ldots, w_n$ represent the numbers of votes held by the n individuals and where q is the quota necessary to win.

Winning coalition   A coalition of voters that has enough votes to pass an issue, that is, a nonlosing coalition.

EXERCISES

1. Give the cardinality of:

   a. The set $\{1, 3\}$

   b. The set $\{1, 2, 4\}$

   c. The set of edges in a cube

   d. The set of faces in a cube

   e. The set of all subsets of the set $\{1, 2, 3\}$

   f. The set of all subsets of the set $\{1, 2, 4\}$ that contain the element 4

   g. The set of all subsets of the set $\{1, 2, 4\}$ that do not contain the element 4

   h. The set of positive even integers

   i. The set of vertices in a 5-cube

   j. The set of all roots of the equation $x^2 - 4x + 3 = 0$

   k. The set of all points on the graph of the equation $y = 2x + 3$

2. Can you prove that any set S with cardinality n has 2^n subsets (including S itself and the empty set $\varnothing$)?

3. For each of the following weighted voting systems describe the following:

   (i) All the winning coalitions

   (ii) All the minimal winning coalitions

   (iii) All the losing coalitions

   (iv) Any dummy voters

   (v) All of the coalitions with veto power. (A coalition has *veto power* if it is a losing coalition and its complementary coalition is also losing.)

   a. [51: 52, 48]

   b. [2: 1, 1, 1]

   c. [3: 2, 2, 1]

   d. [8: 5, 4, 3]

   e. [51: 45, 43, 8, 4]

   f. [51: 28, 27, 26, 19]

   g. [16: 10, 10, 10, 1]

   h. [21: 10, 10, 10, 10, 1]

   i. [51: 28, 24, 24, 24]

   j. [6: 4, 3, 2, 1]

    k. Nassau County in New York in 1958

    l. Nassau County in New York in 1964

    m. [4: 3, 1, 1, 1, 1]

4. a. List the 16 possible combinations for how four voters 1, 2, 3, and 4 can each vote either yes Y or no N on an issue.

    b. List the 16 subsets of the set {1, 2, 3, 4}.

    c. Show a natural one-to-one correspondence between the lists in **a** and **b**.

    d. In how many of the combinations in **a** is the vote (i) 4 Y to 0 N? (ii) 3 Y to 1 N? (iii) 2 Y to 2 N?

5. Determine the Banzhaf power index for each of weighted voting systems given in Exercise 3.

6. Show that the weighted voting systems for Nassau County in New York in 1970 and 1976 had no dummies.

7. Consider a nine-person committee in which each member has one vote and a simple majority wins (for example, the U.S. Supreme Court).

    a. How is the power distributed among the nine voters?

    b. If a coalition of five of these voters made a secret pact to first vote only among themselves with majority rule, and then to vote all the same way in the full committee, then how is power distributed among the nine members?

    c. If a three-person subgroup of the five-person coalition in **b** were also to form an agreement such as in **b**, then how would the power be distributed among the nine committee members?

8. Show that the voting scheme for the United Nations Security Council can be expressed as the weighted voting system [39: 7, 7, 7, 7, 7, 1, 1, 1, 1, 1, 1, 1, 1, 1, 1].

9. In Australia the six states have one vote apiece, the federal government has two votes, and majority rules. Furthermore, the federal government can break a four-four tie in its favor. Can you express this scheme as a seven-voter weighted voting system?

10. In some international boxing competitions the winner of a match is determined as follows: There is a primary panel of five judges; if they rule 5 to 0 or 4 to 1 in favor of one competitor, then he is declared the winner. However, a ruling of 3 to 2 by the first panel is considered indecisive, and the matter is then referred to a second panel of five judges. This secondary panel can reverse a 3 to 2 vote by the primary panel only when they vote 1 to 4 or 0 to 5. But if the secondary panel votes 5 to 0, 4 to 1, 3 to 2, or 2 to 3, then the 3 to 2 decision of the primary panel is upheld. Can this be written as one weighted voting game involving all 10 judges?

11. Certain well-defined voting systems cannot be represented as weighted voting systems. Can you prove that there is no way to pick positive numbers q, w_1, w_2, w_3, w_4 such that $[q: w_1, w_2, w_3, w_4]$ gives the four-person voting system in which the winning coalitions are {1, 2}, {3, 4}, {1, 2, 3}, {1, 2, 4}, {1, 3, 4}, {2, 3, 4}, and {1, 2, 3, 4}? Recall that for every coalition S and every voter i in S we must have $\Sigma_{i \text{ in } S} w_i \geq q$ when S is winning and $\Sigma_{i \text{ in } S} w_i < q$ when S is losing. (Note that this voting scheme is "improper" in the sense that the two *disjoint* coalitions {1, 2} and {3, 4} are both winning.)

12. Consider the seven-person voting system in which the voters are indicated by 1, 2, 3, 4, 5, 6, and 7, and the minimal winning coalitions are $\{1, 2, 3\}$, $\{3, 4, 5\}$, $\{1, 5, 6\}$, $\{1, 4, 7\}$, $\{2, 5, 7\}$, $\{3, 6, 7\}$, and $\{2, 4, 6\}$. Any superset of a winning coalition is also winning. In the following figure, the seven voters correspond to seven points, and the seven minimal winning coalitions correspond to the six lines and the one circle. Can you prove that this voting scheme cannot be expressed as a weighted-voting system? (Note that this voting scheme is "constant-sum" in the sense that the complementary coalition $\{1, 2, 3, 4, 5, 6, 7\} - S$ is losing whenever the coalition S is winning.)

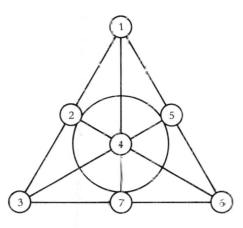

13. In the early 1980s Canada adopted a method for amending the Canadian constitution that gives most of the power to the federal government. Another proposal under consideration for many years would also have involved a vote by the 10 provinces. This proposal can be described briefly in terms of minimal veto-power coalitions as follows: Veto power would be held by Ontario, by Quebec, by any 3 of the 4 Atlantic provinces, by the 3 prairie provinces together, and by British Columbia along with any one of the prairie provinces. Show that this voting scheme for the 10 provinces cannot be expressed as a weighted voting system.

14. Discuss why the British Parliament in early 1974 was "unstable." The parties and their number of seats was as follows:

Party	Seats
Labor	301
Conservatives	296
Liberals	14
Irish Unionists	11
Scottish Nationalists	7
Welsh Plaid Cymru	2
Irish Catholics	1
Others	3
Total	635

15. The Vice President of the United States is allowed to vote to break ties in the U.S. Senate. Discuss how his Banzhaf power index compares with that of an individual senator.

16. Consider some country or city that has a representative form of government consisting of more than two parties. Consider the parties to be the individual voting units and present this government as a weighted voting system. Then compute the Banzhaf power index for each party.

Game Theory: The Mathematics of Competition

Conflict is a central theme in human history and literature. It arises naturally whenever two or more individuals try to control the outcome of events. People compete in such situations because they have both freedom of choice and different values.

Game theory is a serious mathematical subject created to study situations involving conflict and cooperation. It is a new approach in that it brings scientific methods and the powerful tools of mathematics to bear on the topic. Game theory began in earnest in 1944 with the publication of *Theory of Games and Economic Behavior* by John von Neumann and Oskar Morgenstern (see Box 11.1).

A game situation arises when two or more individuals, called **players,** are each able to act freely and to select from a list of available options. These options are referred to as **strategies.** These choices in turn lead to various outcomes, called **payoffs.** Each player has various

preferences among the resulting rewards or penalties. The theory of games, then, is concerned with notions such as selection of optimal strategies, equilibrium outcomes, bargaining and negotiations, coalition formation and stability, equitable allocations, costs or benefits, and the resolution of conflict. This subject deals with the rules of the game, individual and coalition values, side payments and repeated play, as well as various kinds of uncertainty and chance events. Game theory differs from the traditional subjects of statistics and probability in that it treats two or more individuals with different goals or objectives.

Many confrontations are primarily noncooperative, for example, those between combatants in warfare or competitors in sports. In these encounters, the adversaries' ultimate objectives are typically at cross-purposes: a gain for one means a loss for the other. Other social activities, such as those in economics or poli-

BOX 11.1 Historical Highlights

As early as the seventeenth century, such outstanding scientists as Christian Huygens (1629–1695) and Gottfried W. Leibnitz (1646–1716) proposed the creation of a discipline that would make use of the scientific method to study human conflict and interactions. Throughout the nineteenth century, several leading economists created simple mathematical examples to analyze particular illustrations of competitive encounters. The first general mathematical theorem in this subject was proved by the distinguished logician Ernst Zermelo (1871–1956) in 1912. It stated that any finite game with *perfect information* such as checkers or chess has an optimal solution in *pure* strategies; that is, no randomization or secrecy is necessary. A game is said to have perfect information if at each stage of the play, every player is aware of all past moves by himself and others as well as all future choices that are allowed. This theorem is an example of an *existence theorem:* it demonstrates that there must be a best way to play such a game, but it does not provide a detailed plan for actually playing a complex game in order to achieve victory.

The famous mathematician R. E. Emile Borel (1871–1956) introduced the notion of a *mixed,* or randomized, strategy when he investigated some elementary duels around 1920. The fact that every two-person zero-sum game must have optimal mixed strategies and an expected value for the game was proved by John von Neumann (1903–1957) in 1928. Von Neumann's result was extended to the existence of equilibrium outcomes in mixed strategies for multiperson general-sum games by John F. Nash Jr. (1931–) in 1951.

Modern game theory dates from the publication in 1944 of *Theory of Games and Economic Behavior* by the Hungarian-American mathematician John von Neumann and the Austrian-American economist Oskar Morgenstern (1902–1977). They introduced the first general model and solution concept for multiperson cooperative games. Several other suggestions for a solution to such coalitional games have since been proposed. These include the value concept of Lloyd S. Shapley (1923–) which relates to fair allocation and economic prices and serves as well as an index of voting power.

The French artist Georges Mathieu designed a medal for the Paris Musée de la Monnaire in 1971 to honor game theory. It was the seventeenth medal to "commemorate 18 stages in the development of Western consciousness." The first was for the Edict of Milan in 313 AD. Game theory also has a mascot, the tiger, arising from the Princeton University tiger and the Russian abbreviation of the term "game theory" (ТЕОРИЯ ИГР).

tics, typically have a large cooperative component. However, most human interactions involve a delicate mix of cooperative and noncooperative behavior. For example, people in business cooperate to maintain a healthy economy even as they compete for shares in the marketplace.

In the following sections we will present five simple examples of noncooperative two-person games. The first three will be games of complete conflict, the last two will be games of partial conflict.

A Location Game

Two young entrepreneurs, Henry and Lisa, plan to locate a new restaurant at a main-route intersection in the nearby mountains. They can agree on all aspects except one. Lisa likes

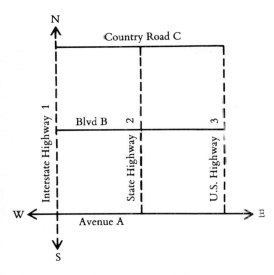

Figure 11.1 The road map for the location problem

low elevations, whereas Henry wants greater heights — the higher up, the better. In this one regard, their preferences are diametrically opposed.

The layout for their location problem is shown in Figure 11.1. You can see that three routes, Avenue A, Boulevard B, and Country Road C, run in the east-west direction and that three highways, numbered 1, 2, and 3, run in the north-south direction. The altitudes at the nine corresponding intersections are given in the following table in thousands of feet and are illustrated in Figure 11.2.

Routes	Highways		
	1	2	3
A	10	4	6
B	6	5	9
C	2	3	7

Henry and Lisa agree to turn their decision into a competitive game, as follows: Henry

will select one of the three routes, A, B, or C, and Lisa will simultaneously pick one of the three highways, 1, 2, or 3. The restaurant will then be located at the resulting intersection.

Henry is rather pessimistic and considers the lowest altitude along each of the routes A, B, and C. He gets the numbers 4, 5, and 2, which are the lowest respective row minima, indicated in the right-hand column of the next table. He notes that the highest of these values is 5. He can elect the corresponding route, B, and guarantee himself an altitude of at least 5000 feet.

Lisa does a worst-case analysis and lists the highest — for her, the worst — elevations for each highway. These numbers, 10, 5, and 9,

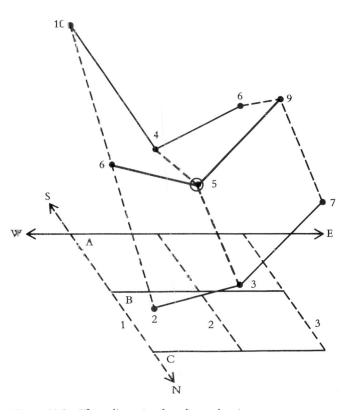

Figure 11.2 Three-dimensional road map showing Henry's and Lisa's selections (color). Underneath, the slanted map shows the map in Figure 11.1 in perspective.

are the column maxima and are listed in the bottom line of this table. The best of these outcomes from her point of view is 5. If she picks State Highway 2, then she is assured of an elevation of no more than 5000 feet.

| | | Lisa | | | |
| | | Highways | | | Row |
	Routes	1	2	3	minimum
	A	10	4	6	4
Henry	B	6	5	9	5
	C	2	3	7	2
	Column maximum	10	5	9	

So Henry has a choice that will result in 5 or higher, and Lisa can choose so as to hold him down to 5 or less. The resulting height 5 at the intersection of route B and highway 2 is, simultaneously, the lowest value along Boulevard B and the highest on State Highway 2. Such an outcome is called a **saddle point,** or **mountain pass,** for a game. It appears obvious that the resolution of this contest is for Henry to pick B and Lisa to elect 2; then each settles for the resulting elevation of 5. This number 5 is called the **value** of the game. The selections B and 2 are the **optimal (pure) strategies,** and, along with the value, are referred to as the **solution** of the game.

There is no need for secrecy in the case of a game with a saddle point. Even if Henry were to reveal his choice of B in advance, Lisa would be unable to use this knowledge to exploit him.

A Duel Game

We now turn to a more interesting game, one that does not have a saddle point in the initial (pure) strategies. Consider the strategic en-

counter that takes place between the pitcher and batter in the game of baseball. First, let's assume that the pitcher has enough control to throw the ball to whatever part of the strike zone he wishes. To simplify matters, let's suppose that he will select one of three pitches:

- HI: a fastball high and inside
- MD: a fastball down the middle
- LO: a fastball low and outside

These alternatives, HI, MD, and LO, are the pitcher's three strategies and are depicted in Figure 11.3.

Next, let's assume that a particular batter is known to average

- .300 against high inside pitches
- .400 against pitches over the middle
- .200 against low outside pitches

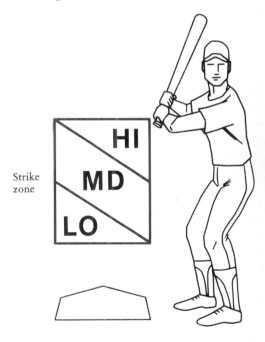

Figure 11.3 The three strike zones, which correspond to the pitcher's three strategies.

A batting average of .300 means that the batter will hit safely about 3 times out of 10.

However, the hitter also has the option of outguessing the pitcher. He can choose to guess HI and step back from the plate as he swings. In this case his average becomes

- .400 against HI

- .200 against MD

- .000 against LO

Similarly, he can guess LO and average

- .000 against HI

- .300 against MD

- .400 against LO

So the hitter also has the three strategies that are designated HI, MD, and LO.

This information can be summarized in the following table, called the **game matrix:**

		Pitcher		
		HI	MD	LO
	HI	.4	.2	.0
Batter	MD	.3	.4	.2
	LO	.0	.3	.4

We assume that all this information is known to both players. The numbers in this table represent the probability that the batter will hit safely; they can serve as a measure of his likely reward. Whereas the batter's objective is to maximize the resulting payoff, the pitcher has the opposite goal. He aims to minimize the likelihood that the batter will hit safely. What is good for the batter is bad for the pitcher, and vice versa. They are complete adversaries. This is an example of a **strictly competitive,** or antagonistic, game. It is also

an example of a **constant-sum** game, because the sum of payoffs to the different players is always the same. In fact, because the batter's expected gain is the same as the pitcher's expected loss, and vice versa, this is a **zero-sum** game.

Clearly, this model is an oversimplification of real baseball. We have excluded from our analysis pitches outside the strike zone, foul balls, bases on balls, and other possible strategies and outcomes. We could enlarge the game to make it more realistic, but that would only complicate our analysis.

To solve this game we must answer the question, Which choices are best for the two players? One approach to this problem is to reason along the following line:

Batter (to himself): If I select MD, which has the best overall average for me, I am protected against the zero payoff in all cases.

Pitcher (to himself): If the batter does choose MD, then I should throw LO and limit him to a .200 average.

Batter: The pitcher knows that I am thinking MD, and he is thus considering a LO pitch. Therefore, I should really guess LO and hit .400.

Pitcher: But I know he is considering LO; so, on second thought, I should surprise him and aim HI.

Batter: In light of the preceding chain of thought, I should expect a HI pitch and swing accordingly.

Pitcher: In light of his last reasoning about HI, I should really throw LO.

Batter: Then I should guess LO.

Pitcher: So I should pitch HI.

Batter: I'd better guess HI after all.

Pitcher: I will then surprise him with LO.

This form of cyclical reasoning can continue on without end. It provides no resolution to the decision problem. Clearly, there is no one pitch that is best in all instances. In fact, the batter can do better than to rely on the safe strategy MD, despite the overall protection it seems to provide.

The answer to the players' problems lies in the notion of a **mixed strategy.** Both players must maintain an element of surprise and must not give away their choices in advance. One good way they can guarantee surprise is to select their strategies randomly. That is, the pitcher should select his pitches at random, and the batter should do so as well. This can be done in a way that ensures an **optimal mixed strategy** for each player. In short, the players can improve their average performance over the long run, if they play the odds and randomize properly.

Let's look more closely at the pitcher's problem. He must determine the best probabilities for each of his three pitches. These probabilities tell the pitcher the ratio of different types of pitches he should select. For example, if he determined a probability of 0.3 for LO, then three-tenths of his pitches should be LO.

An old baseball adage says, "Pitch to the corners." That means that a pitcher should avoid throwing down the middle. (A mathematical analysis verifies this adage in this illustration, showing that the pitcher should indeed assign probability zero to the MD.) So the pitcher should randomize only between HI and LO.

Next, consider the three outcomes that may occur, depending upon whether the batter guesses HI, MD, or LO:

1. If the batter guesses HI, then a randomized mix of HI and LO pitches will cause the batter to average somewhere between .400 and .000. This is indicated by the line labeled HI between the heights .400 and .000 in Figure 11.4. The height of the sloping line indicates the expected batting average of the hitter

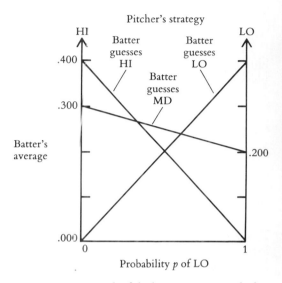

Figure 11.4 A graph of the batting averages, which depend on the batter's guesses and the pitcher's strategy.

when he guesses HI, and the distance along the horizontal axis indicates the probability p with which the pitcher throws LO.

2. If the batter guesses MD, his average will fall somewhere between .300 and .200, as shown in Figure 11.4.

3. Similarly, if the batter guesses LO, his average will be between .000 and .400, as also shown in Figure 11.4.

The pitcher wants to keep the batting average as low as possible against all three batting eventualities simultaneously. The pitcher is therefore concerned with the highest points in the graph, which are indicated by the heavier curve in Figure 11.5. He should select his probability p so as to obtain the lowest point on this top curve.

The mathematical solution to this problem gives the value $p = 0.6$. The pitcher should throw LO with a probability $p = 0.6$ and HI with probability $1 - p = 0.4$. In other words, six-tenths of his pitches will be LO and four-tenths will be HI. This will result in holding the batter to an expected average of .240 or less.

A similar mathematical analysis would show that the hitter has an optimal mixed strategy if he guesses

- HI with probability 0
- MD with probability $q = 0.8$
- LO with probability $1 - q = 0.2$

This will guarantee him an average of .240 over the long run.

The number .240 is the **value** of this game. The pair of optimal mixed strategies, $p = 0.6$ for the pitcher and $q = 0.8$ for the hitter, along with the resulting value .240, gives us the *solution* for this game.

It is usually essential for the players to act in an unpredictable manner — they must maintain secrecy. If the baseball pitcher were to "telegraph" his pitches and the batter were able to detect these signals in advance and react in time, then the batter could raise his batting

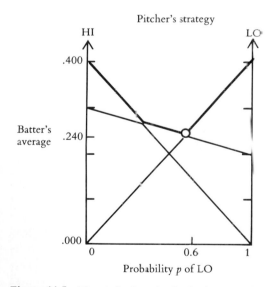

Figure 11.5 The pitcher's optimal mixed strategy is to select p so as to obtain the lowest point (circle) on the top curve (color).

average. On the other hand, if the hitter were not mixing his guesses in an optimal way, then the pitcher could make use of this information to improve his own performance by throwing to the zone that the batter was least likely to guess.

The element of surprise is essential in many other encounters. Examples of the use of mixed strategies include various inspection procedures and auditing schemes. These should employ randomness to keep potential cheaters off guard as well as for statistical purposes. The notion of a "bluff", as in the game of poker, is also a viable strategy. For bluffing, game theory assigns an optimal probability with which one should bluff, given a particular situation. For example, in labor negotiations, a threat to strike is effective only if it is believed to some extent.

Inspection Games

Individual investigators or regulatory agencies often monitor certain accounts or actions to check faults, errors, or illegal activities. The investigators include bank auditors, customs agents, insurance investigators, and quality-control experts. The National Bureau of Standards is responsible for monitoring proper measuring instruments and for maintaining reliable standards. The Nuclear Regulatory Agency demands an accounting of dangerous (and expensive) nuclear materials as part of its safeguards program. The Internal Revenue Service wishes to identify those cheating on their taxes. The military or intelligence service may wish to sneak in or intercept a weapon or secret agent that is in the midst of many decoys. Because it is prohibitively expensive to check out all cases, statistical methods must be used to check for violators. Many such encounters can be modeled as a competitive game, with the goal to obtain optimal mixed strategies for the inspector and the violator.

Suppose you have a choice of either parking

illegally on the street or else parking in a lot and paying. The outcome depends on whether the police happen to be patrolling the area at that particular time. Parking illegally is free if the police are not patrolling, but you receive a $20 parking ticket if they are. It costs only $8 to park in the lot. However, you may be peeved to find that you have paid to park in the lot when the police did not patrol that day. You are willing to assess this outcome as costing $16 ($8 for parking plus $8 for your inconvenience and grief). We summarize the costs in the following table:

		Police	
		Patrol	Not patrol
You	Park illegally	−20	0
	Park at the lot	−8	−16

It seems reasonable to assume that the police will be ranking these outcomes in the opposite way. After all, they prefer most to give out tickets and least to see you get away with your crime. But they would prefer not to have to patrol in the event that you do decide to park in the lot. So let us assume that the police view their payoffs as the negatives of yours. We then have a zero-sum game.

We can compute your optimal mixed strategy $(1 - p, p)$ as follows, where p is the probability you will park in the lot and pay the $8. First, consider the case in which the police do patrol. Your expected loss is then given by

$$E = (-20)(1 - p) + (-8p) = -20 + 12p$$

In this equation, each of the two possible losses is multiplied by its probability. The two results are then added to give the average. In the second case, where the police do not patrol, your expected value is

$$E = (0)(1 - p) + (-16p) = -16p$$

These two equations in two unknowns can be solved simultaneously to obtain

$$p = 5/7, \ 1 - p = 2/7, \text{ and } E = -80/7$$

The $p = 5/7$ indicates you park at the lot five out of seven times and pay an *average* amount of $11.43 each time.

One can similarly calculate the optimal mixed strategy $(1 - q, q)$ for the police in this zero-sum game by solving the simultaneous equations

$$E = (-20)(1 - q) + 0q = -20 + 20q$$
$$E = (-8)(1 - q) + (-16q) = -8 - 8q$$

to obtain

$$q = 3/7, \ 1 - q = 4/7, \text{ and } E = -80/7$$

So the police will patrol 4 out of 7 times. Both players' calculations led to the same game value −80/7, which follows from the **minimax theorem,** a famous result of game theory.

The Prisoners' Dilemma

We will now look at a game that has come to be known as the **prisoners' dilemma.** This elementary two-person game provides a simple explanation of the forces at work behind arms races, price wars, costly advertising campaigns, and many other similar escalations. This game is not strictly competitive (that is, constant-sum), as were our three previous examples. There is some mutual gain to be realized by both players if they can cooperate. But when this game is played in a noncooperative setting, the players' own self-interests lead them to a less-than-optimal payoff. The prisoners' dilemma illustrates a kind of social paradox that we frequently confront in the course of our everyday lives.

The term *prisoners' dilemma* was first assigned to this game by the Princeton mathematician Albert W. Tucker (1905 –) in 1950. The game involves the scenario of two suspects in crime who are held *incommunicado*. Each is given one of two choices: to steadfastly maintain the pair's innocence or to sign a confession accusing the partner of committing the crime. It is usually in the individual's self-interest to confess, yet when both confess, they each reach a bad outcome. What is good for the prisoners as a pair — steadfast denial by both — is frustrated by their pursuit of individual rewards.

We can use this simple model for another crucial international problem: the arms race. Assume that there are two nations called Red and Blue. Each of these superpowers can independently select one of two policies:

(A) Noncooperation: heavily arm in preparation for any possible war contingency.

(D) Cooperation: disarm, or at least agree to a partial ban on armaments.

There are four possible outcomes:

(D, D) Both Red and Blue choose to disarm. Viewed as a whole, this is the most preferred social outcome for them, even in light of certain risks.

(A, A) Both nations select to arm, which is taken as the worst possibility from the global perspective.

(A, D) Red decides to arm, whereas Blue elects to disarm. This amounts to unilateral disarmament by Blue, which is the most preferred of all outcomes to Red but the least desired by Blue.

(D, A) Red disarms, whereas Blue arms. This is considered the worst result for Red and the best outcome for Blue.

This situation can be modeled by means of the game matrix

		Blue	
		A	D
Red	A	Arms race	Favors red
	D	Favors blue	Disarm

Here, Red's choice amounts to picking one of the two rows, whereas Blue's options correspond to the two columns. It may prove helpful to assign numerical payoffs to the four outcomes, as follows:

		Blue	
		A	D
Red	A	(2, 2)	(5, 0)
	D	(0, 5)	(4, 4)

For example, the pair of numbers (0, 5) in the second row and first column signifies a payoff of 0 to the row player Red and a payoff of 5 for the column player Blue. The least desired outcome is 0; the most preferred is 5. Only the relative magnitude, not the actual values, of these payoffs is essential for our analysis. The numbers are not intended to measure the absolute worth of unilateral disarmament or of a disarmed world compared with an armed one. They are used only to suggest the preferences of the players: we simply assume that a player prefers a larger numerical payoff to a smaller one.

Let's examine this arms race more closely. Should Red select strategy A or D? Red can see what will happen if Blue selects his first column A: red will then receive a payoff of 2 for arming and 0 for disarming, so he will arm (first row). Similarly, Red notes the consequences if Blue were to select his second column D. In this case, Red will receive 5 for arming or 4 for disarming. Again Red decides in favor of his first row A. In either case, Red's first row gives him the more desired result. We say that the payoffs to Red in the first row **dominate** those in the second. There is always some advantage to Red to arm, whether Blue arms or disarms.

A similar argument leads Blue also to choose *A*, that is, to pursue a policy of arming. When each nation strives to maximize its own payoff independently, the pair is driven into the outcome (*A, A*), with payoffs (2, 2). The better outcome (*D, D*), with payoffs (4, 4), appears unobtainable when this game is played noncooperatively.

The outcome (*A, A*) is said to be in **equilibrium** because if either nation alone were to deviate from its choice of *A*, then it would be punished with the lower payoff 0, rather than 2. The forces involved prohibit only one nation from moving away from equilibrium (*A, A*).

Even if both nations agree in advance to jointly pursue the globally optimal solution (*D, D*), this outcome is unstable because if either nation alone reneges on the agreement and secretly arms, it will benefit. Each would thus be tempted to go back on its word and select *A*. After all, if you have no confidence in the trustworthiness of your opponent, you may be well advised to cover yourself against such a defection.

In real life, however, people often manage to avoid the noncooperative outcome in the prisoners' dilemma. The game is usually played within a larger context, where other incentives are at work. Moreover, the game is typically played on a repeating basis — it is not a one-time affair. Elements such as reputation and trust also play a role. The players realize the mutual advantages in cooperation and may arrive at this point by slowly phasing down over time. They may also resort to other helpful measures, such as better communications channels, more reliable inspection procedures, truly binding agreements, or promptly enforced penalties for violators (see Box 11.2).

The prisoners' dilemma nicely pinpoints the dynamics behind a frequently occurring social paradox. The resulting standoff, or noncooperative equilibrium outcome, is not as satisfactory a solution as were the optimal strategies derived in our previous constant-sum games. The cooperative outcome, or one that does not take into account immediate self-interest, is obviously the preferred solution in the long run.

The Game of Chicken

Let us look at one more two-person game of partial conflict, known as **chicken,** which leads to troublesome outcomes. Two drivers are approaching each other at high speeds. Each must decide at the last minute whether to swerve to the right or not swerve. There are several possible consequences:

1. Neither driver swerves, and the cars collide head-on. We assign this least preferred outcome a value of 0.

2. Both players swerve. Each loses some prestige by backing off at the brink, but they do remain alive. We give this outcome an intermediate value of 3.

3. One of the drivers swerves and badly loses face, whereas the other does not swerve and is viewed as the winner. Let us select the numerical payoffs of 1 for swerving and 5 for not swerving in this case.

These options can be summarized in the following game matrix:

		Driver 2	
		Swerve	Not swerve
Driver 1	Swerve	(3, 3)	(1, 5)
	Not swerve	(5, 1)	(0, 0)

BOX 11.2 Repeated Play of Prisoners' Dilemma

Robert Axelrod, professor of mathematics at the University of Michigan, is a well-known authority on the problem of the prisoners' dilemma. In the following interview he discusses the effects of playing the prisoners' dilemma game repeatedly.

If you play the game only once, there's no future to your interaction and so you might as well take the short-term gains. In one play there's no chance to reward or punish a defection by the other player: there's no hope that you'll get a mutual cooperation going.

A very important feature of the evolution of cooperation is that there is a long-term relationship. When the prisoners' dilemma is iterated, it gives the players an opportunity to base their current choices on the previous interactions they've had. So if the other player seems willing to cooperate, you can cooperate yourself. If the other player has rarely been willing to cooperate, then it probably doesn't pay to cooperate. Repeating the game allows you to do things like base your own strategy on reciprocity. And therefore it allows you to try to mold the other player's behavior, to encourage him to cooperate with you.

I got the idea of a tournament because I was interested in determining a good way of playing prisoners' dilemma, because it captures some important features of the real world. No one seems to know exactly what's the best strategy. So I invited experts from a variety of fields — people who had written about prisoners' dilemma or game theory — and asked them what strategy they would use in this game. Then I played each with the other to see how well they would do. So I had something analogous to a computer chess tournament.

I was really surprised by the way it came out because the simplest of all the strategies submitted was the one that did best. That was "tit for tat" by Anatol Rapoport. This rule simply says to cooperate on the first move and then do whatever the other player did on the previous move. If the other player cooperated, you cooperate. If the other player defected, you defect. It works best for several reasons.

What the analysis shows is that an effective strategy is not to start defecting: never be the first to defect. But if the other side defects, it pays to be provokable. It also pays to be forgiving after you've been provoked, so as to keep the conflict as short as possible. It pays to respond promptly if someone does something you don't like.

I titled my book *The Evolution of Cooperation* to capture the analogy from biological evolution. It's an evolutionary study of cooperation, asking how it could get started in a world where there isn't any, how it can sustain itself, and how it can grow after it gets started. People are likely to continue to use strategies that are effective and to drop or change them if a strategy they use is not very effective. So in fact things tend to evolve toward more effective strategies.

If both players persist in their attempts to obtain the maximum payoff 5, then the resulting outcome is mutual disaster: the lowest payoff 0 for each. It is surely better for both drivers to simultaneously back down and obtain 3 each. But neither opponent wants to be in the position of being intimidated into swerving (for a payoff of 1) while the other does not give in (and appears as the winner with a payoff of 5).

Many superpower conflicts, prolonged labor disputes, and other power confrontations have elements in common with the game of chicken. We may find some comfort, however, in knowing that of the 78 essentially different two-by-two games of partial conflict, only chicken and the prisoners' dilemma give rise to such disturbing results, where players following their immediate self-interests leads to such suboptimal results for the society as a group.

REVIEW VOCABULARY

Chicken A common two-person symmetric game in which each player has two strategies: to swerve to avoid a collision or confrontation or else to continue straight ahead and cause a collision if the opponent has not chosen to swerve in the meantime. If both players refuse to swerve, disaster results; if only one backs down, the other wins.

Equilibrium A set of strategies, one for each player, is in equilibrium if no one player can unilaterally alter his or her strategy to obtain a better payoff.

Game matrix A rectangular array of numbers. The rows and columns correspond to the strategies for two players, respectively, and the numerical entries represent the resulting payoffs when particular strategies are selected.

Minimax theorem The fundamental theorem for two-person, zero-sum games, stating that there always exist optimal mixed (randomized) strategies that enable both players to obtain their optimal expected value in the game.

Mixed strategies A strategy chosen in a probabilistic manner from a list of options (called pure strategies).

Optimal (strategy) A particular strategy for a player (pure or mixed) that guarantees that the resulting payoff is the best one that this player can expect to achieve against all possible choices by the opposition.

Payoffs The potential outcomes of a game, typically expressed as numbers.

Players The participants who make strategic choices in a competitive encounter.

Prisoner's dilemma A frequently occurring two-person symmetric game in which each player has two strategies: cooperate or defect. The best outcome for the pair taken together occurs when they both cooperate. In the case where one cooperates and the other defects, the resulting payoffs are the worst and best possible, respectively.

Pure strategy Each possible way for a player to play through a game is called a strategy, or pure strategy, to distinguish it from a mixed strategy.

Saddle point (mountain pass) In a two-person, zero-sum game, a pair of strategies, one for each player, that are in equilibrium. Neither player alone can change strategy and achieve a higher payoff. The strategies and resulting value at a saddle point provide a solution for the game.

Solution An optimal strategy for each player, along with the resulting value of the game.

Strategy One of the possible ways to play a game; strategies are mixed or pure, depending on whether they are selected in a probabilistic manner (mixed) or not (pure).

Value The payoffs of a game, usually expressed numerically, that result when the players play optimally.

Zero-sum A game in which the payoff to one player is the opposite (or negative) of the payoff to the opposing player.

EXERCISES

1. Consider the following two-person, zero-sum games, where the payoffs represent gains to the row player and losses to the column player.

(i) $\begin{bmatrix} 6 & 5 \\ 4 & 2 \end{bmatrix}$

(ii) $\begin{bmatrix} 3 & 5 \\ 5 & 4 \end{bmatrix}$

(iii) $\begin{bmatrix} -2 & 3 \\ 1 & -2 \end{bmatrix}$

(iv) $\begin{bmatrix} -1 & 3 \\ 2 & 0 \end{bmatrix}$

(v) $\begin{bmatrix} 0 & 3 \\ -5 & 1 \\ 1 & 6 \end{bmatrix}$

(vi) $\begin{bmatrix} -10 & -17 & -30 \\ -15 & -15 & -25 \\ -20 & -20 & -20 \end{bmatrix}$

(vii) $\begin{bmatrix} 6 & 5 & 8 & 5 \\ 1 & 4 & 2 & -1 \\ 8 & 5 & 7 & 5 \\ 0 & 2 & 6 & 2 \end{bmatrix}$

a. Which of these games have saddle points?

b. Find the solutions for those games given in **a.**

c. List any bad strategies in the above games, that is, ones the players should avoid because the resulting payoffs are dominated by the payoffs for some alternate strategy.

2. Solve the game of batter versus pitcher in baseball, where the pitcher can throw a fastball or a curve and the batter can guess fastball or curve. The batter's batting averages are as follows:

		Pitcher	
		Fastball	Curve
Batter	Fastball	.300	.200
	Curve	.100	.600

3. When it is third down and short yardage to go for a first down in American football, the quarterback can decide to run the ball or pass it. Similarly, the other team can commit itself to defend more heavily against a run or a pass. This can be modeled as a two-by-two matrix game where the payoffs are the probabilities of obtaining a first down. Find the solution for this game.

		Defense	
		Run	Pass
Offense	Run	.5	.8
	Pass	.7	.2

4. A businessman has the choice of either not cheating on his income tax or cheating and making $1000 if not audited. If caught cheating he will pay a fine of $2000 in addition to the $1000 he owes. He feels good if he does not cheat and is not audited (worth $100). If he does not cheat and is audited, he evaluates this at $-$100 (for the lost day). If he is willing to assume that this is a two-person, zero-sum game between himself and the tax agency, then what are the optimal strategies for each player and the expected value of the game?

5. Consider the game played between the opposing goalie and a soccer player who after a penalty is allowed a free kick. The kicker can elect to kick toward one of the two corners of the net or else aim for the center of the goal. The goalie can decide to commit in advance (after the kicker's decision) to either one of the sides or else remain in the center until he sees the direction of the kick. This zero-sum game can be represented as follows, where the payoffs are the probability of scoring a goal:

		Goalie		
		Breaks left	Remains center	Breaks right
Kicker	Kicks left	.5	.9	.9
	Kicks center	1	0	1
	Kicks right	.9	.9	.5

If we assume that decisions between the left or right side are made symmetrically (i.e., with equal probabilities), then this game can be represented by a two-by-two matrix as follows, where $.7 = (\frac{1}{2})(.5) + (\frac{1}{2})(.9)$:

		Goalie	
		Remains center	Breaks side
Kicker	Kicks center	0	1
	Kicks side	.9	.7

Can you find the optimal strategies for the kicker and goalie and the value of this game?

6. You plan to manufacture a new product for sale next year, and you can decide to make either a small quantity, in anticipation of a poor economy and few sales, or a large output, hoping for brisk sales. Your expected profits are indicated in the following table:

| | | Economy | |
		Poor	Good
Quantity	Small	$500,000	$300,000
	Large	$100,000	$900,000

If you want to avoid risk and believe that the economy is playing an optimal mixed strategy against you in a two-person, zero-sum game, then what is your optimal mixed strategy and expected value? Discuss some alternative ways that you may go about making your decision.

7. On an overcast morning, deciding whether to carry your umbrella can be viewed as a game between yourself and nature as follows:

| | | Weather | |
		Rain	No rain
You	Carry umbrella	Stay dry	Lug umbrella
	Leave it home	Get wet	Hands free

Let's assume that you are willing to assign the following numerical payoffs to these outcomes and that you are also willing to make decisions on the basis of expected values (i.e., average payoffs):

$$(\text{Carry umbrella, rain}) = -2$$
$$(\text{Carry umbrella, no rain}) = -1$$
$$(\text{Leave it home, rain}) = -5$$
$$(\text{Leave it home, no rain}) = 3$$

a. If the weather forecast says there is a 50% chance of rain, should you carry your umbrella or not? What if you believe there is a 75% chance of rain?

b. If you are conservative and wish to protect against the worst case, what pure strategy should you pick?

c. If you are rather paranoid and believe that nature will pick an optimal strategy for this two-person, zero-sum game, then what strategy should you choose?

d. Another approach to this decision problem is to assign payoffs to represent what your *regret* will be after you know nature's decision. In this case, each such payoff is the best payoff you could have received under that state of nature, minus the corresponding payoff in the previous table:

		Weather	
		Rain	No rain
You	Carry umbrella	$0 = (-2) - (-2)$	$4 = 3 - (-1)$
	Leave it home	$3 = (-2) - (-5)$	$0 = 3 - 3$

What strategy should you select if you wish to minimize your maximum possible regret?

8. Consider the following miniature poker game with two players, 1 and 2. Each antes $1. Each player is dealt either a high card H or a low card L, with probability one-half. Player 1 then folds or bets $1. If 1 bets, then player 2 either folds, calls, or raises $1. Finally, if 2 raises, 1 either folds or calls.

Most choices by the players are rather obvious, at least to anyone who has played poker: if either player holds H, that player always bets or raises if he or she gets the choice. The question remains of how often one should bluff, that is, continue to play while holding a low card in the hope that one's opponent also holds a low card. This case arises 25% of the time.

This poker game can be represented by the following matrix game where the payoffs are the expected winnings for player 1 and the dominated strategies have been eliminated:

		Player 2 (when holding L)		
		Folds	Calls	Raises
Player 1 (when holding L)	Folds initially	-0.25	0	0.25
	Bets first and folds later	0	0	-0.25
	Bets first and calls later	-0.25	-0.25	0

a. Are there any strategies in this matrix game that a player should avoid playing?

b. Solve this game.

c. Which player is in the more favored position?

9. Consider the following two-person, non-zero-sum games and discuss the player's possible behavior when these games are played in a noncooperative manner (i.e., with no prior communication or agreements). The first payoff is for the row player; the second, for the column player.

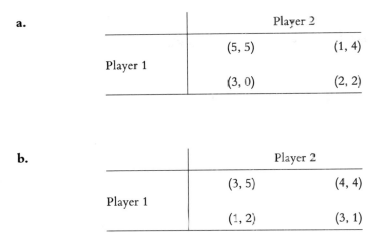

a.

	Player 2	
Player 1	(5, 5)	(1, 4)
	(3, 0)	(2, 2)

b.

	Player 2	
Player 1	(3, 5)	(4, 4)
	(1, 2)	(3, 1)

c. Battle of the sexes:

He buys a ticket for:		*She buys a ticket for:*	
		Boxing	Ballet
	Boxing	(4, 0)	(0, 0)
	Ballet	(1, 1)	(0, 4)

10. For an interesting class project, discuss strategies and expected outcomes in the case of a three-person pistol duel. Assume that three players 1, 2, and 3 are an equal distance apart and that they shoot with accuracies 0.8, 0.6, and 0.4, respectively. Before *each* individual shoots, the players decide by lot which one of the three will take the next shot. When a player is hit he is out of the contest; the lone surviving player is the winner.

a. Discuss this game when it is played in a noncooperative manner.

b. Discuss this game when it is played in a cooperative manner; when coalitions of two can form against the third player, at least until he is eliminated.

C H A P T E R | **12**

Fair Division and Apportionment

Achieving fairness is an important aspect of decision making. How can a group of students divide up a pizza so that each perceives his or her share as fair? How should an estate be divided equitably among the heirs? Can the head of a department fairly allocate such tasks as teaching assignments and committee activities? Can the number of legislators in a representative assembly be assigned in an impartial way? The goal in such fair-allocation problems is to have all persons feel that they obtain a fair and unbiased share of the available benefits or losses, in light of the various limitations present.

In practice we often appeal to someone with authority or experience to answer such questions. Mother will surely divide the cake into fair portions. Managers, we assume, will assign work in a just manner. Labor and management will call in an expert arbitrator to detail the resolution of a dispute. Failing an agreement by a husband and a wife, a judge may dictate the division of the assets in divorce proceedings.

Another common approach to fairness is first to devise some appropriate measure of inequity and then attempt to find an allocation that minimizes inequity—either the largest individual inequity or the total group inequity. Still other approaches take a more statistical view of the problem and seek to achieve even distributions in some average sense, for example, over repeated allocations.

Our approach in this chapter differs from all these. We wish to arrive at methods for fair division that involve only the participants themselves and that satisfy individuals with different value systems. We do not want to assume that each person assigns the same worth or utility to objects or tasks, nor do we want some third party to guess what another's values might or should be. We also want to avoid talking about equity in terms of statistical notions. For example, it is hardly fair to suggest that we flip a coin, giving you all the cake if it is heads and me the whole cake if it is tails, even though we each have an "expected value of half a cake."

In general, a **fair-division problem** consists of n individuals, called **players,** whom we indicate by the numbers $1, 2, \ldots, i, \ldots, n - 1, n$. They must partition some set S of goods (or losses) into n disjoint parts $S_1, S_2, \ldots, S_i, \ldots, S_{n-1}, S_n$. The objective is to find subsets S_i so that person i considers his or her share S_i as fair in his or her own personal value system.

We will illustrate some fair-division schemes for three different cases. First, we will examine the **continuous case,** when the S to be divided is *finely divisible,* like a cake. Second, we will consider the more difficult **discrete case,** where S consists of various *indivisible objects.* For example, an estate's house, furniture, silverware and china, art works, and automobiles are objects that cannot be further subdivided. Finally, we will describe a discrete type allocation known as the **apportionment problem.** It arises when the individual shares must be integers, for example, in assigning the number of representatives each state will be allocated in the U.S. Congress.

The Continuous Case: Two Players

A traditional technique for dividing a finely divisible object S in a fair manner between two players 1 and 2 is "one cuts, the other chooses":

1. Player 1 divides the set S into two pieces S_1 and S_2.

2. Player 2 picks either piece, S_1 or S_2.

3. Player 1 is given the piece not selected by player 2.

Figure 12.1 The New Jersey General Assembly in session in Trenton, the capital. Fair representation is at the heart of the notion of democracy. [New Jersey General Assembly/D. Geoffrey Hobbs.]

Because the roles of players 1 and 2, the cutter and the chooser, are asymmetrical, a player may prefer one role, say, the chooser, over the other. It is not uncommon to flip a coin at the start to decide whether player 1 or 2 will be the cutter.

This method (with or without the coin flip) assures that each of the players can realize a fair piece, provided that two axioms are met:

1. Each player is able to divide the set S into two parts so that *either* one of the pieces is acceptable by that player as fair.

2. Given *any* division of S into two parts, each player will find at least one of the pieces acceptable.

These assumptions apply to the positions of cutter and chooser, respectively. Axiom 1 guarantees that the cutter will be satisfied with whatever piece is remaining, whereas axiom 2 assures the chooser of finding at least one piece agreeable, no matter how the cutter divides object S.

If these assumptions are not in fact realized, then a fair result may not be reached using this method. For example, if one player will accept only the full cake as fair and the other wants a (nonempty) piece, then no method can arrive at a fair split. Any fair-division scheme must presume some reasonability conditions similar to axioms 1 and 2.

It is important to note that our axioms and our method assume neither that the players can assign numerical measures to pieces of S nor that they must delineate all acceptable pieces before the cut is made. The method requires only that the operational requirements in axioms 1 and 2 are met and each player is able to decide whether a given piece is fair. This is an example of applying mathematical concepts to practical problems without explicit use of numbers.

The Continuous Case: Multiple Players

A number of fair-division schemes have been proposed for more than two players, along with corresponding axioms to assure that acceptable distributions can always be reached. One such "divide and choose" approach for the case of n participants is called the **last-diminisher** method. Let's see how this method works for seven students who want to divide a chocolate cake:

1. Student 1 cuts any piece he or she views as fair from the cake.

2. Student 2 can "pass" on this piece, or diminish the piece cut if he or she views it as too large.

3. Students 3, 4, 5, 6, and 7, each have the right — but not the obligation — to further diminish the remaining piece as their turns come.

Figure 12.2 A fair-division scheme can help us decide how to best divide an uncut cake among any number of individuals. [Photo by Travis Amos.]

4. The piece is assigned to the *last* player who elected to diminish it, who then exits from the game. This is student 1 if all the other players passed, that is, if no one chose to challenge the piece as unfair.

5. The above steps are repeated with the remaining six players, using the original cake less the piece that was assigned to the last diminisher.

6. This process is repeated for five players, then four players, then three, until only two players remain in the game.

7. The final two players can continue in the same manner or decide to use the "cut and choose" method.

The reader may want to list the assumptions under which this method will guarantee that a fair division can always result. For example, it might be essential that the values of all the deleted parts always add up to the total value of the cake, that is, that the act of dismembering the whole into many pieces has not lessened any student's appetite for the part he or she eventually receives. In the case of a crumbly cake and actual slicing, this assumption may well fail. On the other hand, if the object being divided is land, the preliminary cuts may only be made by marking on a map, which would hardly devalue the land itself. As in all mathematical models, the conclusions may not apply if the assumptions are not satisfied.

The Discrete Case

In many fair-division problems, including some inheritances, some of the objects to be allocated cannot be further subdivided into smaller parts. One approach in such cases is to attempt to assign numerical values, such as dollar amounts, to the objects and then divide the total sum into fair ratios. The final allocation can then be achieved by assigning either the objects themselves or the dollar equivalents. This will typically require monetary side payments between the players. A great number of auction and bidding schemes have been introduced to force participants to honestly reveal their individual monetary values for specific objects.

In the case of two players dividing one object into equal shares, we can arrive at a fair division as follows:

1. Players A and B enter sealed bids of amounts a and b, representing their respective honest evaluations of the object.

2. The object is awarded to the higher bidder, say B, whose bid was $b > a$.

3. Then the higher bidder B pays the lower bidder A the amount of $a/2 + (b - a)/4$.

As a consequence, player B realizes an amount that B values as

$$b - \left(\frac{a}{2} + \frac{b-a}{4}\right) = \frac{b}{2} + \frac{b-a}{4}$$

It follows that each player receives half of his or her own evaluation (which is his or her fair share) plus a surplus of $(b - a)/4$. Paradoxically, when the participants in a fair division assign different values, one can arrive at a split that gives each of them *more* than a fair share.

For example, consider the case where Alice and Barbara inherit a house; they bid $100,000 and $150,000, respectively. They see Alice's fair share as $50,000 and Barbara's as $75,000. Our method assigns the house to Barbara and has her pay $62,500 to Alice. Barbara ends up with a house worth $150,000 to her, less $62,500 paid out, for a net gain of $87,500. Similarly, Alice receives her fair share ($50,000) plus $12,500, for a total of $62,500. So each receives her fair share plus $12,500.

This method differs from the method commonly practiced, which is to award the object to the higher bidder and give half of this top bid to the losing bidder. In this case, Barbara would have paid Alice $75,000 in return for the house.

Table 12.1 shows the calculation for a more complicated example. Four heirs with equal shares, Alice, Bob, Carol, and Don, will bid for three objects in the estate, a house, a summer cabin, and a boat. We assume that each heir enters a secret bid (in dollars) on each of the objects, as indicated in the top part of Table 12.1.

Apportionment

Another important class of fair-division problems is the *apportionment problem*. It arises when we are required to round fractions so that their sum is maintained at some given constant value (see Boxes 12.1 and 12.2). This constraint appears in many situations, for example, in cases where the fractions correspond to

TABLE 12.1 Fair division of an estate

	Alice	Bob	Carol	Don
House	120,000	200,000	140,000	180,000
Cabin	60,000	40,000	90,000	50,000
Boat	30,000	24,000	20,000	20,000
Fair division:				
Sum of the bids:	210,000	264,000	250,000	250,000
Fair shares:	52,500	66,000	62,500	62,500
Objects awarded:	Boat	House	Cabin	—
at the high bids:	30,000	200,000	90,000	0
Remaining claims:	22,500	−134,000	−27,500	62,500
Total surplus:	76,500			
Share of surplus:	19,125	19,125	19,125	19,125
Final settlements:	Boat + 41,625	House − 114,875	Cabin − 8,375	81,625

BOX 12.1 Number Systems

The concept of number is one of the most fundamental notions in mathematics. One starts with the **natural numbers** 1, 2, 3, 4, 5, 6, 7, . . . , which are also called the **whole** numbers, the **counting** numbers, or the **positive integers.** The famous nineteenth-century German mathematician Leopold Kronecker (1823–1891) attests to the fundamental role of the natural numbers in his much quoted quip: "God made the integers: all else is the work of man."

One important extension of the natural numbers is the **positive fractions.** These are ratios p/q of natural numbers p and q. The number p is called the **numerator** and q the **denominator** of the fraction p/q. These ratios are also referred to as **positive rational numbers.** Any such fraction p/q can be divided by another fraction r/s according to the rule

$$\frac{p}{q} \div \frac{r}{s} = \frac{p \times s}{q \times r}$$

and the quotient will be a rational number. In other words, the positive fractions are said to be closed under the operation of division, as well as under addition and multiplication.

Fractions can also be expressed in *decimal notation,* as in these examples:

$$\frac{1}{4} = \frac{25}{100} = 0.25$$

$$\frac{5}{4} = 1\frac{1}{4} = 1\frac{25}{100} = 1.25$$

$$\frac{8}{3} = 2.666666 \quad . \, .$$

$$\frac{7}{12} = 0.583333 \, . \, . \, .$$

$$\frac{1}{7} = 0.142857142857 \, . \, . \, .$$

Note that the decimal representations of some fractions, such as one-fourth (0.25), terminate, whereas others, such as eight-thirds (2.6666 . . .), go on without end. However, all rational numbers expressed in their decimal form will either terminate or else become infinitely repeating. For example, one-seventh will continue to repeat the block of digits 142857 unendingly.

Use of the Hindu-Arabic number system, along with the decimal notion, was a great leap forward in the history of mathematics. It greatly simplified performing the elementary operations of arithmetic.

However, as even the ancient Greeks knew, fractions are not sufficient to represent all "real" measurements. For example, some ratios that arise naturally cannot be represented as rational numbers. The ratio of the diagonal of a square to one of its sides is $\sqrt{2} = 1.4142146.$ . . . The ratio of the circumference of a circle to its diameter, denoted by the Greek letter pi, is $\pi = 3.1415927$ . . . . These are examples of irrational numbers. They cannot be represented either as terminating decimals or as repeating decimals.

BOX 12.2 Rounding Fractions

In many practical applications of arithmetic we are forced to *round* fractions. To "round off" a fraction is to approximate it by another fraction that uses fewer digits in its decimal representation. For many purposes, it may be sufficient to approximate the fraction $\frac{1}{3}$ by 0.33333 or $\frac{1}{7}$ by 0.14286.

For example, if your savings account at the local bank grows at the annual rate of $6\frac{2}{3}\% = 0.066666$ . . , compounded quarterly, then the interest that accrues after three months on an investment of $1000 is

$$\frac{1}{4} \times 6\frac{2}{3}\% \times (\$1000) = \frac{1}{4} \times \frac{20}{3} \times \frac{1}{100} \times \$1000$$

$$= \frac{5}{3} \times \$10$$

$$= \frac{\$50}{3}$$

$$= \$16.666666 \, . \, . \, .$$

The bank will typically credit $16.66 to your account and keep the remaining fraction of a cent for itself. Customers rarely complain when the bank keeps their two-thirds of a penny, although such fractions from many different accounts may add up to a nontrivial sum for the bank. On the other hand, if the bank were computing interest on a loan, it would normally round up and charge you $16.67 in interest for the quarter.

In most practical applications, the rounding of fractions is done routinely and produces trivial results. However, when one is adding or multiplying a very large string of numbers, the accumulation of many small errors may eventually create a significant error. In most cases where it may matter (for example, in computer programs), one can usually incorporate additional decimal places into the calculations to reduce the potential final error to a negligible level.

In some situations, however, the rounding of fractions can have major impact. For example, the election of Rutherford B. Hayes to President of the United States in 1876 depended, among other things, on the way in which fractions had been rounded. Had a different procedure been employed for rounding fractions in the U.S. Congress and the Electoral College, including the method they were supposed to be using according to the law at the time, Samuel J. Tilden would have been declared President instead.

The method used for allocating fractions may at times prove significant, especially if we are concerned with fairness. The surprising conclusion is that there is no one method for apportioning fractions that is entirely free of all undesirable outcomes. Every method must on occasion demonstrate some counterintuitive anomalies or undesirable behavior.

numbers of persons (who must be assigned in whole numbers) or in statistical tables where we may wish to round percentages while we maintain the sum at 100%.

For example, consider a university with 20,000 students and six colleges. The student populations are displayed in Table 12.2.

If we use the normal rounding procedure — rounding fractions up when they are greater than one-half and down when they are less than one-half — to round these percentages to one decimal place or to integer values, the total percentages do not sum to 100% (see Table 12.3).

TABLE 12.2 **University enrollment by colleges**

College	Number of students	Percent
Arts & Sciences	6716	33.580
Engineering	4832	24.160
Agriculture	4093	20.465
Business	3211	16.055
Law	852	4.260
Architecture	296	1.480
Total	20,000	100.000

TABLE 12.3 **Percentages rounded two different ways**

College	Rounded percent	Rounded percent
Arts & Sciences	33.6	34
Engineering	24.2	24
Agriculture	20.5	20
Business	16.1	16
Law	4.3	4
Architecture	1.5	1
Total	100.2	99

An alternate apportionment will realize 100% for these fractions (see Table 12.4).

TABLE 12.4 Percentages apportioned

College	Rounded percent	Rounded percent
Arts & Sciences	33.6	34
Engineering	24.1	24
Agriculture	20.5	20
Business	16.0	16
Law	4.3	4
Architecture	1.5	2
Total	100.0	100

Figure 12.3 A university must resolve many apportionment problems, such as arise in class scheduling. [Photo by Robert Cohen. © 1987 Photographic Services, University of Delaware. All rights reserved.]

Apportionment problems arise frequently in allocation situations. A typical school district, for example, must schedule courses to classrooms; place teachers in schools; assign students to schools, to classrooms, or to different sections of a course; and assign school buses to various routes or schools. The U.S. Navy must allocate personnel with various qualifications and ranks to a large variety of positions and tasks. A traffic engineer has to assign subway cars or buses to different trains or routes to best meet the expected demand. Planners everywhere must apportion their limited resources to optimize the quality of service.

The apportionment problem frequently arises in determining political representation in democratic institutions. In this case, a fair representative share may be a fraction, whereas the representatives themselves are individuals with one vote apiece. In some national governing bodies with a representative form of government, the number of seats in the parliament that are assigned to a particular party is directly proportional to the number of votes the party received, with the total size of the assembly held constant. A party's proportional share, which we call its **quota,** typically will not be a

whole number; so it must be replaced by a nearby integer value.

For example, if a party obtains 41.23% of the votes in a national election for a parliament with 120 seats, its quota is given by the number q in the formula

$$\frac{q}{120} = 41.23\%$$

or

$$q = (120)(0.4123)$$
$$= 49.476$$

However, the number of individual representatives assigned to this party must be a whole number, such as 49 or 50, or perhaps even some other integer value.

The apportionment problem also arises in a federal form of government, where the number of representatives from a given region is proportional to the population of that region. For example, the U.S. House of Representatives in recent years has had 435 voting members, where the number from each state is intended to be proportional to the state's population. The population of California, for example, was 23,668,562, according to the 1980 census, whereas the population of all 50 states at that same time was 225,867,174. So California's *quota* is given by

$$\frac{q}{435} = \frac{23,668,562}{225,867,174}$$

which yields

$$q = 45.5835 \ldots$$

The current method for apportioning seats in the House, called the Hill-Huntington method, assigns California 45 seats.

The U.S. House of Representatives is one of the best-known and most frequently studied cases of political apportionment. After each decennial census, this matter is typically a topic of intense debate. Some half-dozen different methods have been seriously considered

for use in the U.S. Congress, and four of these have been implemented at various times. These are referred to in U.S. political history as the methods of Hamilton, of Jefferson, of Webster and Willcox, and of Hill and Huntington; each is named after famous U.S. statesmen or distinguished mathematical scientists who argued for its use. John Quincy Adams also proposed a method, but it was never used. The first presidential veto in U.S. history occurred when George Washington vetoed an apportionment bill advocated by Alexander Hamilton in favor of one supported by Thomas Jefferson.

The fascinating history of apportionment in the U.S. Congress is told in a delightful and important book, *Fair Representation: Meeting the Ideal of One Man, One Vote,* by Michel L. Balinski and H. Peyton Young (Yale University Press, New Haven, 1982). After describing the classical methods of apportionment in some detail, the authors recommend as most appro-

Figure 12.4   The United States House of Representatives in session in Washington, D.C.

priate for political apportionment the method advocated by Senator Daniel Webster in the nineteenth century and by statistician Walter F. Willcox throughout most of this century. The authors' suggestion differs from the currently used Hill-Huntington method, which has been the law of the land since 1941.

It should be noted that the Electoral College, which formally elects the President of the United States, gives to each state a number of votes equal to its number of congressmen, including both senators and all House members. Thus, the method of apportionment used for Congress can also affect who is elected President, as was the case in 1876.

Undesirable Outcomes

Given several different methods for apportioning fractions, we would naturally ask whether one method is better than the others. Is there any method that is completely fair in the sense that it satisfies all of the criteria we most desire? Perhaps surprisingly, the answer is no.

Every *known* apportionment method will, in some particular cases, demonstrate some undesirable property. Furthermore, research done in the 1970s by Balinski and Young has proved that *any possible* apportionment method, whether known or not yet discovered, must in fact produce some unpleasant result in some instances. These difficulties are unavoidable. In particular, we can list three properties we would want to hold for any fair apportionment method, and then prove that *no* method can possibly exist that always satisfies all three conditions simultaneously. Any attempt to avoid these difficulties is doomed to failure. Balinsky and Young's result is another good illustration of an *impossibility theorem*, which states that some desired outcome cannot in general be realized (see Box 9.2 in Chapter 9). Such discoveries may well appear, at least initially, as paradoxical or counterin-

tuitive. Let's look at each of three desired conditions in turn, seeing how it may be violated.

The apportionment methods of Jefferson and of Adams violate what is known as the **quota condition** — a standard requiring that the apportioned assignment be one of the two whole numbers nearest to the true quota. For example, the census of 1830 listed New York State's population as 1,918,578 and the total population of the United States as 11,931,000. At that time, the House of Representatives had 240 seats, so the quota for New York State was determined by the formula

$$q = \frac{1,918,578}{11,931,000} \times 240 = 38.593 \ldots$$

The Jefferson method of apportionment, which is biased in favor of large states, gave New York an apportionment of 40, whereas the Adams method, which is biased in favor of small states, would have assigned only 37 seats. Both these methods violated the quota condition because they produced an apportionment

Figure 12.5 Thomas Jefferson favored a method of apportionment biased in favor of states with large populations. [The Bowdoin College Museum of Art.]

differing by a full integer or more from the quota. Both the Webster-Willcox method and the Hill-Huntington method would have assigned 39 seats to New York in 1830, satisfying the quota condition. Even these two methods can also violate the quota condition in other cases, although they are statistically less likely to do so.

The apportionment method of Hamilton will always satisfy the quota condition. However, it often demonstrates an undesirable property called the **Alabama paradox.** Using the census of 1880, the Hamilton method would have given the state of Alabama 8 seats if the U.S. House of Representatives had a total of 299 seats, whereas this method would assign Alabama only 7 seats if the House size were raised to 300. Thus, an increase in the House size would cause Alabama to lose a seat. Both Texas and Illinois were to gain a seat. (A method that produces this result is said to violate a condition called **house monotonicity.**)

In an attempt to discover an apportionment method that would never violate either the quota condition or the condition of house monotonicity, Balinski and Young invented the **quota method.** It is a variant of the Jefferson method, which assigns seats one at a time, but it is modified so that the quota condition is verified at each step.

Unfortunately, the quota method may violate a third desirable property, called **population monotonicity:** using this method, a state may increase its population and yet end up with fewer seats! For example, using the quota method, five states with the populations

$$122 \quad 35 \quad 17 \quad 16 \quad 10$$

would receive

$$4 \quad 2 \quad 0 \quad 0 \quad 0$$

seats, respectively, in a House of 6 seats. However, if the populations were changed to

$$122 \quad 39 \quad 17 \quad 16 \quad 10$$

then the resulting apportionment by the quota method would be

$$4 \quad 1 \quad 1 \quad 0 \quad 0$$

You can see that the second state increased in population but nonetheless lost a seat.

Balinski and Young showed that no apportionment method will *always* satisfy the three desirable properties of quota condition, house monotonicity, and population monotonicity. The selection of an apportionment method is necessarily a compromise between these three drawbacks, in which one must pick the drawback one is willing to accept.

Class Scheduling

We will now look at some simple illustrations of the apportionment problem in practical applications. These examples will also indicate how some of the undesirable properties occur in practice.

Consider a small senior high school with only one mathematics teacher who teaches five classes each day. One hundred students preregister to take one of three mathematics courses: 51 students sign up for tenth-grade geometry, 30 for eleventh-grade algebra, and 19 for twelfth-grade calculus. This yields the following apportionment problem: Given a total of five sections, how many sections of each subject should be offered?

If there are 100 students and five sections, the average class size will be 20. However, we cannot realize this average in each class because 20 does not divide evenly into the numbers 51, 30, and 19. Instead, we compute the ideal fair share of sections for each subject — the quotas. For example, the quota for geometry will be $q = \frac{51}{100} \times 5 = 2.55$. Similarly, the quotas for algebra and calculus are 1.50 and

TABLE 12.5 Apportioning course sections

Grade	Course	Number of students	Quota q	Integer part	Fractional part	Fraction rounded	Apportionment	Course average
10	Geometry	51	2.55	2	0.55	1	3	17
11	Algebra	30	1.50	1	0.50	0	1	30
12	Calculus	19	0.95	0	0.95	1	1	19
Totals		100	5.00	3	2.00	2	5	20

0.95, respectively. Table 12.5 summarizes this information.

The apportionment problem consists in rounding these three quotas to whole numbers while maintaining the sum at 5. One very natural way to do this is called the method of **largest fractions,** or the Hamilton method. First, assign to each subject the largest integer in its quota. These integer parts of the quotas are called the *lower quotas.* For this example, they are the numbers 2, 1, and 0, respectively, and are shown in the fifth column of Table 12.5. They account for only three of the five sections. Second, assign the remaining two sections to those subjects with the largest fractional parts in their quotas. Because the calculus course has the largest remainder, with fractional part 0.95, and geometry has the next highest, with 0.55, the resulting apportionment is 3, 1, and 1. The resulting average class sizes for the three individual subjects are then 17, 30, 19, respectively.

Suppose now that when the term begins the actual enrollments in the three math courses

have changed to 52, 33, and 15. The new quotas, integer parts, fractional parts, and apportionment are given in Table 12.6.

Comparing Tables 12.5 and 12.6, you observe that even though the number of students

Figure 12.6 The apportionment method of largest fractions, also known as the Hamilton method, was named for Alexander Hamilton. [*Alexander Hamilton,* John Trumbull; National Gallery of Art, Washington; Andrew W. Mellon Collection.]

TABLE 12.6 Apportioning course sections revised enrolllments

Grade	Course	Number of students	Quota q	Integer part	Fractional part	Fraction rounded	Apportionment	Course average
10	Geometry	52	2.60	2	0.50	0	2	26
11	Algebra	33	1.65	1	0.65	1	2	16.5
12	Calculus	15	0.75	0	0.75	1	1	15
Totals		100	5.00	3	2.00	2	5	20

taking geometry has increased from 51 to 52, the number of geometry sections decreased from 3 to 2 because one section formerly allocated to geometry has been switched to algebra, which had a still larger increase in enrollment. This particular anomaly, the failure of **quota monotonicity,** is inherent in any reasonable apportionment method: one group's quota may increase, only to have its apportionment decrease. This unpleasant outcome is distinct from the properties we called house monotonicity and population monotonicity.

Except for Hamilton's method, all the apportionment methods we have discussed are known as **divisor methods.** Any divisor method consists of two parts. First, it focuses on a particular number d called a **divisor** and determines how often d divides into the respective populations. The resulting quotients will have an integer part, plus a fractional remainder. Second, a divisor method must specify the conditions under which each remainder is to be rounded down or rounded up. Jefferson's method discards all such fractional parts, whereas the Adams method rounds any remaining positive fraction upward. On the other hand, the method of Webster-Willcox rounds fractions down or up, according to whether they are less than or greater than one-half.

Intuitively, we see that the divisor corresponds to some desirable unit size. For example, the U.S. Constitution requires that congressional districts average at least 30,000 people. In our classroom example, the divisor may correspond to some ideal ratio of students to teachers or to some desired minimal (or maximal) class size. Such divisors need not be integer values.

Let's return to our classroom example and consider a divisor equal to the average class size of 20. The number of times that 20 divides the preregistration figures of 52, 33, and 15 is given in Table 12.7, along with the total number of sections apportioned, using this divisor, according to the methods of Jefferson, Webster-Willcox, and Adams. Because none of these totals of 3, 6, and 6 meets the requirement of precisely five sections, none of these apportionment schemes provides a suitable answer for this particular divisor.

So we will consider other divisors. For example, a divisor of 26 yields an Adams apportionment of two section each of geometry and algebra, and one of calculus (see Table 12.8). This divisor of 26 can be viewed as the maximum class size for any single section, although it does not necessarily follow that we must assign precisely 26 students to each of the two sections in geometry.

Going in the other direction, a divisor of 16 (see Table 12.9) results in a Jefferson apportionment of 3, 2, and 0. In this case, the divisor can be viewed as the minimal allowable class size, although here we are likely to overrule this minimum requirement of 16 per section and schedule a class for the 15 students wanting calculus, even though the added section

TABLE 12.7 Apportioning classes with divisor 20

Course	Number of students	Quotient		Apportionment		
		Integer	Fraction	Jefferson	Webster	Adams
Geometry	52	2	$\frac{12}{20}$	2	3	3
Algebra	33	1	$\frac{13}{20}$	1	2	2
Calculus	15	0	$\frac{15}{20}$	0	1	1
Totals	100	3	2	3	6	6

TABLE 12.8 Apportioning classes with divisor 26

Course	Number of students	Quotient		Apportionment		
		Integer	Fraction	Jefferson	Webster	Adams
Geometry	52	2	$\frac{0}{26}$	2	2	2
Algebra	33	1	$\frac{7}{25}$	1	1	2
Calculus	15	0	$\frac{15}{26}$	0	1	1
Totals	100	3	$\frac{22}{26}$	3	4	5

TABLE 12.9 Apportioning classes with divisor 16

Course	Number of students	Quotient		Apportionment		
		Integer	Fraction	Jefferson	Webster	Adams
Geometry	52	3	$\frac{4}{16}$	3	3	4
Algebra	33	2	$\frac{1}{16}$	2	2	3
Calculus	15	0	$\frac{15}{16}$	0	1	1
Totals	100	5	$\frac{20}{16}$	5	6	8

would have to be taken away from one of the other courses. A similar minimal condition arises in the U.S. House of Representatives because the Constitution requires at least one seat for each state. Of course, apportionment methods can be redefined to take account of various minimum or maximum conditions.

Finally, a divisor of 22 yields the apportionment shown in Table 12.10, where the fraction $\frac{11}{22}$ has been rounded upward under the Webster-Willcox method. Thus, for a class of 22 — or 21 if you wish — the Webster-Willcox method assigns two sections each of geometry and algebra and one to calculus.

Salary Increments

We conclude with a financial example that illustrates the Alabama paradox. A chairman of

TABLE 12.10 Apportioning classes with divisor 22

Course	Number of students	Quotient		Apportionment		
		Integer	Fraction	Jefferson	Webster	Adams
Geometry	52	2	$\frac{8}{22}$	2	2	3
Algebra	33	1	$\frac{11}{22}$	1	2	2
Calculus	15	0	$\frac{15}{22}$	0	1	1
Totals	100	3	$\frac{34}{22}$	3	5	6

TABLE 12.11 Apportioning salaries

Professor	Current salaries	Increased by 5%	Rounded to 100s	Rounded to 1000s
A	43,100	45,255	45,200	45,000
B	42,150	44,257	44,300	44,000
C	10,000	10,500	10,500	11,000
Totals	95,250	100,012	100,000	100,000

TABLE 12.12 Apportioning larger salaries

Professor	Current salaries	Increased by 6%	Rounded to 100s	Rounded to 1000s
A	43,100	45,686	45,700	46,000
B	42,150	44,679	44,700	45,000
C	10,000	10,600	10,600	10,000
Totals	95,250	100,965	101,000	101,000

a small college mathematics department supervises three other faculty, whose current salaries are $43,100, $42,150, and $10,000. (Because the third person works only part time, his salary is low.) The chairman is instructed by his dean to award salary increments for the coming year at approximately 5%, up to a total salary pool of $100,000, and to round all salaries to multiples of $1000. So, the chairman increases everyone's salary proportionally by 5% and using the Hamilton method rounds to multiples of $100 and $1000, so as to sum to exactly $100,000. These figures are shown in Table 12.11.

The chairman is displeased with the result, however, because he had to round the first two professor's salaries downward. In an attempt to alleviate the problem, the dean responds by allocating an additional $1000 to the salary pool, bringing it up to a total of $101,000. The chairman then recomputes his figures, using 6% instead of 5% (see Table 12.12).

In this case, as the salary pool increased from $100,000 to $101,000, the third professor's salary decreased from $11,000 to $10,000. The added condition that salaries be in multiples of 1000 caused him to receive no increase whatsoever, even though his 6% fair share of $600 brings him closer to $11,000 than to $10,000. This illustration of the Alabama paradox plagues the Hamilton apportionment method and many of its variants.

In recent years mathematics has been employed to solve many problems of a highly human nature, such as concerns about equitable allocation. In this chapter we have described a few of the many possible schemes a group can use to arrive at a fair distribution of goods or services. On the other hand, we have also indicated how some fair-division problems, namely, the apportionment problem, cannot always be solved in a completely satisfactory manner.

REVIEW VOCABULARY

Adams method The apportionment scheme of John Quincy Adams is the divisor method that rounds all positive fractions upward to the next integer.

Alabama paradox An apportionment method exhibits the Alabama paradox if an increase in the size of a legislative body can cause an individual state to lose a representative.

Apportionment problem The apportionment problem arises when one must round a list of fractions to integers in a way that preserves the sum of the original fractions. It occurs in assigning representatives to parties or states in an assembly, in assigning teachers to courses in classrooms, in rounding percentages to integers that sum to 100% in a statistical table, and in many other instances.

Continuous (divisible) case A fair-division problem in which the object to be divided has no indivisible components and can be finely divided into parts; examples include time, land, money, or sand.

Discrete (indivisible) case A fair-division problem in which some parts of the objects to be divided, such as the house and cars of an estate, cannot be finely divided into arbitrarily small parts in any manner.

Divisor method An apportionment scheme in which the resulting apportionment is obtained in two steps. First, the populations are divided by a suitable constant value d, called a *divisor*, to obtain integer values plus fractional remainders. Second, a rule is given to determine whether the remainders will be dropped or rounded up in order to arrive at the final apportionment. The methods of Adams, Jefferson, and Webster-Willcox are divisor methods, as is the Huntington-Hill method currently used by the U.S. Congress for the House of Representatives.

Fair-division problem To divide up some gains or losses into n separate parts so that each of n people considers the part he or she receives as a fair allocation.

Fair-division scheme A method for solving a fair-division problem. Each participant in the procedure must have a way to realize a piece that he or she views as fair in his or her own value system. Any such scheme must be based on certain reasonability assumptions in order to guarantee that a fair division will exist.

Hamilton method (method of largest fractions) The apportionment method of Alexander Hamilton assigns to each state the integer part (lower quota) of its quota and then adds 1 (to get upper quota) for those states

with the largest fractional parts in their quota. The numbers of 1s so added is enough to reach the given House size.

Jefferson method The apportionment scheme of Thomas Jefferson is the divisor method that rounds all fractions down to zero and takes only the integer parts as the apportionment.

Monotone A function of a variable is monotone if this function will not decrease in value whenever the variable increases in value. Several types of monotonicity are desirable for apportionment schemes: (1) An apportionment method is *house monotone* (avoids the Alabama paradox) if no state can lose a seat in the house when the size of the house increases. (2) An apportionment method is *population monotone* when no state can lose a seat when only its population increases. (3) An apportionment method is *quota monotone* when no state can lose a seat whenever its quota increases. Property 3 is rarely achieved in an apportionment scheme. Properties 1 and 2 hold for all divisor methods, although property 2 fails for the popular Hamilton method.

Quota A state's quota in an apportionment problem is given by the formula $q = hp/P$, where h is the size of the house (the total number of representatives), p is the population of the state, and P is the total population of all the states. The quota represents the state's fair share, but it is typically a fraction that must be rounded to an integer value. The largest integer less than or equal to q is called the *lower quota*; the smallest integer greater than or equal to q is called the *upper quota*.

Quota condition A condition satisfied by an apportionment method if it assigns each state an apportionment (integer value) that satisfies the lower quota and the upper quota conditions.

Webster-Willcox method The apportionment scheme of statesman Daniel Webster and statistician Walter F. Willcox is the divisor method that rounds fractions greater than or equal to $\frac{1}{2}$ upward and fractions less than $\frac{1}{2}$ downward.

EXERCISES

1. a. Give a precise definition of "grade point average" (GPA), as it is computed at your college.

b. In order to graduate summa cum laude at Mount LAX College, students need to have a 3.75 GPA in their last two years. If Mary has 228 quality points for 61 credit hours, will she receive this honor?

c. In order to graduate from Mud Tech, where "+" and "−" after grades count in the averages, a student must have an overall GPA of 2.000. If John has 253.95 quality points for 127 credit hours, will he graduate on time?

2. a. Entering the last day of the 1941 baseball season, Boston Red Sox slugger Ted Williams had 179 hits in 448 times at bat. If Williams did not play that day, would he have averaged .400 for the year?

b. Williams got 6 hits in 8 appearances during a double header that day to be the last batter in more than 45 years to exceed .400. What was his batting average when rounded to three decimal places?

3. a. Who won the American League batting championship in 1945 when George Stirnweiss of New York got 195 hits in 632 times at bat and Tony Cuccinello of Chicago went 124 for 402?

b. Who won the American League batting championship in 1949 when Ted Williams of Boston got 194 hits in 556 times at bat and George Kell of Detroit went 179 for 522?

c. Who won the American League batting championship in 1970 when Carl Yastrzemski of Boston got 186 hits in 566 times at bat and Alex Johnson of California went 202 for 614?

d. Who won the National League batting championship in 1931 when Bill Terry of New York hit 213 out of 611, Jim Bottomly of St. Louis hit 133 out of 382, and Chick Hafey of St. Louis hit 157 out of 450?

4. If Sheila bids $1.50 and Jean bids $1.25 for one frisbee, how would you reach a fair division of this object?

5. a. Describe a fair division for three heirs A, B, and C who inherit a house in the city, a small farm, and a valuable sculpture and who submit sealed bids (in dollars) on these objects as follows:

	A	B	C
House	145,000	149,999	165,000
Farm	135,000	130,001	128,000
Sculpture	110,000	80,000	127,000

b. Describe a fair division for the three heirs A, B, and C if their shares are $\frac{5}{10}$, $\frac{3}{10}$, and $\frac{2}{10}$, respectively.

6. Consider the following scheme used to divide a large submarine sandwich among three children:

STEP 1. Mother passes the knife slowly from left to right over the top of the sandwich. The first child who says "Stop" is given the piece to the left of the knife and leaves the contest.

STEP 2. Mother continues to move the knife. The next child to say "Stop" receives the piece to the left and exits the game.

STEP 3. The piece to the right of the knife in step 2 is given to the remaining child.

a. Is this a reasonable fair-division scheme?

b. What assumptions need to be made in order for this to be a fair method?

c. Can this approach be extended to more than three children?

7. Pablo Picasso left his enormous estate to six heirs with shares as follows:

Person	Relationship	Share
Jacqueline	Wife	$\frac{11}{32}$
Claude	Child	$\frac{3}{32}$
Poloma	Child	$\frac{3}{32}$
Maya	Child	$\frac{3}{32}$
Bernard	Grandson	$\frac{6}{32}$
Marina	Granddaughter	$\frac{6}{32}$

Can you suggest a reasonable fair-division scheme for the heirs that takes into account the huge number of objects involved?

8. It makes for interesting class projects to implement the continuous and discrete fair division schemes discussed in this section in the classroom using a (nonhomogeneous or irregular shaped) cake, a large pizza, a can of mixed nuts, a bag of mixed candies, or an assortment of candy bars.

9. Consider a small college with 3 divisions as follows:

Arts	690
Science	435
Business	375
Total	1500

a. How would you apportion the 5 seats in the student senate to the 3 divisions?

b. If the student numbers the next year are 555, 465, and 480, respectively, how would you apportion the 5 seats?

c. Has any division gained students but lost representation in going from **a** to **b**?

10. Reconsider the college example given in the text (p. 236).

College	Number	Percent	Percent rounded
Arts & Sciences	6716	33.580	34
Engineering	4832	24.160	24
Agriculture	4093	20.465	20
Business	3211	16.055	16
Law	852	4.260	4
Architecture	296	1.480	2
Total	20,000	100.000	100

The last column gives the rounded percentages apportioned according to the Hamilton method of largest fractions.

a. How would the last column appear if one used the apportionment methods of (i) Jefferson, (ii) Webster-Willcox, and (iii) Adams?

b. Give a value of a divisor d that works for each of the three methods in **a.**

11. Consider a state with 11 counties with populations (in thousands) of 8785, 126, 125, 124, 123, 122, 121, 120, 119, 118, and 117.

a. What is the best apportionment for an assembly with 100 seats?

b. What would the apportionment be when you use the methods of (i) Hamilton, (ii) Jefferson, (iii) Webster-Willcox, and (iv) Adams?

c. For which methods in **b** is the quota condition violated? That is, when does some state receive more than its upper quota or less than its lower quota?

12. Consider a state with 6 counties with populations of 9215, 159, 158, 157, 156, and 155. Answer all the questions in the previous exercise.

13. Consider a country with 6 states with populations: 27,774, 25,178, 19,947, 14,614, 9225, and 3292 and a house size of 36 seats. Find the apportionment using the methods of (i) Hamilton, (ii) Jefferson, (iii) Webster-Willcox, and (iv) Adams. (You may wish to use a hand calculator.)

14. For an interesting project, take some national assembly in which the number of representatives a party receives is proportional to the number of votes it receives in an election, then calculate the number of seats each party receives according to the four different apportionment schemes listed in Exercise 12. Israel, Japan, and several European countries have such systems.

15. Apportionment methods can also be characterized in terms of minimizing the distance (or the inequity) between the quota point $q = (q_1, q_2, \ldots, q_s)$ and the integer apportionment point $a = (a_1, a_2, \ldots, a_s)$, where $a_1 + a_2 + \cdots + a_s = h = q_1 + q_2 + \cdots + q_s$ for a house size of h seats and s states.

a. Can you show that the apportionment scheme that minimizes the Euclidean distance

$$\min_a \left[\sum_{i=1}^{s} (a_i - q_i)^2 \right]^{1/2}$$

gives the Hamilton method?

b. Can you show that the apportionment scheme that minimizes the Manhatten distance

$$\min_a \left[\sum_{i=1}^{s} |a_i - q_i| \right]$$

gives the Hamilton method?

16. a. What is the cardinality of the set {1, 2, 3, . . .} of natural numbers?

b. What is the cardinality of the set {2, 4, 6, . . .} of even positive integers?

c. Which of the sets in **a** and **b** has the larger number of elements? In what sense?

d. Georg Cantor (1845–1918) argued that the set {1, 2, 3, . . .} of natural numbers has as many elements as the set of positive rational numbers by showing a one-to-one correspondence between the set {1, 2, 3, . . .} and all the fractions:

$$
\begin{array}{cccccc}
\dfrac{1}{1} & \dfrac{1}{2} & \dfrac{1}{3} & \cdots & \dfrac{1}{p} & \cdots \\[2ex]
\dfrac{2}{1} & \dfrac{2}{2} & \dfrac{2}{3} & \cdots & \dfrac{2}{p} & \cdots \\[2ex]
\dfrac{3}{1} & \dfrac{3}{2} & \dfrac{3}{3} & \cdots & \dfrac{3}{p} & \cdots \\[2ex]
\vdots & \vdots & \vdots & & \vdots & \\[2ex]
\dfrac{q}{1} & \dfrac{q}{2} & \dfrac{q}{3} & \cdots & \dfrac{q}{p} & \cdots \\[2ex]
\vdots & \vdots & \vdots & & \vdots &
\end{array}
$$

Can you show such a relationship?

PART IV

ON SIZE AND SHAPE

The geometric regularity of naturally occurring phenomena is strikingly shown in the hexagonal tiling of this magnified moth's head. [Photo by Don Davis/Smithsonian.]

Mathematics is the study of patterns and relationships. In some applications of mathematics, this truth is obscured by the details of the problem being solved. Consider as examples the spiral growth structure of a sunflower; the spiral distribution of stars, gas, and dust in the Andromeda galaxy; the intricate symmetries of old pottery designs; and the patterns of growth exhibited by human populations or by bank accounts.

Mathematicians instinctively search for geometrical and numerical patterns and for symmetry. Their discoveries of patterns and symmetries often enable us to better understand practical problems. Geometry grew as a body of knowledge because of people's need to explain and control the world around them. The set of tools created has enabled mankind to range further and further from daily experience.

From determining the circumference of the earth to digging straight tunnels, the ancient Greeks used the tools of geometry to make seemingly impossible measurements. More modern astronomers and mathematicians succeeded in uncovering the secrets of planetary motion. And new geometries, which have led to a much deeper understanding of the nature of mathematics itself, paved the way for the discoveries of Einstein, the basis of our present view of the workings of our universe.

Perhaps no other subject has intrigued people through the centuries as much as geometry. We perceive the symmetries and patterns of nature and search to understand the intrinsic order and beauty we observe.

CHAPTER 13

Patterns

Patterns abound in nature. It has long been observed by botanists that particular numbers occur with many natural spirals, such as in the growth of leaves around a plant stem, the spirals on the surface of a pinecone or pineapple, and the spirals formed by the florets at the center of any daisylike flower. These numbers are called the Fibonacci numbers: 1, 1, 2, 3, 5, 8, 13, 21, 34, 55, 89, 144, 233, 377, ... This Fibonacci sequence begins with the numbers 1, 1 and obtains each next term by adding the two preceding terms.

Look at the sunflower and seashell in Figure 13.1. The sunflower seeds and the successive sections of the chambered nautilus appear along a "growth spiral." You see a set of spirals running in a counterclockwise direction, and another set in the clockwise direction. It is (just barely) possible to count the number of spirals in both directions; in the sunflower there are 55 in one and 89 in the other direction. Why are these numbers of spirals the same as two that appear next to each other in the purely mathematical Fibonacci sequence?

Symmetry

The spiral distribution of the seeds in a sunflower head or the spiralling of leaves around a plant stem represents an aspect of symmetry, that of *regularity*. Another aspect of symmetry is *balance*. Spirals, although regular, are certainly not balanced; indeed, they seem to be perpetually rotating, or just ready to rotate. Balanced symmetry is more common in such man-made objects as pottery, wallpaper, and buildings.

Mathematicians describe a variety of kinds of balance by using the geometric tool of rigid motion, or **isometry,** which means "of the same measure." A rigid motion is a transformation in the plane (or in space) in which the original figure and its image are congruent.

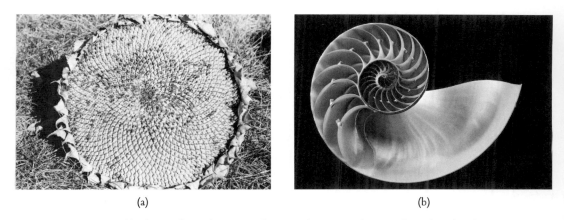

(a)   (b)

Figure 13.1 (a) This sunflower has 55 spirals in one direction and 89 spirals in the other direction. (b) A chambered nautilus shell. [Photo by Nancy Rodger.]

There are four of these rigid motions in the plane. We will illustrate them by selecting from among the many patterns made by people throughout the world those of the Kuba people of Zaire in Central Africa (Figure 13.2), who are noted for their fascination with pattern and symmetry (see Box 13.1).

Figure 13.2 The Kuba people of Zaire (colored area) are fascinated with pattern and symmetry.

The simplest isometry is reflection in a line, or a mirror image, and is called *bilateral symmetry:* the figure looks the same on both sides of a line, except that the two sides are mirror images of each other. Figure 13.3a shows an infinite strip that can be reflected in a vertical line through or between the V's in the design. Figure 13.3b has a horizontal line of reflection symmetry along the middle of the strip.

A second familiar rigid motion is *rotation.* A figure has *rotational symmetry* if it has a center about which it can be rotated by a certain angle without changing its appearance. A circle is such a figure—any rotation at all about its

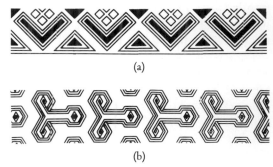

(a)

(b)

Figure 13.3 Two Kuba patterns having bilateral symmetry, one with vertical lines of symmetry (a) and the other with a horizontal line of symmetry (b).

BOX 13.1 The Patterns Created by the Kuba People

Close to 100 Kuba patterns can be recognized. It was considered an achievement by the Kuba people to invent a new pattern. Every Kuba king had to create a new pattern at the outset of his reign. The pattern was displayed on his drum throughout his reign and, in the case of some kings, on his dynastic statue.

When missionaries first showed a motorcycle to a Kuba king in the 1920s, he showed little interest in it. But the king was so enthralled by the novel pattern the tire tracks made in the sand that he had it copied and gave it his name.

(Adapted from Jan Vansina, *The Children of Woot,* University of Wisconsin Press, 1978, p. 221.)

The pattern made by tire tracks fascinated the Kuba people. [Travis Amos.]

---

center leaves it unchanged. A square is another, because a rotation of 90° about its center leaves it unchanged. Figure 13.4 shows an infinite strip that is unchanged by a 180° rotation about any point at the center of the small crosshatched regions.

The third plane isometry is *translation,* which is a slide in a certain direction by a certain distance. Only infinite figures can be translated without changing their appearance. The infinite Kuba pattern in Figure 13.5 shows translation symmetry but no other kind. Similarly, an infinite chessboard could be moved by sliding (translating) all the squares two squares to the left; we cannot detect move-

ment in either the Kuba pattern or the squares. Unlike the Kuba pattern, however, the infinite chessboard has other symmetries — it could be rotated by 90° about the center of any square or could be reflected in a line containing the diagonal of any square.

The fourth plane isometry is less familiar. It is called *glide reflection* because it is composed of a glide (i.e., translation) by a certain distance along some line, followed by a reflection in that line. An infinite row of alternating p's and b's has glide reflection:

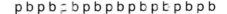

Figure 13.4 A Kuba pattern having rotational symmetry.

Figure 13.5 A Kuba pattern having translational symmetry.

If the **p**'s are translated as far as the **b**'s and are then reflected in a suitable horizontal line (so that each **p** becomes **b** and vice versa), the row of **p**'s and **b**'s will not appear to have changed. Figure 13.6 shows a Kuba carved pattern whose only symmetry (except for translation) is glide reflection.

Figure 13.6   A Kuba pattern having glide reflection symmetry.

It is a remarkable fact that every rigid motion of the plane, no matter how complicated, can be built from one of these four symmetries. For example, if we take the plane of the blackboard, represented by some cardboard triangle placed on it, send the triangle off to NASA and then to Mars, then bring it back via the moon and place it anywhere on the original blackboard, the resulting position could have been obtained from the original one simply by applying some one of these four simple rigid motions. In that sense, the rest of the trip was completely superfluous.

The fact that every rigid motion can be built from one of these four symmetries leads to a simple classification of all patterns, according to which combinations of the four fundamental rigid motions leave a pattern invariant. This kind of classification was originally developed by crystallographers in the nineteenth century who wanted to be able to classify and recognize the three-dimensional patterns associated with crystal structure. They proved that there are exactly 230 such pattern types in three dimensions.

In the patterns found by archeologists on pottery or cloth, there are fewer categories because patterns in the plane are not as complex as the three-dimensional patterns formed by crystals (see Boxes 13.2 and 13.3). In fact, there are only 17 two-dimensional plane patterns (patterns that extend in all directions in a single plane) and only 7 one-dimensional borders or strips (patterns that extend in only

---

BOX 13.2   Archeologist Comments on Symmetry in Ancient Design Themes

Dorothy Washburn is an archeologist at the University of Rochester. She comments on how various cultures have made use of symmetry in their designs:

> It's important to think about how archeology and anthropology go together. We're both studying human behavior, but more specifically we're really studying pattern behavior. Human beings do things in a very consistent fashion and over the years, within a given cultural group, their activities are really nonrandom. In an archeological context we can see repetition of patterns in their material culture. One of the interesting things for us and the material (baskets, cloth, pottery) we're studying here is that the patterns and designs are very consistent. And we've never been able to see that before, except through symmetry.
>
> Early descriptions of design have been largely idiosyncratic and dealt just with the individual type of material—the type of textiles or the type of pottery. But now we can look at the way the motifs in the design are arranged in material from all different cultures throughout the world—regardless of whether they're contemporary cultures or past cultures—and see how these designs are put together and how they've changed through time and space.

One of the most interesting things we've found is that a given cultural group, a tribe or a band unit, will choose just a few symmetries to structure their designs. And this is very consistent. I've tested this through studies of California Indian baskets. Last summer I went to Zaire, where we did similar work with the Kuba cloth weavers. We found a consistency in the archeological record among the Anasazi, one of the prehistoric traditions of the American Southwest, which also is the same kind of consistency in choices and preferences that people are using today.

One of the interesting things about archeology is that we want not just to excavate sites, but to excavate in order to learn about ethnic groups and tribes as cultural units. Otherwise we're excavating a site here and a site there, and it's just piecemeal. But if we try to reconstruct culture as a whole, then we need to use other analytical methods. Now, if we can show through ethnographic and modern-day analysis that a tribal group is very consistent in structuring their designs, then we can go, for example, to look for those consistencies in pots or tilings or textiles within sites in a valley system. If we find a whole group of sites in a valley system, for example, the whole Ica Valley system in Peru, then we can conclude that they were a unified cultural group. It is a really important thing—a major step forward for the archeologist—to move beyond excavating sites to excavating whole cultures and cultural units.

BOX 13.3 Design Motifs Signal Change in Cultural History

Archeologist Dorothy Washburn reflects on the significance of design as an indicator of a culture's rich history:

For material culture that we're dealing with, the patterns are finite designs, one-dimensional designs, or two-dimensional designs on flat surfaces. Pots naturally have a curved surface, but we can imagine that we can flatten them out. It's easier to see it, perhaps, if we think of tilings on bathroom floor tiles, wallpaper, or textiles. These are all flat, two-dimensional surfaces on which patterns are found.

We find that although some people do produce random patterns, by far the largest number of decorative designs that cultural groups produce are based on symmetry. The point is that anytime you repeat a motif in a systematic fashion, you're using one of the four symmetry motions to make that repetition.

Let's take an example from material found on Crete. The site of Knossos had 3000 years of uninterrupted prehistory—wonderful stratigraphy that we didn't have in the Southwest. For 1500 years 2 symmetries and 2 design patterns were used consistently throughout the stratigraphy. Then suddenly a whole series of 10 different symmetries were used. The design motifs were the same, but it was the rearrangement of these motifs into the different designs which clued us that something really interesting was happening.

That interesting thing was, in fact, the beginning of trade in the Aegean. We could see that in the stratigraphic record simply by looking at the introduction of new symmetries from cultures outside that island. The increase in the number and variety of symmetries indicated that trade was coming in. So it was an incredibly sensitive marker of change.

one direction, for example, along the border of a cloth or around the rim of a bowl, imagined as unrolled into the plane).

Examples of the seven possible borders are shown in Figure 13.7. Each is imagined as extending infinitely far to the left and right. These all appear in the pottery of San Ildefonso pueblo in what is now New Mexico. Let us go through them one by one to see how each is really different from the rest.

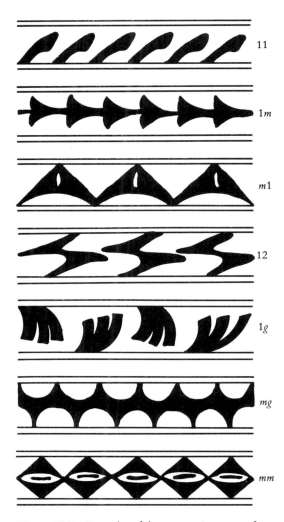

Figure 13.7 Examples of the seven strip patterns from the pottery of San Ildefonso pueblo, New Mexico.

On the first pattern there is no rigid motion except *translation,* which takes the pattern onto itself. (We imagine the individual black figures to be exactly alike, so that each can indeed be translated to the one next to it without appearing to have moved.) The second pattern adds a mirror reflection along the horizontal axis of the pattern to the translation, which moves the pattern to the left or right. A reflection in this central line interchanges the top and bottom halves of the arrowheads, so that the pattern does not appear to have moved. In the third pattern the mirror reflections are in vertical lines, perpendicular to the direction of the strip.

The fourth pattern has no reflections, but there are half-turns (180° rotations) about suitable points in the center of, or between, the individual black figures. The rotations leave them, and the whole strip, unchanged in appearance. In the fifth figure there is a glide reflection, translating each figure hanging from above (or below) to the next one below (or above) and then reflecting in a horizontal line. Except for translations (by twice the distance of a figure), there are no rigid motions of this strip except glide reflections.

The sixth pattern combines several motions: reflections (in vertical lines), glide reflections, and half-turns. Finally, the seventh pattern combines all possible rigid motions of a strip: translations, reflections in both horizontal and vertical lines, glide reflections, and half-turns.

With these models you can imitate the work of an archeologist or a wallpaper designer by trying to identify strip patterns seen along the tops of buildings or recorded by car tires on a dirt road. Every repeated strip pattern you see will surely be one of the seven types.

A similar classification of two-dimensional patterns yields exactly 17 such patterns. Examples of each of these types are shown in Figure 13.8. Again, if you are energetic, you can try to match up whatever two-dimensional pattern you find with one of these 17.

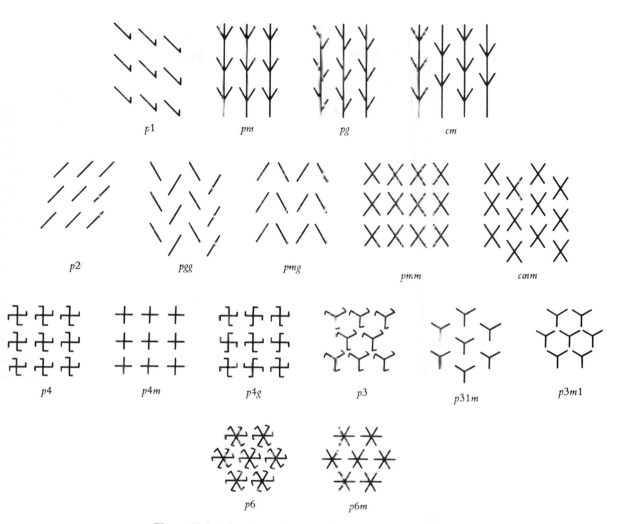

Figure 13.8 The 17 two-dimensional crystallographic patterns, with the crystallographer's notation.

Patterns and Tilings

When we contemplate nature or look at a human face, we can discern patterns. From the bilateral symmetry of our faces to the growth of petals in a chrysanthemum, nature has a penchant for patterns.

We have seen that people, too, have a penchant for patterns. When our ancestors used stones to cover the floors and walls of their houses, they selected shapes and colors to form pleasing designs. We can see the artistic impulse at work in mosaics from Roman dwellings to Moslem religious buildings (see Figure 13.9). The same intricacy and complexity arise in other decorative arts — on carpets, fabrics, and baskets.

Such patterns have one feature in common: they use repeated shapes to cover a plane surface without gaps or overlaps. If we think of

Figure 13.9 These mosaics from the Hakim Bey Mosque, Konya, reveal our long fascination with patterns. [Bildarchiv Foto Marburg/Art Resource.]

the shapes as tiles, we can call the pattern a *tiling,* or *tessellation.* Even when efficiency is more important than esthetics, designers value clever tiling patterns. In manufacturing, for example, stamping the components from a sheet of metal is most economical if the shapes of the components fit together without gaps — in other words, if the shapes form a tiling.

Regular Tilings

The simplest tilings are those using only one size and shape of tile. The easiest of these **monohedral** tilings to imagine and construct are those made from squares. Apart from varying the size of the square, which would change the scale but not the pattern of the tiling, we

can get different tilings for different offsets of one row of squares from the next. However, only one tiling is **edge-to-edge,** that is, the edge of a tile coincides entirely with the edge of a bordering tile.

Any tiling by squares can be refined to one by triangles by drawing a diagonal of each square. These triangles are **isosceles,** that is, two of their three sides are the same length. Because **equilateral** triangles, with three equal sides, can be arranged in rows by alternately inverting triangles, it is easy to offset these rows by different amounts to produce many different tiling patterns. However, as with squares, there is only one pattern of equilateral triangles that forms an edge-to-edge tiling. For simplicity, from now on we will consider only edge-to-edge tilings.

What about tiles with more than four sides? We are interested especially in *regular polygons,* plane figures with straight sides of equal lengths and equal angles. A square is a regular polygon with four sides and four interior angles of 90° each; an equilateral triangle has 60° interior angles. A polygon with five sides is a pentagon, one with six sides is a hexagon, and one with *n* sides is an *n*-gon.

What Other Regular Tilings Are There?

There is a very common edge-to-edge tiling with regular hexagons. However, if we look for one with regular pentagons, we won't be able to find one. How do we know whether we're just not being clever enough or there really isn't one to be found? This is the kind of question mathematics is uniquely equipped to answer. In the other sciences, phenomena may exist even though we have not observed them; such was the case for bacteria before the invention of the microscope. In the case of an edge-to-edge tiling with regular pentagons, we can conclude with certainty that there isn't one.

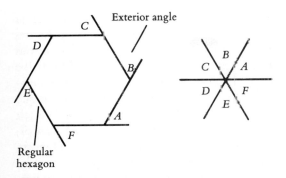

Regular
hexagon

Figure 13.10   The exterior angles of a regular hexagon, like those of any regular polygon, add up to 360°.

The key to our proof is the size of the interior angle of a regular polygon. Proceeding around the polygon in the same direction, we see that each interior angle is paired with an exterior angle. If we bring all the exterior angles together at a single point, they will add up to 360° (see Figure 13.10). If the polygon has n sides, then each exterior angle must measure $360/n$ degrees. The exterior angle plus its corresponding interior angle make up a straight line, or 180°; therefore, the interior angle must be 180° minus the exterior angle, or

$$\text{Interior angle} = 180° - \frac{360°}{n}$$

$$= \frac{n \times 180°}{n} - \frac{2 \times 180°}{n}$$

$$= \frac{(n-2)180°}{n}$$

This formula — which we have manipulated slightly into a form that will be useful to us later — gives the size of the interior angle of a regular polygon of n sides. We can check that for $n = 3$ it gives 60° and for $n = 4$ it gives 90°, as we might expect. For $n = 5$, the case of the regular pentagon, it gives 72°.

In the tilings by regular triangles, squares, and hexagons, we notice that at any vertex

(endpoint where edges meet), the same number of polygons come together: six in the case of triangles and four and three, respectively for the other two tilings. Moreover, the polygons that come together must have interior angles at that vertex that add up to exactly 360°. If they added up to more, there would be overlapping; if less, there would be gaps (see Figure 13.11). In either case, we would not have a tiling.

To explore possible tilings with a regular n-gon, let's suppose that m of them meet at each vertex. We will see what values of m are possible for a given value of n. We have just observed that m interior angles of the n-gon must add up to exactly 360°. Using our formula for the size of the interior angle, we have

$$\frac{m(n-2)180°}{n} = 360°$$

or, dividing by 180°,

$$\frac{m(n-2)}{n} = 2$$

and multiplying by n we get

$$m(n-2) = 2n$$

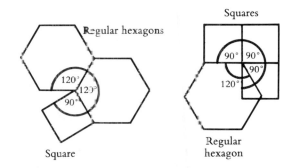

Figure 13.11   Polygons that come together at a vertex in a tiling must have interior angles that add up to 360° — no less, no more.

so that dividing by $(n - 2)$ gives

$$m = \frac{2n}{n - 2}$$

In other words, once we specify n, the value of m is automatically determined. We check it for $n = 3$: we get $m = 6$, as we know was the case for our tiling by equilateral triangles. For $n = 4$, we get $m = 4$, which corresponds to our pattern of squares. But for $n = 5$, we get $m = \frac{10}{3}$. What can this mean? This means that there is no whole-number solution for m — there is no whole number of pentagons we can arrange at a vertex to add up to exactly $360°$. In other words, we have shown beyond the shadow of a doubt that it is *impossible* to have an edge-to-edge tiling of the plane with regular pentagons.

If we continue trying larger values for n, we see that for $n = 6$ we get $m = 3$, corresponding to our tiling with hexagons. For $n = 7$, $m = \frac{14}{5}$: again we face impossibility. What about octagons? For $n = 8$, $m = \frac{16}{6}$. In fact, no larger value of n will give a whole number for m. We can see why this is so by looking closely at the equation for m: we're dividing $2n$ by a little less than n, so that for any n beyond 6, m will always be between 2 and 3. This fact may be more apparent from rewriting the expression for m as

$$m = \frac{2n}{n - 2} = 2 + \frac{4}{n - 2}$$

from which we see that m is more than 2 and that we get lower values of m for higher values of n. Thus, only three regular polygons — the triangle, the square, and the hexagon — can tile the plane edge-to-edge.

The follow-up question, of course, is which combinations of regular polygons can tile the plane edge-to-edge? The particular arrangement of polygons around a vertex is called a *vertex figure*. A systematic tiling that uses a mix of regular polygons in which all vertex figures are alike is called *semiregular*. As before, our technique of adding up angles at a vertex (to be $360°$) eliminates some impossible combinations, such as "square, hexagon, hexagon." Having found the possible arrangements, we must confirm the actual existence of each tiling by constructing it. For example, even though a possible arrangement of regular polygons around a point is "triangle, square, square, hexagon," it is not possible to construct a tiling that has that vertex figure at every vertex. The result of this investigation is that in a semiregular tiling no polygon can have more than 12 sides. In fact, polygons with 5, 7, 9, 10, or 11 sides also do not occur in a semiregular tiling.

Nonperiodic Tilings

All the patterns we've exhibited and discussed so far have been **periodic** tilings. If we outline any portion of the tiling on a transparency, it is possible to slide the transparency a certain distance horizontally, without rotating it, until the outline exactly matches another section of the tiling; we can achieve the same result by moving the transparency some second direction (possibly vertically) a certain (possibly different) distance.

There are infinitely many periodic tilings and infinitely many nonperiodic ones. As an example of a nonperiodic tiling, we can consider again a non-edge-to-edge tiling by squares, where the second row from the bottom is offset half a square to the right from the bottom row, the third row a third of a square more, and so forth (see Figure 13.12a). An elegant theorem in number theory assures us that $1/2 + 1/3 + 1/4 + \cdots + 1/n$ never adds up to a whole number, so this type of tiling never repeats. For an edge-to-edge example, take the usual edge-to-edge square til-

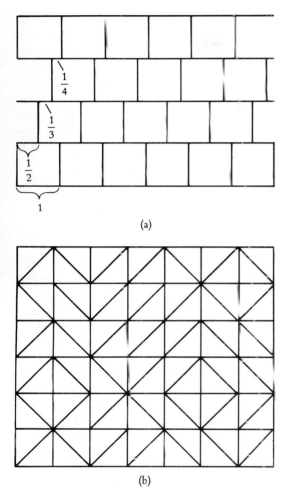

Figure 13.12 (a) A non-edge-to-edge tiling by squares that is not periodic. (b) An edge-to-edge tiling by right triangles that is not periodic.

whatsoever. This is one instance in which mathematicians have not yet been able to provide either a counterexample or an impossibility proof.

The Penrose Tiles

For a long time, mathematicians believed that the corresponding fact would be true for any *set* of tile shapes: if you could construct a nonperiodic tiling with the shapes, you could construct a periodic one. But in 1964 the first set of tiles was found that permits *only* nonperiodic tiling. No wonder it took so long to discover that set — it contains 20,000 different shapes! Over the next several years smaller sets of this type were discovered, with as few as 100 shapes. But it was still amazing when in the 1970s Roger Penrose, a mathematical physicist at Oxford, found a new set that would do the job — with just 2 tiles! (See Figure 13.13.)

Penrose calls his tiles "darts" and "kites." It is easy to specify their construction, as one of each can be obtained from a single rhombus. A **rhombus** is a planar figure with four equal

ing and divide each square into two right triangles by adding at random either a rising or a falling diagonal. The chance of this procedure resulting in a periodic tiling by right triangles is 0 (see Figure 13.12b).

As far as mathematicians know, every shape that can be used to construct a monohedral nonperiodic tiling can also be used to form a periodic one. It is an open question whether this assertion is true for every shape of tile

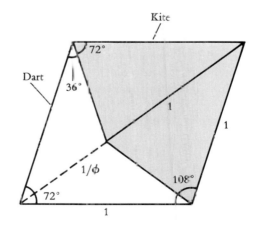

Figure 13.13 Construction of Penrose's "dart" and "kite" (colored area). $\phi = (1 + \sqrt{5})/2$.

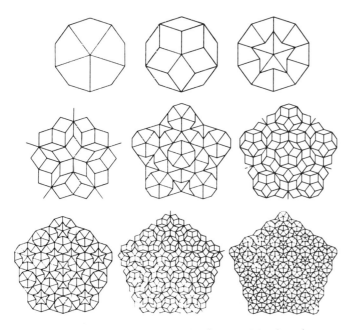

Figure 13.14 Successive decomposition (i.e., the cutting up of large tiles into smaller ones) of patches of tiles of a Penrose nonperiodic tiling. Note that the final pattern (lower right) exhibits a 10-gon pattern. [Courtesy of Roger Penrose.]

sides and equal opposite interior angles. The particular rhombus from which the Penrose tiles are constructed has interior angles 72° and 108°. If we divide the longer diagonal according to the golden mean—the proportion 1 to $2/(1 + \sqrt{5}) \approx 1.6$, so dear to the Greeks—and connect the dividing point to the remaining corners, we split the rhombus into a dart and a kite (color).

Although tilings with Penrose's pieces cannot be periodic, they do possess a type of symmetry called **fivefold** and **tenfold symmetry.** Consider any one of the decagons (10-gons) of Figure 13.14. If we rotate the pattern around its center through one-fifth of a turn, the pattern looks the same as before.

Where does this rotational symmetry come from? The rhombus we split up has the angles shown in Figure 13.13. Except in the recess of the dart and matching part of the kite, all the

internal angles of the kite and dart are either 72° or 36°. Now, 72° goes into 360° five times, and 36° goes ten times. If we recall that it is the interior angles that matter in arranging polygons around a point, we see that fivefold or tenfold symmetry could conceivably result from using such tiles. (By no means does this prove that all Penrose patterns must have fivefold or tenfold symmetry. They do, but it is not easy to show.)

Shechtman's Crystals and Barlow's Law

Although Penrose's discovery was a big hit among geometers and in recreational mathematics circles in the mid-1970s, few people thought that his work might have practical significance. Yet in 1982 scientists at the U.S. National Bureau of Standards discovered unexpected fivefold symmetry while looking for new ultrastrong alloys of aluminum.

Manganese doesn't ordinarily alloy with aluminum, but the experimenters were able to produce small crystals of alloy by cooling mixtures of the two metals at a rate of millions of degrees per second. Following routine procedures, chemist Daniel Shechtman began a series of tests to determine the atomic structure of the special crystals. But there was nothing routine about what he found: the atomic structures of the manganese-aluminum crystals were so startling that it took Shechtman three years to convince his colleagues they were real.

Why did he encounter such resistance? His patterns—and the crystals that produced them—defied one of the fundamental laws of crystallography. Like our discovery that the plane cannot be tiled by regular pentagons, **Barlow's law,** also called the **crystallographic restriction,** says that no crystal can have fivefold symmetry. Rotation around the axes of a crystal may produce identical patterns

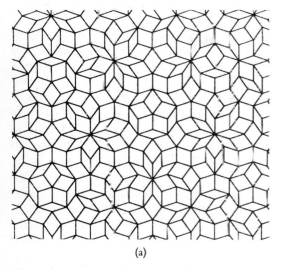

(a)

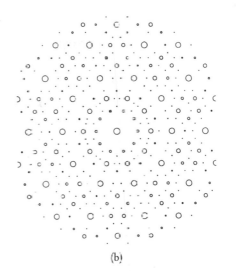

(b)

Figure 13.15 (a) A nonperiodic Penrose tiling. (b) A three-dimensional crystallike structure based on this tiling. (c) An actual crystal pattern observed by chemist Daniel Shechtman in a special manganese-aluminum alloy. Note the similarity to the pattern in (a). (d) A single-grain crystal of an aluminum-lithium-copper Shechtman alloy. Its fivefold symmetry is apparent. [(a), (b), and (c) are from D. R. Nelson and B. I. Halperin, *Science,* 229:233–238 (1985).]

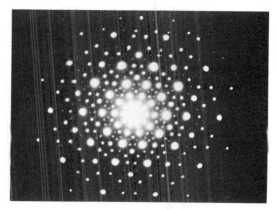

(c)

of atoms with a half-turn, third-turn, quarter-turn, or sixth-turn, but repetitions occurring at fifths of a turn are ruled out by a mathematical argument similar to the one we used for the regular pentagons.

Yet fivefold symmetry is exactly what Shechtman and his colleagues observed in the new alloy crystals! Other scientists soon spotted the connection with Penrose tilings and created models using three-dimensional analogues of the Penrose tiles to form crystallike structures. Like the two-dimensional Penrose patterns, these have orderly fivefold symmetry but are nonperiodic.

Figure 13.15b shows what a three-dimensional crystallike structure would look like if it were based on the Penrose tiles. Figure 13.15c is the actual pattern Shechtman observed in the alloy. You can see how closely they match. They seem to contradict Barlow's law, which,

(d)

like our proof of impossibility of a regular pentagonal tiling, was a mathematical theorem—a fact we can be certain of. How then could it be contradicted by Shechtman's alloy?

Barlow's law, as a mathematical theorem, shows that fivefold symmetry is impossible in a periodic tiling of the plane or of space. Chemists, for good theoretical and experimental reasons, believe that crystals are modeled well by three-dimensional tilings. An array of atoms with no symmetry whatever would not be considered a crystal. Until Penrose's discovery, however, no one realized that nonperiodic tilings—or arrays of atoms—can have the regularity of fivefold symmetry. Chemists could simply say that Shechtman's alloys aren't crystals. In the classical sense they aren't, but in other respects they do resemble crystals. It is scientifically more fruitful to extend the concept of crystal to include them rather than rule them out. We can then go on to consider what other marvels are possible in crystallography.

Once again, as so often in history, pure mathematical research anticipated scientific applications. Penrose's discovery, once just a delightful piece of recreational mathematics, is now prompting a major reexamination of the theory of crystals. Mathematics is fundamental to our understanding of scale, growth, and pattern, but it is even more important as a tool for explaining what we don't understand.

REVIEW VOCABULARY

Barlow's law, or **crystallographic restriction**   A law of crystallography that states that a crystal may have only rotational symmetries that are twofold, threefold, fourfold, or sixfold.

Bilateral symmetry   Ordinary mirror (i.e., reflection) symmetry, as seen in the letter A.

Edge-to-edge tiling   A tiling in which bordering tiles meet only along full edges of each.

Equilateral triangle   A triangle with all three sides equal.

Exterior angle   The angle outside a polygon formed by one side and the extension of an adjacent side.

Fibonacci numbers   The numbers in the sequence 1, 1, 2, 3, 5, 8, 13, 21, 34, . . . (each number is obtained by adding the two preceding numbers).

Fivefold (or tenfold) symmetry   Rotational symmetry by a fifth (or a tenth) of a turn about some point.

Glide reflection   A combination of translation (= glide) and reflection in a line parallel to the translation direction. Example: . . . p b p b p b . . .

Interior angle   The angle inside a polygon formed by two adjacent sides.

Isometry   Another word for **rigid motion.** Angles and distances, and consequently shape and size, remain unchanged by an isometry. (For plane figures there are only four possible isometries: reflection, rotation, translation, and glide reflection.)

Isosceles triangle   A triangle with two equal sides.

Kuba   A group of people in central Africa (Zaire), known for their pattern-making skills.

Monohedral tiling   A tiling with only one size and shape of tile (the tile is allowed to occur also in "turned-over," or mirror-image, form).

n-gon   A polygon with n sides.

Periodic tiling   A tiling that repeats at fixed intervals in two different directions, possibly horizontal and vertical.

Rotational symmetry   A figure has rotational symmetry if a rotation about its "center" leaves it looking the same, as shown by the letter S.

Translation symmetry   An infinite figure has translation symmetry if it can be translated (slid without turning) along itself without appearing to have changed. Example: . . . A A A A A A . .

EXERCISES

1. The figure below shows a computer-generated "sunflower," plotted in the way the sunflower is believed to lay out its seeds. (You can see the spirals radiating out from the center in two directions, clockwise and counterclockwise.)

a. How many spirals are there in each of the two directions?

b. Are these two numbers (in your answer to **a**) adjacent numbers in the Fibonacci sequence?

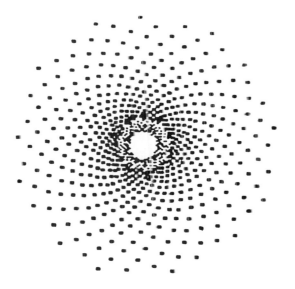

2. Which of the 26 capital letters of the alphabet have

a. reflection symmetry?

b. rotational symmetry?

(Assume that each letter is drawn in the most symmetric way. For example, the upper and lower loops of "B" should be the same size.)

3. In *The Complete Walker III* (p. 505), Colin Fletcher's answer to "What games should I take on a backpacking trip?" is the game he calls "Colinvert": "You strive to find words with meaningful mirror (or half-turn) images." Some of the words he found are

MOM   WOW   pod   MUd   bUM

    a. Which of his words reflect into themselves?

    b. Which of his words rotate into themselves?

    c. Find some more words or phrases of these various types.

4. The eight one-dimensional patterns shown below all appear on the brass straps for a single lamp from nineteenth-century Benin. Identify each of these patterns as one of the seven types shown in Figure 13.7 and described in the text: 11, 1*m*, *m*1, 12, 1*g*, *mg*, or *mm*.

(a)   (b)   (c)   (d)   (e)   (f)   (g)   (h)

5. In each of the four examples below, two adjacent triangles of an infinite strip are shown. For each example,

    a. determine a motion (translation, reflection, rotation, or glide reflection) that takes the first (= left) triangle to the second (= right) one.

    b. draw the next four triangles of the infinite strip that would result if the second triangle were moved to the next space by another motion of the same kind, and so on.

    c. identify the resulting strip as one of the seven possible strip patterns.

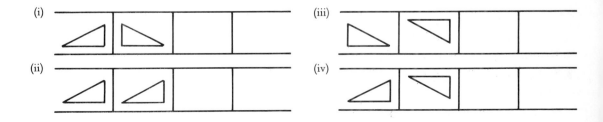

TABLE 13.1 Frequency of strip designs on Mesa Verde pottery and Begho smoking pipes

	Mesa Verde		Begho	
Strip type	Number of examples	Percentage of total	Number of examples	Percentage of total
11	7		4	
1m	5		9	
m1	12		22	
12	93		19	
1g	11		2	
mg	27		9	
mm	19		165	
Totals	174		230	

6. Table 13.1 shows comparative data about the frequency of occurrence of strip designs of various types on pottery (Mesa Verde, USA) and smoking pipes (Begho, Ghana, Africa) from two different continents.

a. Fill in the two columns of "percentage of total."

b. Which types of motions appear to be preferred for designs from each of the two localities?

c. State some other general conclusions from the data of this table.

d. On the evidence of the table alone, in which locality is each of the strips shown below most likely to have been found? Explain.

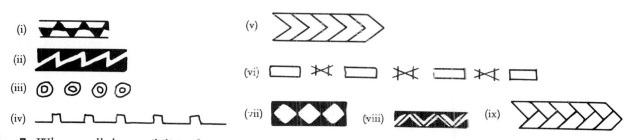

7. What are all the possibilities for regular polygons meeting at a point in a tiling, in terms of the numbers of each kind? There are 17. Four of these split into 2 kinds of each if we consider the order of the polygons around the vertex. Once you have drawn all the 21 possible vertex figures, can you argue further to show that only 11 of these give rise to tilings? These 11 include the 3 regular tilings and the 8 Archimedean (or semi-regular) tilings. (This is a long project. The details can be found in *Geometry: An Investigative Approach* by P. G. O'Daffer and S. R. Clemens, pp. 91–101.)

8. What tessellations with regular polygons are possible on a sphere? (To see an example of a semiregular tessellation of the sphere, find a soccer ball.)

| **14**

Growth and Form

Fantasy films have made us familiar with assorted giant creatures, including King Kong, Godzilla, and the 50-foot-high grasshoppers in "The Beginning of the End." We also find supergiants in literature, such as the giant of "Jack and the Beanstalk," Giant Pope and Giant Pagan of *The Pilgrim's Progress,* and the Brobdingnagians of *Gulliver's Travels.*

Much as we appreciate those stories, even from an early age we don't really believe in monsters and giants. Could such beings ever exist? What adaptations might they undergo? These questions pose the *problem of scale.* Certainly, there have been large land mammals (mammoths) and huge sea mammals (the blue whale)—not to mention the dinosaurs. But the tallest humans have been only 9 to 10 feet tall, the largest mammoth was 16 feet at the shoulder (about twice as tall as an elephant), and even the tallest dinosaurs weren't much taller than giraffes.

But what about supergiants and utterly huge monsters? That they have never existed suggests physical limits to size. In fact, with a few simple principles of geometry, we can show that such size is impossible. The creatures and objects in our world could not exist, unchanged in shape, on a scale vastly different from what we observe (see Box 14.1).

Geometric Similarity

The basic and powerful mathematical idea that we will use is that of geometric similarity. By geometric standards, two objects are similar if they have the same shape, regardless of the materials they are made of. They may even be of different sizes. Corresponding angles, however, must be equal, and corresponding line segments must all have the same factor of proportionality.

Figure 14.1   Could King Kong actually exist? [The Museum of Modern Art/Film Stills Archive.]

For example, when a photo is enlarged, it is enlarged by the same factor in both the horizontal and vertical directions. Even an oblique (i.e., nonhorizontal, nonvertical) line segment between two points on the original will be

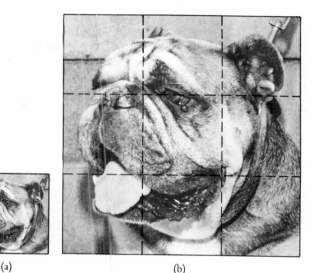

(a)   (b)

Figure 14.2   Two geometrically similar photographs. [Photo by Travis Amos.]

enlarged by that same factor. We call this enlargement factor the **scaling factor.** In the photos in Figure 14.2, the scaling factor is 3: the enlargement is 3 times as wide and 3 times as high as the original. We notice, however, that the enlargement can be divided into 3 × 3 = 9 rectangles, each the size of the original. Hence, the enlargement has $3^2 = 9$ times the area of the original. More generally, if the scaling factor is M (instead of 3), the resulting

BOX 14.1 On Growth and Form

D'Arcy Thompson (1860–1948) was the author of *On Growth and Form* (1917), a pioneering essay about the way things grow and the shapes they take. J. T. Bonner, coauthor of *On Size and Life* (1983), calls Thompson's work "good literature as well as good science; it is a discourse on science as though it were a humanity."

Thompson's father was a classicist, and Thompson himself became president of the Classical Associations of England and Wales and of Scotland. He translated Aristotle's *Historia Animalium* and wrote books on the birds and fishes mentioned in Greek literature. Natural history was his real love; *On Growth and Form* was the first book to describe in quantitative terms the processes of growth and shaping of biological forms, including many of the aspects we discuss in this chapter.

enlargement will have an area $M \times M = M^2$ ("*M* squared") times the area of the original. Thus, the *area* of a scaled-up object increases as the *square* of the scaling factor.

What about enlarging three-dimensional objects? If we take a cube and enlarge it by a scaling factor of 3, it becomes 3 times as long, 3 times as high, and 3 times as deep as the original (see Figure 14.3).

We observe that the area of each face (side) of the enlarged cube is $3^2 = 9$ times as large as that of a face or our original cube. Since this fact is true for all six faces, the total surface area of the enlargement is 9 times as much as the original. More generally, for objects of any shape whatever, the total *surface area* of a scaled-up object increases as the *square* of the scaling factor.

What about volume? The enlarged cube has 3 layers, each with $3 \times 3 = 9$ little cubes. Thus, the total volume is $3 \times 3 \times 3 = 3^3 = 27$ times as much as the original cube. In general, the *volume* of a scaled-up object increases as the *cube* of the scaling factor. Thus, for an object scaled up with a scaling factor of *M*, the enlargement will have M^3 ("*M* cubed") times the volume of the original. This relationship, like the relationship between surface area and M^2, holds even for irregularly shaped objects, such as science fiction monsters. (Box 14.2 discusses the correct comparative and relational terms for describing increases and decreases.)

Scaling Real Objects

Real three-dimensional objects, of course, are made of matter. Matter is affected by other matter via forces, according to physical laws. For example, you react to the earth by staying close to it, and you exert a (much smaller) gravitational attraction on the earth that tends to keep it close to you. Gravitational force, which we perceive as weight, is the most easily observed influence on the size and shape of objects.

In Figure 14.4, we consider two cubes. The first is a cube of steel 1 foot on a side. The bottom face supports the weight of the entire cube. The *pressure* exerted on the bottom face is equal to the weight of the cube divided by the area of the bottom face. A 1-foot steel cube weighs 492 pounds, and the area is 1 square foot; the pressure exerted is therefore 492 pounds per square foot.

The second cube in Figure 14.4 is made of the same steel but is 2 feet on a side. The area of

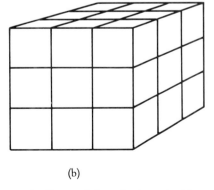

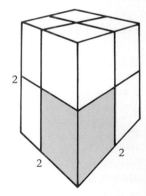

Figure 14.3 Cube (b) is made by enlarging cube (a) by a factor of 3.

Figure 14.4 A cube of side *n* and a cube of side 2*n*.

BOX 14.2 The Language of Growth, Enlargement, and Decrease

Erin's Own Irish sphagnum moss peat. It enriches your soil and makes your growing easier. Compressed to $2\frac{1}{2}$ times normal volume."—Printed on sacks of the product (*New Scientist* (May 8, 1986): 80)

Poor or imprecise use of language can often confuse a reader or listener. There is particular occasion for such confusion in discussions of growth or decrease, when the speaker or writer may exaggerate.

Let us take a photograph with area 1 and enlarge it 3 times, so that it is 3 times as wide and 3 times as high. The enlargement has area 9. What are some ways you might compare this area to the original area? Fill in the blank in the sentence: "The enlargement's area is _____ the area of the original."

"Nine times," "nine times as large as," "nine times as great as" are all simple, descriptive, correct, and unconfusing alternatives. However, "nine times greater than," "nine times more than," "nine times larger than" are confusing alternatives. These phrases sound even grander, because they use the comparative form "more" in addition to "times." "More," "larger," and "greater" generally refer to quantity over and above the original; so these phrases may be interpreted as meaning "ten times as large" (the original plus nine times more). The problem is easiest to see if we ask what "one times greater than" could possibly mean.

This problem is not cured by speaking of percent instead. In all the phrases we have considered, "nine times" could be replaced by "900%" and the same criticism would hold. Percent has even further hazards for the careless user of language. An increase from one to nine is an increase of 800%, not 900%. In discussions of percent we also need to distinguish percent from percentage points: if support for the President has decreased from 60% to 30%, it has dropped 30 percentage points but decreased 50% (because the drop of 30 percentage points is 60% of the original 60 percentage points).

There are also perils associated with speaking of decreases. For a reduction from nine to one, the reduced quantity may be described incorrectly as "nine times less than" or "900% less than" instead of "one-ninth as much." Being aware of correct usage helps to avoid the kind of confusion generated by these incorrect phrases.

---

the bottom face will increase as the square of the scaling factor, so it will be $2^2 \times 1 = 4$ square feet. The volume goes up as the cube of the scaling factor, so this larger cube has a volume of $2^3 \times 1 = 8$ cubic feet. Because both cubes have the same **density** (ratio of mass to volume), the larger cube has 8 times the mass of steel as the smaller. Hence it weighs 8 times as much as the smaller cube, or $8 \times 492 = 3936$ pounds.

When we divide this weight by the area (4 square feet) of the bottom face, we see that the pressure exerted on the bottom face is 984 pounds per foot, or *twice* the pressure on the bottom face of the original cube. This makes sense because over each 1 square foot stand 2 cubic feet of steel.

If we scale the original cube of steel up to a cube 3 feet on a side, we observe *triple* the pressure on the bottom face. In general, the pressure on the bottom face will be M times as much if the scaling factor is M.

How High Can a Mountain Be?

At some scale the pressure owing to the weight of the cube will exceed the steel's ability to withstand that pressure — and the steel will crumble under its own weight. That point for steel is reached for a cube slightly over 4 miles on a side — the pressure exerted by the cube's weight exceeds the resistance to crushing (ability to withstand pressure) of steel. A 4-mile-long cube of steel would be more than 20,000 times as long as the original 1-foot-cube; that is, the scaling factor is greater than 20,000.

Unfortunately, bone has nowhere near as high a resistance to crushing as steel. This fact helps to explain why there couldn't be any King Kongs. A creature scaled up by a factor of 20 would weigh $20^3 = 8000$ times as much! The weight increases with the cube of the scaling factor, but the ability to support the weight — as measured by the cross-sectional area of the bones — increases only with its square.

These simple consequences of the geometry of scaling apply to other objects, natural and artificial, not only to supermonsters. For example, we can determine how high the tallest mountains can be. Assuming the mountain is made of granite, which has a resistance to crushing about half that of steel, the tallest mountain can be about 7 miles high. That height is not much greater than the actual tallest mountain, Mt. Everest, which is a little less than 6 miles high. By doing a similar calculation, Galileo arrived at his prediction that the tallest trees cannot exceed 300 feet (see Box 14.3).

The Tension Between Volume and Area

The tension between volume (weight) and area, a major consequence of the scaling problem, also explains why and how a real building or machine must differ from a scale model. The balsa wood or plastic of the model would never be strong enough to use for the real thing; the materials in the scaled-up version must be aluminum, steel, or reinforced concrete. One way to compensate for the problem of scale is to use stronger materials in the scaled-up object (see Box 14.4). *A large change in scale forces a change in either materials or form.* In a change of form, we solve the problem of scale by redesigning the object so that its weight is better distributed.

Let's go back to our original cube. It supports all its weight on its bottom face. In the

---

BOX 14.3 Galileo and the Tallest Trees

Galileo Galilei (1564–1642) was the first to describe the problem of scale, in 1638, in his *Dialogues Concerning Two New Sciences* (in which he also discussed the earth's revolving around the sun). He proposed the solutions of change of form or of materials. We quote D'Arcy Thompson's account from *On Growth and Form* (1917):

> [Galileo] said that if we tried building ships, palaces or temples of enormous size, yards, beams and bolts would cease to hold together; nor can Nature grow a tree nor construct an animal beyond a certain size, while retaining the proportions and employing the material which suffice in the case of a smaller structure. The thing

Color Plate 14 A totally artificial computer-generated fractal mountain and valley created by Benoit B. Mandelbrot and Richard F. Voss. [Copyright 1982 by B. B. Mandelbrot and R. F. Voss.] (See Part IV opener, Box 21.2.)

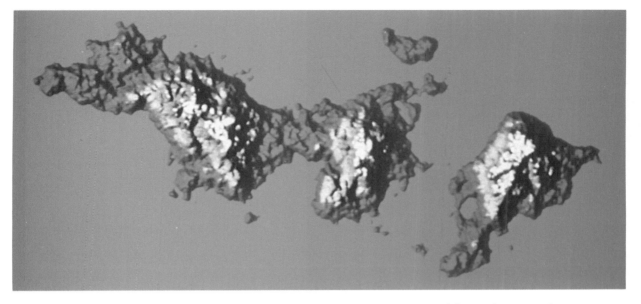

Color Plate 15a Computer-generated fractal islands. [Copyright B. B. Mandelbrot and R. F. Voss.] (See Part IV opener, Box 21.2.)

Color Plate 15b Computer-generated fractal planet as seen from its moon. [Copyright B. B. Mandelbrot and R. F. Voss.] (See Part IV opener, Box 21.2.)

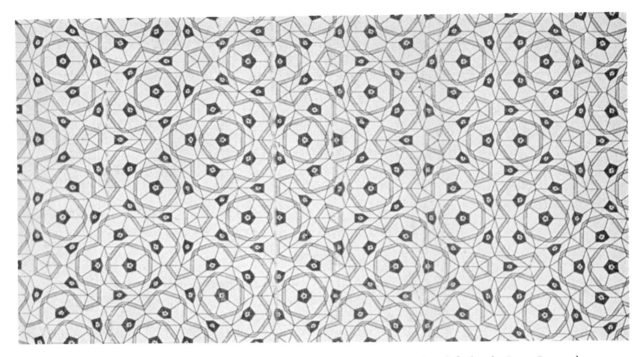

Color Plate 16a A Penrose (aperiodic) tiling made with two rhombs (rhombuses). [Tiling by Roger Penrose.] (See pp. 253–264.)

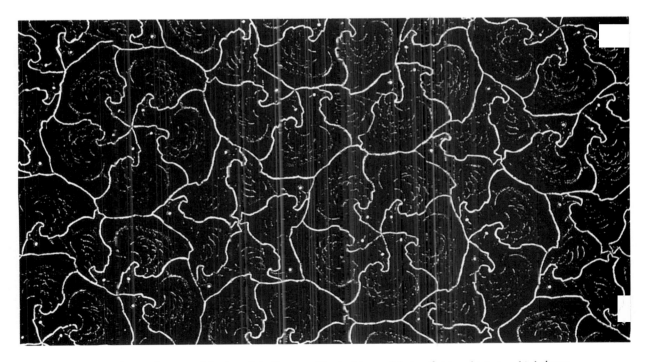

Color Plate 16b A modification of the Penrose tiling by kites and darts refashioned into two bird shapes. [Tiling by Roger Penrose.] (See pp. 263–264.)

Color Plate 16c A five-colored Penrose tiling by kites and darts. [Tiling by Roger Penrose.] (See pp. 263–264.)

Color Plate 17 The cartwheel tiling. [From Roger Penrose.] (See pp. 263–264.)

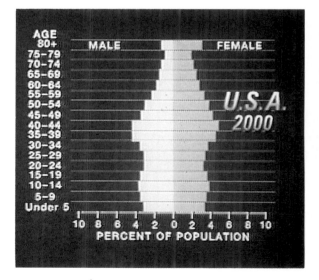

Color Plate 18 Graph of the estimated population of the United States in the year 2000 grouped by age and sex and shown as a percentage of the total population. This population estimate is based on a geometric growth model. (See pp. 290–293.)

Color Plate 19 The ratio of lengths of similar triangles can be used to calculate the height of a pyramid. (See pp. 306–310.)

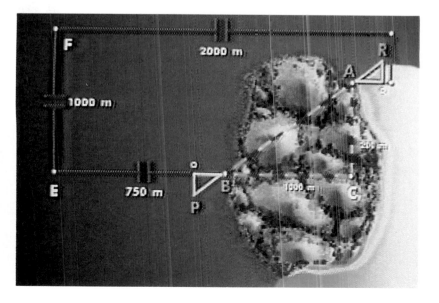

Color Plate 20 The method of similar triangles was probably used 2500 years ago by Greek engineers on the island of Samos to dig a tunnel through Mount Castro to bring water to the capital city. (See pp. 310–311.)

(a)

(b)

(c)

Color Plate 21 (a) M. C. Escher's *Heaven and Hell* (also known as "Angels and Devils"). Photographed from one of Escher's notebooks (in black and white), this example demonstrates a repeating pattern of the Euclidean plane. Note the uniform size of the figures and the central meeting of four angel and four devil wing tips. [© M. C. Escher Heirs c/o Cordon Art—Baarn—Holland.] (See pp. 313, 314–315, 331–332.) (b) Escher's angels and devils pattern carved on an ivory sphere by Masatoshi (in black and white). This mapping of the pattern onto a sphere shows the different effects of a spherical geometry. Note that three angel and three devil wing tips meet in this version. [© M. C. Escher Heirs c/o Cordon Art—Baarn—Holland.] (See pp. 313, 314–315, 334–336.) (c) M. C. Escher's *Circle Limit IV (Heaven and Hell)*. This example shows the repeating angels and devils pattern mapped onto a hyperbolic geometry. At the center three angel and three devil feet meet. As one moves outward the figures get smaller. Note that four angel and four devil wing tips meet. [© M. C. Escher Heirs c/o Cordon Art—Baarn—Holland.] (See pp. 313, 314–315, 332–334.)

Color Plate 22a *Seven Butterflies Pattern.* This pattern is based on a repeating pattern of the Euclidean plane by M. C. Escher in which six butterflies meet at left front wingtips; increasing this number to seven results in a repeating pattern of the hyperbolic plane. A fundamental region for *Seven Butterflies Pattern,* as a colored pattern, consists of 168 butterflies; within any "ring" of butterflies of one color there are three different ways to arrange the other seven colors around the center. [Pattern designed by Douglas Dunham, University of Minnesota, Duluth] (See pp. 313, 314–315, 332–334.)

Color Plate 22b *Six Fish Pattern.* This pattern is based on M. C. Escher's similar pattern of fish, *Circle Limit III.* The difference is that six fish meet at left fin-tips in this pattern (instead of the four fish that meet at left fin-tips in *Circle Limit III*). In *Six Fish Pattern,* as in *Circle Limit III,* 12 fish form a fundamental region for the colored pattern; of course one fish serves as a fundamental region for either pattern if color is disregarded. Also, as in *Circle Limit III,* each white backbone is a circular arc making acute angles with the bounding circle; thus in hyperbolic geometry, such an arc is not a hyperbolic line but is an equidistant curve from the circular arc with the same endpoints that is orthogonal (at right angles) to the bounding circle (and thus is a hyperbolic line). [Pattern designed by Douglas Dunham, University of Minnesota, Duluth.] (See pp. 313, 314–315, 322–334.)

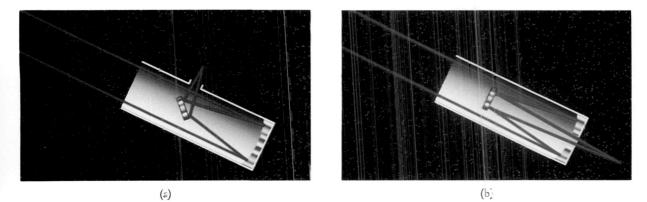

(a)   (b)

Color Plate 23 (a) Schematic diagram showing the light path in a Newtonian reflecting telescope. (See pp. 323–325.) (b) Schematic diagram showing the light path in a Cassegrain reflecting telescope. (See pp. 325–326, 327.)

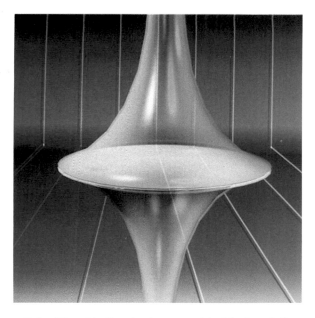

Color Plate 24 Pseudosphere, a model of the hyperbolic plane. (See pp. 332–334.)

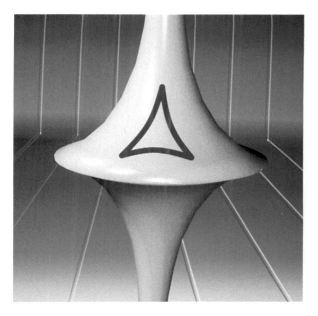

Color Plate 25 In the hyperbolic geometric model, the sum of the angles of any triangle on the surface of a pseudosphere is less than 180°. (See pp. 332–334.)

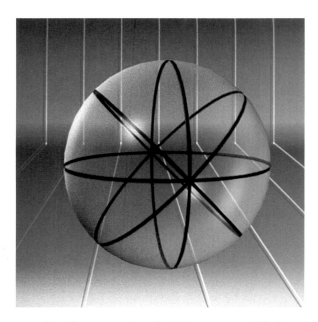

Color Plate 26 In the spherical geometric model, the shortest paths between two points are great circles. Note that every pair of great circles intersect. (See pp. 334–336.)

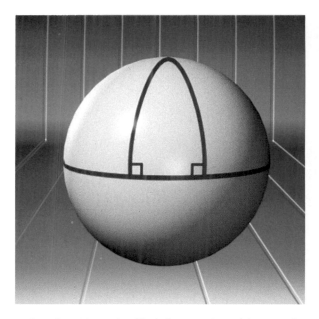

Color Plate 27 In the elliptical geometric model, a triangle on the surface of a sphere can have two or three right angles. In this geometry, the sum of the angles of any triangle is greater than 180°. (See pp. 334–336.)

Even these giant sequoias may grow no taller than their form and materials allow. [Larry Ulrich.]

will fall to pieces of its own weight unless we either change its relative proportions, which will at length cause it to become clumsy, monstrous and inefficient, or else we must find new material, harder and stronger than was used before. Both processes are familiar to us in Nature and in art, and practical applications, un-dreamed of by Galileo, meet us at every turn in this modern age of cement and steel.

In fact, Galileo suggested 300 feet as the limiting height for a tree. Giant sequoias, which grow only on the West Coast of the United States and hence were unknown to Galileo, have been known to grow as high as 360 feet. Galileo's reasoning was correct; the tallest giant sequoias adapt their form in ways that evade the limits of his model.

BOX 14.4 Mile-High Buildings?

Architect Hugh Stubbins discusses skyscrapers:

Scale is a relative thing. If you think of a human being standing beside the Great Pyramid at Giza, you can imagine the scale of a human being against that big mass, or even against the Empire State Building in New York. Now, scale in the past and ancient times was made so it would impress people; they made temples to impress people. Today I don't think we build big buildings to impress people; they're big because they have to be big. The requirement for this building was to have a million and a quarter square feet of usable space. Now, that's a lot of square footage! If the building were high and solid, then nobody would be close to a window. It would be just a cavern of space, which would not be very pleasant to be in. And one of the main purposes of architecture is to develop spaces that are pleasant for people to work in, to live in, and to spend their time in.

The reason for tall office buildings is real estate costs: the cost of property in cities is so high that in order to justify the cost of the land, you have to put a lot of building on it. And if you built it from street to street, the city would be completely filled with nothing but canyons.

Well, I think buildings can get too big. William LaMeasure has devised a building that could be built a mile high. Years ago Frank Lloyd Wright said he had designed a building that would reach a mile high. Well, LaMeasure checked into that and found out that Wright's scheme was absolutely impossible—it could never be built. But LaMeasure is such a genius at engineering that he has devised a mile-high structure that *could* be built. There are four buildings actually, at four corners of a square, which are about 400 feet apart. They go up, and then there are horizontal buildings that run between these four posts. And they're open every other floor so the air can get through it. He says that such a structure can be built 5000 feet high. But who would ever want to do it? I can't imagine anybody wanting to do that. It takes too long to get from the ground up to where you're going. It's a fun thing for an engineer to think about and to try to make happen, but I don't think it'll ever happen.

---

version scaled up by a factor of 3, each small cube of the bottom layer has a bottom face that is supporting that cube's weight *plus* the weight of the other two cubes piled on top of it.

Now, let's redesign the scaled-up cube, concentrating for simplicity only on the front face, with its nine small cubes. We take the three cubes on top and move them to the bottom, alongside the three already there. We take the three cubes on the second level, cut each in half, and put a half cube over each of the six ground-level cubes (see Figure 14.5). We have the same volume and weight that we started with, but now there is less pressure on the bottom face of each small cube. Of course, our new design is not geometrically similar to the object we started with—it's no longer a cube. By changing the proportions, we have given up the precise scaling of geometrical similarity, but we have managed to compensate for the scaling problem.

We observe in nature both strategies for adaptation to scaling: change of materials and change of form. Smaller animals do not even have bony internal skeletons; larger animals do. Animals such as a mouse and an elephant, which are made of much the same materials, differ in shape. If a mouse were scaled up to the size of an elephant, its legs could no longer support it. It would need the disproportionately thicker legs of the elephant, as well as the elephant's thick hide to contain its tissue.

Fatal Falls, Dives, Jumps, and Flights

Area-volume tension has many other practical consequences. It brings our childhood fanta-

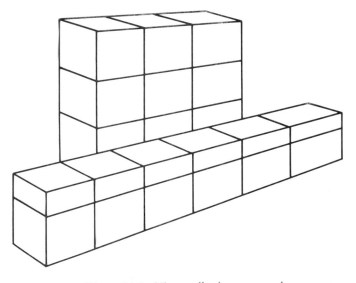

Figure 14.5 Nine small cubes rearranged to support greater weight.

sies into sharp but disappointing terms. We can forget about "leaping tall buildings in a single bound," "soaring like an eagle," diving miles below the sea, and jumping from an airplane without a parachute.

Area-volume tension affects how creatures respond to falling, another of gravity's effects. A mouse may be unharmed by a ten-story fall, a cat by a two-story fall, but a human may well be injured just by falling from its own height.

What is the explanation? The energy acquired in falling is proportional to the weight of the falling object, hence to its volume. This energy must be absorbed either by the object or by what it hits or must be otherwise dissipated at impact — for example, as sound. The fall is absorbed over all or part of the surface area of the object, just as the weight of the cube was distributed over its base. With scaling up, volume (hence weight, hence falling energy) goes up much faster than area. As volume increases, the hazards of falling the same height increase.

Whales can hold their breath and stay under water for as long as 20 minutes. Why can't we? The lung capacity, or volume of oxygen carried, is proportional to the cube of the length of the mammal. But the rate at which oxygen is absorbed and consumed is proportional to the surface area of the lungs, and hence to the square of the length of the mammal. We would therefore expect the limits of duration of dive to be proportional to the length of the animal, although special adaptations of particular species may play a large role. Accordingly, blue whales are about 16 times as long as human adults and therefore hold their breath 16 times as long.

A flea can jump about 2 feet vertically, many times its own height. From a standing start a human can jump at most 5 feet. Many people believe that if a flea were as large as a person, it could jump a thousand feet into the air. Imagining — against our earlier arguments — that there could be so large a flea, we know its limits: a scaled-up flea could jump about the same height as a small flea. A flea jumps by expending energy in flexing its leg muscles. If the jumping muscles form a constant fraction of the animal's body, then the energy developed per ounce of muscle is independent of the size of the animal. Thus, big fleas and little fleas alike could jump about 2 feet.

Wouldn't it be nice to be able to fly? It turns out that the minimum cruising speed for an object flying level is proportional to the square root of wing loading, which is defined as body mass divided by wing area. (We don't discuss *gliding,* in which the object either loses altitude or depends on thermal currents to maintain it.) Body mass is proportional to the cube of the length, and for birds, wing area is proportional to the square of body length. The ratio — the wing loading — is therefore proportional to body length.

It follows that the minimum cruising speed is proportional to the square root of length, or equivalently, to the sixth root of body mass. If a bird or plane is scaled up by a factor of 4, then it must fly $\sqrt{4} = 2$ times as fast. Now, a plane scaled up by 4 will have $4^3 = 64$ times as much weight as the smaller one. Hence it requires $\sqrt{4} \times 4^3 = 128$ times as much power to stay up.

Let's apply the same principle to birds. For a sparrow, the minimum speed is 20 miles per hour. An ostrich is 25 times as long as a sparrow, so the minimum speed for an ostrich would be $\sqrt{25} \times 20 = 100$ miles per hour. Have you seen any flying ostriches lately? Heavy birds fly fast or not at all! Power considerations derived from further dimensional analysis force us to conclude that the maximum possible weight for a flying animal would be about 35 pounds.

Keeping Cool (and Warm)

Area-volume tension is also of crucial importance to a creature's maintenance of thermal

BOX 14.5 "Take That, King Richard!"

Did wearing armor give an advantage to the shorter warrior? [All rights reserved, the Metropolitan Museum of Art.]

Shakespeare, following the propaganda of the Tudor historians, painted Richard III as a humpbacked Machiavellian monster. Did Richard have an advantage in armored combat because he was short? That suggestion was made some years ago by one of the leading modern historians of the Tudor era, Garrett Mattingly.

[B]etween a short man and a tall man, height increases by the linear dimension—from 5 feet 2 inches, say, to 6 feet—while the surface of the body increases as the square. Since it's . . . the surface of the body that the armorer must plate with steel, the armor of a short warrior, like Richard, would be lighter than a tall warrior's by a lot more than the few inches' difference in height would indicate. So Richard's notorious deadliness in battle would have been possible at least in part—or so Mattingly's speculation ran—because his armor, while protecting him as well as the big man's, left him less encumbered.

After the lecture, someone said to Mattingly that he had grasped the right idea—but by the wrong end. Muscle power, the listener claimed, is a matter of bulk—and physical volume goes up by the cube, whereas the surface to be protected goes up by the square. So the large warrior should have more strength left over than the little guy after putting on his armor. And the large warrior, swinging a bigger club, can deliver a far more punishing blow—because the momentum of the club depends on its weight, which goes up with its volume, which means by the cube. Richard was at a terrible disadvantage.

But wait a minute, a second listener said. That's true about the club—but not about the muscles. The strength of a muscle is proportional not to its bulk but to the area of its cross section. And since the cross section of muscles obviously increases by the square, just as the surface of the body does, the big guy, plated out, has no more, or less, advantage over the little guy than if both were naked.

But hang on, a third person—interjected—an engineer. That's right about the muscles, but it's not right about the armor. The weight of the armor increases not simply with the increase in the surface area that it must cover but slightly faster. The reason is that to obtain the same strength with a larger area of metal, there must be reinforcing ribs. Or else the metal must be significantly thicker overall. So maybe Richard had an advantage after all.

To maximize protection within the weight, the armorer adopted two strategies: variable thickness and deflection. The unexpected fact is that armor was made as thin as possible. Thickness reinforcement, and structural stiffening of a large surface were concentrated where opponents' weapons were likely to hit. From these strong, shaped places, the metal tapered away, until the sheet steel was as thin as the lid of a coffee can at the sides of the rib cage beneath the arms, or across the fingers, or at the cheek of a helmet.

(Quoted from Horace Freeland Judson, *The Search for Solutions*. Holt, Rinehart and Winston, pp. 54–56.)

---

equilibrium. Both warm-blooded and cold-blooded animals gain or lose heat from the environment in proportion to body surface area. A warm-blooded animal usually is losing heat; its equilibrium consumption, or food intake needed to maintain body heat, depends primarily on the amount of its surface area, the temperature of its ambient environment, and the insulation provided by its coat or skin. Other factors being equal, a scaled-up mammal scales up its food consumption by surface area (proportional to the square of the scaling factor) not by volume (proportional to its cube).

Cold-blooded animals, such as alligators or perhaps most dinosaurs, have a somewhat different problem. Whereas mammals regulate their metabolism and maintain a constant internal body temperature, cold-blooded animals are at the mercy of their environment. They absorb heat from the environment for energy, and they also dissipate any excess heat (e.g., from metabolism) to keep their temperature below unsafe levels. The amount of heat that must be gained or lost is proportional to total volume, because the entire animal must be warmed or cooled. But the heat is exchanged through the skin, so the rate is proportional to surface area.

Dimetrodon was a mammallike reptile that roamed present-day Texas and Oklahoma 280 million years ago (see Figure 14.6). Over the course of its evolution, it doubled in size and altered in form, evolving a disproportionately larger fan or sail on its back to absorb and dissipate heat. It appears that the size of the fan increased precisely in proportion to the volume of the animal.

Similarity and Growth

Although a large change of scale forces adaptive changes in materials or form, within narrow limits—perhaps up to a factor of 2—

Figure 14.6   *Dimetrodon* may have evolved a sail to absorb and dissipate heat efficiently. [Courtesy, Field Museum of Natural History.]

BOX 14.6 On Being the Right Size

J. B. S. Haldane. [The Bettman Archive/BCC Hulton.]

J. B. S. Haldane (1892–1964) was a biologist who contributed to the theory of evolution by helping to show that natural selection and Mendelian genetics are compatible. He then helped to develop the theory of population genetics.

His life and work accommodated several contradictions. Although he majored in classics and mathematics at Oxford, he worked in genetics and biochemistry. Haldane joined the Communist party in the 1930s, became a leading Communist and contributor to the *Daily Worker,* then left the party in the 1950s. Having been an army captain in World War I who later admitted to enjoying the war, he spent his last years in India espousing nonviolence.

In "On Being the Right Size" (1928), his classic short essay on the problem of scale, Haldane succinctly surveys area-volume tension, flying, the size of eyes, and even — as one might expect from a man with so intense an interest in politics — the best size for human institutions: "I find it no easier to picture a completely socialized British Empire or United States than an elephant turning somersaults or a hippopotamus jumping a hedge."

creatures can grow according to a law of similarity. In other words, they grow in such a way that their shape is preserved. A striking example of such growth is that of the chambered nautilus *(Nautilus pompilius)*. Each new chamber that is added on to the nautilus shell is the same shape as the previous chamber, and the shape of the shell as a whole — an equiangular, or logarithmic, spiral — remains the same (see Figure 13.1b).

Most living things grow over the course of their lives by a factor greater than 2. An adult is not simply a baby scaled up. Relative to the size of the body, a baby's head is much larger than an adult's. Even the proportions of facial features are different: in a baby, the tip of the nose is about halfway down the face; in an adult, the nose is about two-thirds of the way down. The arms of the baby are disproportionately shorter than an adult's. In the growth from baby to

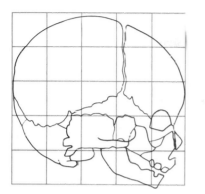

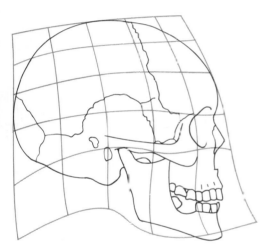

Figure 14.7 Modeling the changes in shape of a human head from infancy to adulthood. [From Richard C. Lewontin, "Adaption." © *Scientific American* 239(3):220, 1978. All rights reserved.]

interlocking scalings. For a simple model of the process in which a baby's head changes shape to grow into an adult head, we can use graph paper: we put a picture of the baby's skull on graph paper, then deform the grid until the pattern matches an adult skull (see Figure 14.7).

It is often valuable to be able to predict what a developing organ will look like in the future. For example, what does a child look like now who was kidnapped at age three, two years ago?

At Face Systems, Inc., in New York, a computer and a more sophisticated version of our graph-paper technique are used to answer such questions. Staff transfer to the computer screen a photo of the missing child, taken right before the disappearance, and a photo of an older sibling or of one of the parents as an older child. Then they superimpose the two images and warp the image of the missing child's face to fit the shape of the other face. The result is a rough idea of what the missing child may look like. As mathematicians and biologists refine their models of how faces change over time, this technique will improve. It may become possible for a child to gain an idea of how he or she may look at age 40 or 65.

Logarithmic Plots and Allometry

If we measure the arm length or head size for humans of different ages and compare it with body height, we observe that humans do not grow in a way that maintains geometric similarity. The arm, which at birth is one-third as long as the body, is by adulthood closer to two-fifths as long. We plot body height on a horizontal axis and arm length on the vertical axis. If we use ordinary graph paper with linear scales (axes marked off evenly in units), we get a curve. If the growth were proportional, that is, according to similarity, we would have got a

adult, different parts of the body scale up, each with a *different* scale factor.

Although the laws for such differential growth are much more complicated than for proportional scaling, more sophisticated mathematics—for example, *differential geometry,* the geometry of curves and surfaces—permits analysis of even these complex and

straight line. Graphing therefore provides a way to test for proportional growth.

Is there an orderly law by which we can relate arm length to height? Let's plot again, this time using **log-log paper** — graph paper on which each axis is marked off evenly, not in units but in orders of magnitude: 1, 10, 100, 1000, and so on (see Figure 14.8). On this graph the data plot closely to a straight line. We can discern two different straight lines: one that fits early development, with slope 1.2, and another that fits development after nine months of age, with slope 1.0. The change from one line to another at that age indicates a change in pattern of growth. The pattern after nine months, characterized by the straight line with slope 1, is indeed proportional growth, sometimes called *isometric growth*. But the earlier growth, described by a straight line with slope other than 1, also follows a definite pattern, called *allometric growth*. Because slope 1.2 is greater than 1, we know that arm length is increasing relatively faster than height.

If we denote arm length by y and height by x, a straight line fit on log-log paper corresponds to the algebraic relation

$$\log_{10} y = B + a \log_{10} x$$

where a is the slope of the line and B is the point where the graph crosses the vertical axis. If we raise 10 to the power of each side, we get

$$y = bx^a$$

where $b = 10^B$. This equation describes a power curve: y as a constant multiple of x raised to a certain power.

For a slope $a = 1$, we get $y = bx$ — a linear relationship describing proportional growth. On ordinary graph paper, proportional growth appears as a straight line, allometric growth as a curve. On log-log paper, both patterns appear as straight lines.

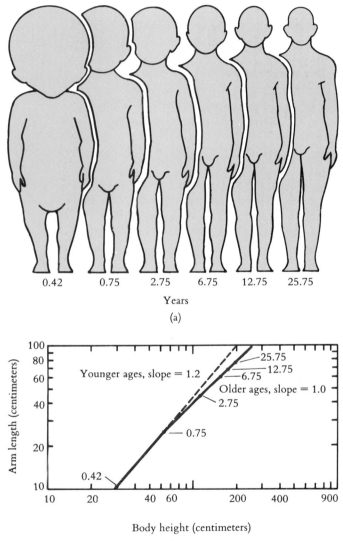

Figure 14.8 (a) The proportions of the human body change with age. (b) A graph of human body growth on log-log paper. The numbers shown beside the points indicate the age in years; they correspond to the stage of human development shown in (a). [From Thomas A. McMahon and John Tyler Bonner, *On Size and Life*, Scientific American Library, 1983.]

REVIEW VOCABULARY

Allometric growth   The growth of one feature at a rate proportional to a power of another.

Allometry   The law of allometric growth.

Density   Weight per unit volume.

Geometric similarity   Two objects are geometrically similar if they have the same shape, regardless of the materials they are made of.

They need not have the same size. Corresponding angles, however, must be equal, and line segments between corresponding points must all have the same factor of proportionality.

Log-log paper   Graph paper with scales marked so that each axis is marked off evenly in orders of magnitude.

EXERCISES

1. Jonathan Swift's Gulliver also traveled to Lilliput, where the Lilliputians were human-shaped, but only about 6 inches tall. In other words, they were geometrically similar in shape to ordinary human beings but only one-twelfth as tall. Discuss:

   a. What would a Lilliputian weigh?

   b. Are Lilliputians ruled out by the size-shape and area-volume considerations in this chapter? If you think they are, what considerations do you find convincing? If not, why not?

2. Steel has a resistance to crushing of 10.5×10^6 pounds per square foot. Use this figure to verify that a cubic steel mountain will crumble if it is larger than 4 miles on a side.

3. How high could a mountain be on the moon, whose gravity is 0.16 times that of earth? On Jupiter (2.6 times that of earth)? On a celestial body with gravity $r\%$ of the earth's?

4. Use algebra to demonstrate that for proportional growth, "area scales as volume to the two-thirds power."

5. Icarus of Greek legend escaped from Crete with his father, Daedalus, with wings made by Daedalus and attached with wax. Against his father's advice, Icarus flew too close to the sun: the wax melted, the wings fell off, and Icarus fell into the sea and drowned. What must have been his minimum cruising speed, assuming he was proportional to a bird?

6. Compare breath-holding times of a child and friends of different heights. Take several observations for each person (let them rest between trials!). Plot your results on ordinary graph paper. Do your experimental results confirm what the theory predict? Why or why not?

7. Recent years have seen the beginnings of true human-powered controlled flight, in the Gossamer Condor and other superlightweight planes. The Gossamer Condor is far longer than an ostrich. How can it fly at a speed of 12 miles per hour? Why can't it fly faster? How can it fly at all?

8. Consider the following data for various mammals:

	Weight (kilograms)	Calories per kilogram
Guinea pig	0.7	223
Rabbit	2	58
Man	70	33
Horse	600	22
Elephant	4,000	13
Whale	150,000	1.7

Graph these data on log-log paper. What can you conclude from your graph? [Note that you may have to "create" your own log scale on the weight axis, as the weights span six order of magnitude. Commercial log paper is usually limited to 2, 3, or perhaps 5 orders of magnitude ("cycles").]

9. Listed in the table below are the winning times in the 1983 World Rowing Championships for rowing shells seating one, two, four, and eight oarsmen. The men's times are for 2000 meters, the women's for 1000 meters. Convert the times to speed. For men and women separately, plot speed versus number of oarsmen on ordinary graph paper, and then on log-log paper. Is the relationship proportional? Allometric?

Fit the best line you can through the log-log data and estimate its slope. Compare your results with the theoretical discussion and data in McMahon and Bonner's *On Size and Life,* Scientific American Books, Inc., 1983, pp. 42–47 and 67.

Event	Number of oarsmen	Men (2000 meters)	Women (1000 meters)
Single sculls	1	6:49.75	3:36.51
Pairs without coxswain	2	6:35.85	—
Fours without coxswain	4	6:14.83	3:26.68
Eights (with coxswain)	8	5:34.39	2:56.22

The Size of Populations

Many of the problems we face relate to population change over time. Each of us has a stake in the problems that are associated with human population growth, such as hunger and disease. Our food supplies are also affected by the growth and behavior of nonhuman populations such as medflies, locusts, and grasshoppers. Even inanimate populations may affect our lives. A "population" of barrels of nuclear waste from a generating plant poses storage and disposal issues, and the growth rate of such a population poses important environmental questions. Similarly, but more favorably, a growing population of dollars in a bank account can provide resources that enrich our lives.

Each of these examples of population change is familiar. But to apply the ideas and methods of mathematics to the study of populations we must clarify and refine even familiar notions. We will therefore pose and answer some fairly precise questions about populations:

1. How are its members described?

2. How big is the population?

3. Where is it located and how is it distributed?

4. How fast is it growing or shrinking?

5. How is its structure or composition changing?

We will use the term **structure** to refer to the divisions of a population into subgroups. For example, human populations are frequently structured according to age. Indeed, U.S. census data are normally presented in terms of age groups. However, human populations can also be structured in many other ways. Indeed, for a specific problem it may be advantageous to structure a population using economic, social, or educational criteria.

In this chapter we will pay particular attention to the size of a population and to the way in which its size changes over time. We will use the term **growth** to refer to both positive and negative changes in population size. Thus, *growth* may indicate either an increase or a decrease.

Geometric Growth and Financial Models

We begin our study with a population whose behavior is relatively simple — the population of dollars in a bank account. In one of the most common financial transactions, we deposit money in a savings or money market account. Our primary concern is with the growth of such savings. Suppose that we deposit $1000 in an account that, we are told, "pays interest at a rate of 10%, compounded and paid annually." Assuming that we make no other deposits or withdrawals, how much is in the account after one, two, or five years?

The $1000 is usually called the *initial balance* or the *principal* of the account. At the end of one year, interest is added. The amount of interest is 10% of the principal, $100 in this case. If $100 is added to the account at the end of the first year, the balance at the beginning of the second year becomes $1100. During the second year the interest is also 10%, so at the end of the second year 10% of $1100, or $110, is added to the account.

Notice that during the second year we earn interest on both the original balance of $1000 and on the $100 interest we earned during the first year. Interest paid on both the principal amount and on the accumulated interest in an account is known as *compound interest*. It is a simple but remarkable idea. Because we receive more interest during the second year than during the first, the account grows more

during the second year. At the beginning of the third year the account contains $1210, so at the end of the third year we receive $121 in interest. Again this is larger than the amount we received at the end of the preceding year. Moreover, the increase during the third year

Third-year interest
− second-year interest = $121 − $110
= $11

is larger than the increase during the second year:

Second-year interest
− first-year interest = $110 − $100
= $10

Thus, not only is the amount in the account increasing each year, but the amounts that are added each year are also increasing. This type of growth is sometimes called **geometric growth.**

There is another way to pay interest, called *simple interest*. Using this method, interest is paid only on the original balance. In our example, we receive $100 interest at the end of the first year, and therefore at the beginning of the second year our account contains $1100. Then we receive another $100 interest at the end of the second year, so at the beginning of the third year our account contains $1200. In fact, at the end of each year we receive just $100 in interest. Clearly this method yields less than the amount we would have if interest were compounded.

Although the simple-interest method is seldom used in today's competitive financial markets, we frequently observe this type of growth, called **simple,** or **arithmetic, growth,** in other contexts. If you withdraw and spend the interest paid each year in an account paying compound interest, then that account effectively pays simple interest. Nuclear waste generated by a power plant gener-

TABLE 15.1 **The growth of $1000: compound interest versus simple interest**

Years	Amount in account from compounded interest	Amount from simple interest
1	1100.00	1100.00
2	1210.00	1200.00
3	1331.00	1300.00
4	1464.10	1400.00
5	1510.51	1500.00
10	2593.74	2000.00
20	6727.50	3000.00
50	117,390.85	6000.00
100	13,780,612.34	11,000.00

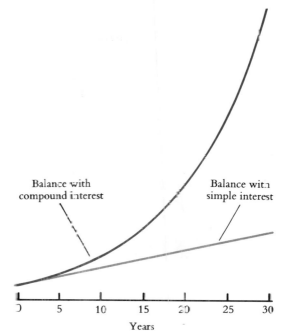

Figure 15.1 The growth of $1000: compound interest and simple interest.

ally follows a simple-growth model because the amount of waste added each year depends on the fixed size of the power plant, not on the growing size of the waste dump.

The amounts in accounts paying interest at the rate of 10% per year with compound and simple interest are shown in Table 15.1 and in the graph in Figure 15.1. The impressive growth associated with compound interest (see Box 15.1) is dramatically illustrated by these figures, as is the distinction between geometric and arithmetic growth. It is this distinction that led the demographer and economist Thomas Malthus (1766–1834) to his famous theory that human populations grow

BOX 15.1 Investing in Manhattan Island

The sums that result from small initial deposits and relatively modest interest rates over long time periods are truly astonishing. The miracle of compound interest is frequently illustrated with the following example:

If Manhattan Island had been purchased from the Indians in 1650 for $1, and if the Indians had immediately deposited that $1 in an account that paid compound interest at the rate of 5% compounded yearly, then that single dollar would now be worth over $11,000,000!

The effect of small changes in interest rates can also be illustrated. If the assumed interest rate were 6% rather than 5% — a 20% increase — then the current value of the account would be over $267,000,000, an increase of greater than 2200% over the amount compounded at the 5% rate.

BOX 15.2 Thomas Malthus, 1766–1834

Thomas Malthus. [The Bettman Archive/ BBC Hulton.]

Thomas Malthus, a nineteenth-century English demographer and economist, based a well-known prediction on his perception of the differences in the rates of growth of two populations, the human population and the "population" of food supplies.

His belief was that human populations increase at a much faster rate than do food supplies. He concluded, however, that over the long run there would be restrictions on the natural growth of human populations too, including war, disease and starvation— hardly an optimistic forecast and doubtless, responsible for the dreary image associated with his views.

geometrically, whereas food supplies grow arithmetically (see Box 15.2).

Suppose we have the option of depositing our $1000 in a bank that pays interest at the rate of 10% per year but compounds the interest quarterly, that is, four times per year. With an interest rate of 10% per year, we are entitled to one-fourth, or 2.5%, each quarter.

Let's see what happens during the first year. At the end of the first quarter, we have the original balance plus $25 interest, so the balance at the beginning of the second quarter is $1025. During the second quarter we receive interest equal to 2.5% of $1025, or $25.63, so our new balance at the end of the second quarter is $1050.63. Continuing in this manner, we find that the balance at the end of the first year is $1103.81.

Note that quarterly compounding (four interest payments per year) gives us a larger balance at the end of the year than we got with interest compounded annually (see Table 15.2). If interest were compounded monthly (12 times per year), the balances would be still

larger. A comparison of yearly, quarterly, monthly, and daily compounding for an interest rate of 10% is shown in Table 15.2.

It is useful to summarize our results before we consider other examples. We will agree that from now on, all interest rates will be expressed as fractions, so that the rate r will mean (rate in %)/100. If an amount P is deposited in an account that pays interest at the rate r, then the amount in the account one year later is

Original principal + interest
$$= P + Pr = P(1 + r)$$

This amount can be viewed as a new starting balance. Hence, in the next year the amount $P(1 + r)$ grows to

$$P(1 + r) + P(1 + r)r$$
$$= P(1 + r)(1 + r) = P(1 + r)^2$$

The pattern can be continued until we reach the following conclusion: *If a principal P is de-*

TABLE 15.2 Comparing compound interest: the value of $1000, under 10% annual interest, if interest is compounded

Years	Compounded yearly	Compounded quarterly	Compounded monthly	Compounded daily	Compounded continuously
1	$1100.00	$1103.81	$1104.71	$1105.16	$1105.17
5	$1610.51	$1638.62	$1645.31	$1648.61	$1648.72
10	$2593.74	$2685.06	$2707.04	$2717.91	$2718.28

BOX 15.3 *e*— A Very Special Number

The results given in Table 15.2 illustrate a general trend: for a fixed interest rate, more frequent compounding results in larger ending balances. Thus, as you move from left to right in any row of the table, the amounts increase. However, the amounts in an account that result from more and more frequent compounding do not grow indefinitely; instead, they get closer and closer to a number that can be determined in advance from the interest rate. This number is shown in the far right column in each row.

What happens when we increase the frequency of compounding? With interest at 10% per year compounded n times per year, the amount we have at the end of one year is $1000(1 + 0.1/n)^n$. To simplify matters, however, we can look at the slightly simpler form $(1 + 1/n)^n$. As n increases, the term $(1 + 1/n)^n$ gets closer and closer to a special number called e. This is illustrated in the following table:

n	$(1 + 1/n)^n$
1	2.0000000
5	2.4883200
10	2.5937425
50	2.6915879
100	2.7048138
1,000	2.7169239
10,000	2.7181459
100,000	2.7182682

The value of e is 2.71823 . . . (the dots indicate that the decimal continues).

Similarly, for any number x, the quantity $(1 + x/n)^n$ gets closer and closer to the number e^x as n becomes large. Thus, as n becomes large, the quantity $(1 + 0.1/n)^n$ gets closer and closer to $e^{0.1} = 1.10517$. . . . No matter how frequently interest is compounded, the original $1000 at the end of one year cannot grow beyond $1105.17. The corresponding values for other years are shown in the last column of Table 15.2.

BOX 15.4 Variations on a Theme

There is nothing special about using a year as the unit of time for when interest is compounded. In fact, we can use any time unit as long as the interest rate is expressed in the same units used to measure time. Given an interest rate for a year, we can determine the amount in an account with interest compounded quarterly, monthly, or in any other period by simply using our formula, with an interest rate adjusted to reflect the new time unit and with the time interval expressed in terms of our new unit of time.

To illustrate this, suppose we have in a bank account a principal of $P = 1000$ with interest at an annual rate of 10%. Using the formula $P(1 + r)^m$ we can determine the amount in the account after 10 years using several different compounding periods:

1. If we take the period of time to be a year, then the interest rate of 10% per year gives $r = 0.1$, and we determine the amount in the account after 10 years to be

$$1000(1 + 0.1)^{10} = 1000(1.1)^{10} = 1000(2.59374) = 2593.74$$

2. If we take the unit of time to be a quarter of a year, then $r = \frac{0.1}{4} = 0.025$, and after 10 years (which is 40 quarters) the account contains

$$1000(1 + 0.025)^{40} = 1000(1.025)^{40} = 1000(2.68506) = 2685.06$$

3. Finally, if compounding is to be done monthly, then our unit of time is one-twelfth of a year and $r = \frac{0.1}{12} = 1000(\frac{12.1}{12})^{120}$. The amount in the account after 10 years, which is 120 months, is

$$1000\left(1 + \frac{0.1}{12}\right)^{120} = 1000\left(\frac{12.1}{12}\right)^{120} = 1000(2.70704) = 2707.04$$

These are the entries in the last row of Table 15.2.

---

posited in an account that pays interest at the rate r per year, compounded annually, then after m years the account contains $P(1 + r)^m$. Box 15.4 explains compounding for other time units.

Growth Models for Biological Populations

We can now use a geometric growth model to make rough estimates about sizes of human populations. Indeed, population projections reported in the popular media are based on

such models; they use the difference between the birth rate b and the death rate d as the growth rate r. A shortcoming of this model is that b and d rarely remain constant for very long, so extreme care must be exercised in making projections.

In the short run, however, predictions based on the model may provide useful information. Let's apply this model to two questions:

• If birth and death rates remain about the same as they are now, how large will the population be a specific number of years from now?

• What will happen if the birth rate increases by a certain percent while the death rate decreases by a different percent?

Suppose that the population of the United States was 230,000,000 in 1987 and that it increases at an average growth rate of 0.9% per year until the year 2000. (This growth rate is roughly what we are seeing in the mid-1980s.) What is the anticipated size of the U.S. population in the year 2000? What will it be if the average growth rate turns out to be 0.6% per year or 1.2% per year?

To answer these questions we apply our geometric model with the initial population size equal to 230,000,000. In the first case $r = 0.009$ (because 0.9% is just under 1%, or 0.01). Using the formula $P(1 + r)^m$, where $m = 13$, we find that the projected population size in the year 2000, 13 years from the year 1987, is

Population in 2000
= (population in 1987) $(1 + \text{growth rate})^{13}$
= $(230,000,000) (1 + 0.009)^{13}$
= $(230,000,000) (1.123531)$
= $258,412,190$

In the same way we find that with a growth rate of 0.6% per year, we predict a population in the year 2000 of 248,600,264, whereas a growth rate of 1.2% per year yields a predicted population of 268,580,513. It is clear that an uncertainty of three-tenths of one percent, or 0.003, in the growth rate of a population has major implications, even over fairly short time horizons. The presence or absence of 10,000,000 people would have a significant impact on our social and economic systems. Indeed, much of the concern over the long-range funding of the Social Security programs results from uncertainties over birth rates.

Growth rates in some third world countries are well above those experienced by the United States and other industrialized nations. It is not uncommon to find growth rates of 3%

per year (and more) in developing nations. With a growth rate of 3%, a population of 100,000,000 in 1986 becomes a population of 151,258,970 by the year 2000, a 50% increase. If such growth rates were to be maintained, the population would double in about 24 years. Projections of this sort are at the root of worldwide concern over our ability to provide sufficient food for all people.

It is now time to think more carefully about the implications of using a geometric growth model to describe the growth of a population. Examining Figure 15.2, we see that for a positive growth rate r, the size of the population increases as time increases, at least for all times shown in the graph. In fact, no matter how long a time span we consider, the population continues to grow. More important, no matter how large a number we specify, the model predicts that at some time the population size will exceed that number.

Such predictions are clearly unreasonable for many situations. For instance, no biological population can continue to increase without limit. Its growth is eventually constrained by the availability of resources such as food, shelter, and psychological and social "space." As the population grows, it eventually reaches a level at which there are no resources for new

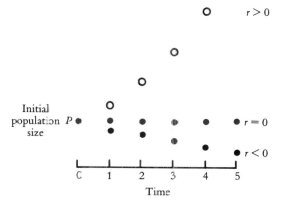

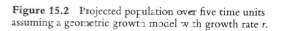

Figure 15.2 Projected population over five time units assuming a geometric growth model with growth rate r.

members. It is clear that a geometric growth model cannot describe forever the growth of such a population.

If a population of size P is growing according to a geometric growth model with growth rate r, then the amount that the population grows in one unit of time is Pr. Hence for a geometric growth model, the **relative growth rate**—the amount a population grows in one unit of time, per unit of population—is $(Pr)/P = r$.

Let's now think about the way actual biological populations behave. Their relative growth rate is likely to depend on the size of the population and may actually decrease as population size increases. For sufficiently large populations, the relative growth rate may be negative. How can we account for this decline in the growth rate? This could occur if as the population size increases, the resources per individual decrease, and thus the energy available for growth and reproduction decreases. There may be a maximum population size that can be supported by the available resources. Such a population size is called the **carrying capacity** of the environment. In one model for growth that takes into account the carrying capacity, we reduce the relative growth rate r by a factor that indicates how close the population size P is to the carrying capacity M:

Relative growth rate

$$= r\left(1 - \frac{\text{population size}}{\text{carrying capacity}}\right)$$

$$= r\left(1 - \frac{P}{M}\right)$$

This method is known as the *logistic model* of population growth.

Indeed, it is easy to see that this model has the properties we want. For small population sizes, that is, for values of P that are small relative to M, the quantity P/M is small; therefore $1 - P/M$ is close to 1 and the relative

growth rate is close to r. As the size of the population increases, the relative growth rate decreases—because the term containing the population P has a negative sign. For a population size equal to the carrying capacity, that is, when $P = M$, the relative growth rate is zero.

If at any time the population size exceeds the carrying capacity, as might happen with a significant decrease in the food supply, then the relative growth rate becomes negative (because $P > M$) and the population size decreases.

The logistic model provides excellent predictions for the growth of some populations, particularly in laboratory environments. For example, the graph in Figure 15.3 shows the growth of a population of fruit flies in a glass enclosure with a limited food supply. On the same coordinate system we show the predictions of an arithmetic population model, a

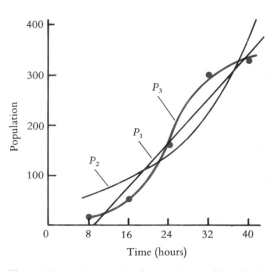

Figure 15.3 The growth of a population of fruit flies: P_1 is the best approximation by a simple growth model; P_2 is the best approximation by a geometric growth model, and P_3 is the best approximation by a logistic model. The red points show the actual values. [Adapted from Daniel Maki and Maynard Thompson, *Mathematical Models and Applications,* Prentice-Hall, 1973, p. 431.]

geometric population model, and a logistic population model where the parameters r and M of the equation have been selected to give the best fit to the data. As the positions of the data points show, predictions based on the logistic model come closest to the actual growth.

Harvesting Renewable Resources

A *renewable natural resource* is, as its name implies, a resource that tends to replenish itself if we allow it to evolve without intervention. Obvious examples are fish, wildlife, and forests. Petroleum and coal are examples of nonrenewable resources. We would like to predict how much of a resource we can harvest and still allow the resource to replenish itself.

We will concentrate on the subpopulation consisting of individuals that have a commercially significant harvest value. In the instance of a forest, the subpopulation might be trees of a commercially useful species and appropriate size. Because trees of different sizes have different commercial value, we will measure the size of the population in common units of equal value. For example, we measure a forest not by counting the trees but by estimating the number of board feet of usable timber. Similarly, we might measure the size of a fish population in terms of pounds rather than numbers of fish.

When a population is measured in this way — in common units of equal value — we say that we are considering the **biomass** of the population. We will measure the size of our population in terms of biomass. For simplicity, we suppose that the size of our population is measured once a year — a common practice with many naturally occurring populations.

We use a figure called a **reproduction curve,** which gives the population size of the next generation if the current population size is known. A typical reproduction curve is shown in Figure 15.4. The size of the popula-

tion in the current year is measured on the horizontal axis: let x be a typical value for this size. Then the size of the population the next year is given by the vertical distance between the horizontal axis and the reproduction curve. This value, when projected onto the vertical axis, is denoted by $f(x)$ or by y. The reproduction curve represents the total change in the population from one year to another; it therefore includes the growth of continuing members plus the addition of new members. Although the precise shape of the reproduction curve will vary from one species to another, reasonable biological conditions result in a curve of the general shape in Figure 15.4.

A few experiments with various choices for x show that whenever the reproduction curve is above the broken line inclined at a 45° angle with the horizontal axis, then the size of next year's population is larger than the size this year. Whenever the reproduction curve is below the broken line, then the population size next year is smaller than this year's. Note that the special population size labeled x_e, the population at which the reproduction curve crosses the broken line, is exactly the same

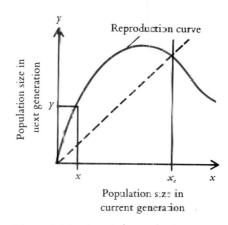

Figure 15.4   A typical reproduction curve.

next year as this year. The population size x_e is known as the **equilibrium,** or *steady state,* population size.

We introduce the term *yield* to refer to the amount harvested at each harvest. For the present discussion we will focus on a **sustained-yield** harvesting policy, that is, a harvesting policy that if continued indefinitely will maintain the same yield. Sustained-yield harvesting policies are obviously important to timber companies and other corporations that plan to use a natural resource over a long period of time. General problems of determining the best or optimal harvesting policy frequently have answers that are sustained-yield policies.

Under a sustained-yield harvesting policy, the population after each year's harvest will maintain the same size. To achieve this stability, the amount harvested must be exactly equal to the amount by which the population naturally increases each year. Now, because the population increases from x to $f(x)$ in one year, the amount of natural growth must be $f(x) - x$. For sustained yields, this must be

equal to h, the amount harvested. Thus, our constraint on harvesting policies leads us to the condition $f(x) - x = h$, or equivalently $x = f(x) - h$ (see Figure 15.5).

A common problem facing a timber company or a fishery is to determine a harvesting policy that results in the largest possible sustainable yield, or **maximum sustainable yield.** In terms of our diagram, the goal is to select x so that the sustainable harvest is as large as possible. The value of the population size that achieves this is shown as x_M in Figure 15.6. At this point, the length of the line between the broken line and the reproduction curve — the line representing the harvest — is as large as possible.

Let's now consider how the costs of harvesting may complicate this analysis. Consistent with our goal of keeping things as simple as possible, we will assume that although the cost of harvesting does not vary from one unit of the population to another, it does depend on the number of units harvested. This assumption incorporates the familiar principle of **economy of scale:** the cost of harvesting one

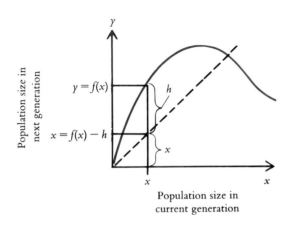

Figure 15.5 The relation between population size x and harvest h for a sustained-yield harvest policy.

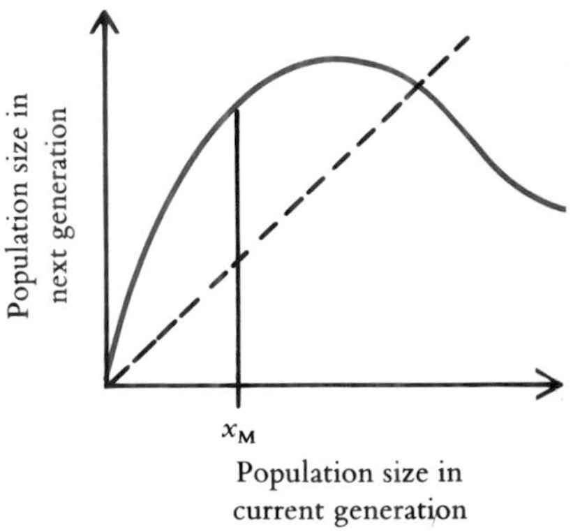

Figure 15.6 Determining the maximum sustainable yield, x_M, using a reproduction curve.

THE SIZE OF POPULATIONS

unit decreases as the size of the population increases. For example, a logger's costs of harvesting one tree of a certain size are less when that tree is harvested from a stand consisting of many harvestable trees than when the tree is harvested from a forest containing only a few such trees.

Figure 15.7 shows a curve that relates the cost of harvesting a single unit to the size of the population. The size of a population from which one unit is harvested is shown on the horizontal axis; the cost of harvesting a single unit is measured on the vertical axis. The fact that the curve slopes downward and to the right is a reflection of the basic principle that the cost of harvesting a single individual is less in a large population than in a small population.

We need some way to indicate symbolically the relation between our unit cost and population size. If our population is of size x, then the cost c of harvesting one unit depends on x. Our formula is therefore $c = g(x)$. The curve in Figure 15.7 gives us information on the relation between x and $g(x)$, in that it is the graph of the function g.

Now, if we harvest h units from a population of initial size x, then the associated cost is the sum of the cost of harvesting the first unit

$g(x)$; the cost of harvesting the next, this time from a population of size $x - 1$, $g(x - 1)$; the cost of harvesting the next, $g(x - 2)$; and so on, up to the cost of harvesting the hth unit. This total cost, which depends on x and h, will be denoted by G.

Let p denote the price we can obtain for one harvested unit. Then our net profit P from harvesting h units from a population consisting of x units is

$$\text{Net profit} = \text{value of harvested units} - \text{cost of harvest}$$

$$P = ph - G$$

$$= p(f(x) - x) - G$$

An optimal harvesting policy will depend on the relation between price and costs. There are two cases. First, if the price we receive for a harvested unit is less than the cost of harvesting that unit for all population sizes, then it is impossible to make a positive net profit, and the best we can do is to have a net profit of zero. This is achieved by harvesting no units from our population.

The second case is of greater interest: we have values of the population size for which the cost of harvesting one unit from the population is less than the price we receive for that unit. Now it is possible to generate a positive net profit. Moreover, there must be some population size, call it x_Q, that gives a maximum net profit. Because the unit costs of harvest decrease as the population size increases, the population size x_Q is actually larger than the population size x_M, the maximum sustainable yield, as shown on the curves in Figures 15.8a and 15.8b.

Notice that our solution of the net-profit-maximization problem results in maintaining the population. It is never in our best economic interest to harvest the entire population (see Boxes 15.5 and 15.6).

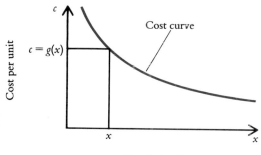

Figure 15.7 The cost of harvesting one unit from a population of size x.

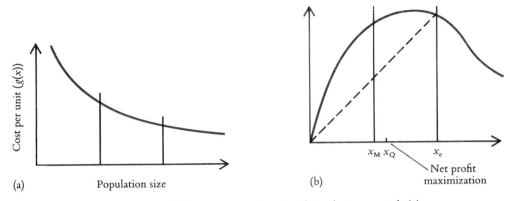

Figure 15.8   (a) The cost curve gives the relation between x and $g(x)$.
(b) The reproduction curve for the harvest cost in (a), showing x_Q.

BOX 15.5 Extinction of the Passenger Pigeon

Although once numbering in the billions, passenger pigeons are now extinct. [Courtesy, Field Museum of Natural History, Chicago.]

Historically, the utilization of a renewable resource has followed a characteristic pattern. First comes a stage of expanding harvests, perhaps based on a new use of the resource or on new harvesting technology. This is followed by concern for overutilization. Conservation measures are then adopted and the industry either stabilizes or collapses.

In some cases, the population has actually become extinct. For example, the passenger pigeon was once considered to be the world's most abundant land bird. Over one century ago, they numbered from three to five billion, traveling and nesting in huge flocks, mostly in eastern North America. But by 1914, the last remaining passenger pigeon had died at the Cincinnati Zoo.

The demise of the passenger pigeon can clearly be traced to expanding harvests, brought about by new technology—in this case, the development of the eastern railroad network and the telegraph. To understand how these developments were able to severely diminish such an abundant species, we can look at the ecological characteristics that were once key to the passenger pigeon's earlier success: colonization and nomadism.

Although they numbered in the billions, passenger pigeons were not a solitary species. They nested in colonies containing *millions* of pairs (the entire population consisted of perhaps fewer than a dozen flocks). The flocks were sometimes so immense that they were reported to have obscured the sun. (The largest flight ever recorded was estimated to contain 2.23 billion birds.) By traveling and nesting in such large groups, pigeons were able to literally "shield" themselves from predators. No matter where they nested, there were not enough local predators to significantly reduce their numbers. This concept is known as "predator satiation."

Passenger pigeons fed on large crops of nuts found in the deciduous forests of eastern (and occasionally midwestern) North America. Because the location of crops large enough to accommodate their numbers varied from year to year, passenger pigeons rarely nested in the same place two years in a row, with nesting sights ranging from New York and Pennsylvania to Michigan or Wisconsin. Thus, it was difficult to predict their location from one nesting season to the next.

By all accounts, the final decline of the passenger pigeon was rapid. The arrival of flocks of passenger pigeons had always meant food to the local people, but it was probably not until pigeons were actually harvested for *market* that the population began to markedly diminish. (Market harvesting began before 1800, but was not a major industry until 1840.)

While we can never know for certain, it is believed that the technological developments of the nineteenth century—namely, the railroad and the telegraph—increased the efficiency and scope of market harvesting to the point where it was ultimately responsible for the extinction of the passenger pigeon. By the time of the Civil War, the railroad network through America, east of the Mississippi, was complete. This network allowed the professional pigeoners, who numbered about 1000 in their heyday, rapid access to all major nesting colonies. It also provided a fast means of shipping barrels of pigeons to the big city markets in the east and midwest.

The telegraph was able to keep professional pigeoners informed of the locations of nesting colonies. In fact, the entire operation was organized so efficiently that word of any pigeon nestings spread rapidly for hundreds of miles. Since the railroads benefitted from the pigeon harvest, it is likely that they, too, helped to see that this information was transmitted.

The fact that passenger pigeons nested in gigantic colonies—which at one time, had assured their safety from predators—now made them especially accessible to harvesting for market. People did not understand that such an abundant resource could ever be severely diminished. They also did not allow for undisturbed nesting sites so that the pigeons could replenish their numbers. Instead, harvests at the nesting sights were so efficient and complete that there were no successful nesting colonies for a period of over 10 years. The last known colonial nesting attempt occurred in 1887 in Wisconsin, but the site was rapidly abandoned by the birds, probably because of disturbances.

If the harvest had *not* occurred at the nesting colonies, it is unlikely that the adult population could ever have been exterminated. Or, if only the adults had been harvested, the species might have survived. But because the fat nestlings were especially prized, and because many birds were driven away from nesting sites by the violent hunting methods sometimes used (shooting, setting trees on fire), *the adults could not replace themselves,* and the fate of the passenger pigeon was sealed.

Passenger pigeons were once a renewable resource, but within a period of about 20 years — twice an individual's lifetime — they became extinct. Certain other species, such as bison, have also been reduced to numbers below a level of economic significance.

BOX 15.6 Why Eliminate a Renewable Resource?

We might well ask why anyone would want to eliminate a renewable resource. In at least some instances, such as the case of the now extinct passenger pigeon, populations have been completely harvested. We can now apply our approach to see if we can understand why.

Sustained-yield policies operate under the assumption that some of the revenues will be received at a fairly distant future time. It is reasonable that the value of these revenues should be discounted to reflect the loss of income that would be earned if the funds were available for investment today. Using the appropriate financial term, we should consider the *present value* of revenues to be received in the future rather than their current value.

To be specific, if funds could be invested at an interest rate of r per year, the present value P of an amount A to be received n years in the future is related to A by the formula $A = P(1 + r)^n$. Hence $P = A/(1 + r)^n$. In this model, our goal is to maximize the sum of the present values of all future receipts of a sustained-yield harvesting policy. We refer to this sum as the present value of our return. The optimal harvesting policy will depend on the price p per unit harvested, the cost function G (or the equivalent g), and the interest rate r.

Again there are several cases to consider. In the first case, suppose that the cost of harvesting $g(x)$ exceeds the price for all population sizes x. Then it is impossible to have a positive net profit, and the best we can do is have a return of zero. The optimal policy in this case is to harvest nothing.

For the second case, suppose that there is a population size x' such that $g(x') = p$. Then there is a population size between x' and x_e (the steady state population size) for which the present value of the total return is maximized. In particular, the population that is maintained has a positive size — the population is not completely harvested.

For the third case, suppose that p is larger than $g(x)$ for all population sizes x. Then the conclusion depends on the value of the interest rate r. If r is small, then the situation is the same as in the second case. On the other hand, if r is large, then it may be that the optimal harvesting policy is to harvest the entire population immediately. This conclusion certainly corresponds to our intuition: if the price is high enough and if the proceeds can be invested at a sufficiently high rate of return (interest), then the most profitable course of action is to harvest everything and invest the proceeds.

REVIEW VOCABULARY

Biomass A measure of a population in common units of equal value.

Carrying capacity The maximum population size that can be supported by the available resources.

Compound interest The method of paying interest on both the principal amount and on the accumulated interest in an account.

Equilibrium The population size for which the size next year will be exactly the same as the size this year.

Maximum sustainable yield The largest sustainable yield.

Relative growth rate The amount a population grows in a unit of time, per unit of population.

Renewable natural resource A resource that tends to replenish itself if it is allowed to

evolve without external intervention; examples are fish, forests, wildlife.

Reproduction curve A curve that shows "population size in the next generation" plotted against "population size in the current generation."

Simple or arithmetic growth Any growth process that has the pattern of simple interest.

Simple interest The method of paying interest on only the principal amount in an account.

Steady state The population size for which the size next year will be exactly the same as the size this year.

Sustained yield A harvesting policy that can be continued indefinitely while maintaining the same yield.

Yield The amount harvested at each harvest.

EXERCISES

1. The population of Egypt was 48,503,000 in 1985, and the growth rate at that time was 2.5% per year.

a. If this growth rate continues until the end of the century, what will be the size of the population of Egypt at that time?

b. If the growth rate changes to 2% in 1990 and to 1.5% in 1995, what will be the size of the population of Egypt at the end of the century?

c. If the growth rate changes to 3% in 1990 and to 3.5% in 1995, what will be the size of the population at the end of the century?

2. Suppose that a country now has a population of 10,000,000.

a. If the population grows at a rate of 1% per year for 10 years and then grows at a rate of 4% for another 10 years, what is the final population size?

b. What is the final population size if the population grows at a rate of 4% per year for 10 years, and then grows at a rate of 1% per year for another 10 years?

c. Can you now propose a general conclusion about the effects on final population size of the order in which high and low growth years occur in a population?

3. In 1985 the population of Brazil was 135,564,000 and the growth rate was 2.26% per year. In the same year the population of the United States was 238,740,000 and the growth rate was 0.89% per year. If these growth rates were to continue into the future, when would the population of Brazil be the same as the population of the United States?

Hint: Begin by finding the population sizes at 10, 20, 30, 40, and 50 years in the future. If the population of Brazil is less than that of the United States at time T_1 and greater than that of the United States at T_2, then the two populations must be equal at some time between T_1 and T_2.

4. Here is a rule of thumb for finding the time it takes a population to double: If r is the rate of increase measured in percent, then the time t for a population to double is approximately $72/r$. Verify this rule for growth rates of 5%, 6%, 7%, 8%, 9%, and 10%.

5. Here is another example of geometric growth: You take a job for two weeks (10 days), and you have the option of being paid $100 per day or $1 for the first day, $2 for the second day, $4 for the third day, and so on (your pay each day is twice that of the preceding day). Which method of payment will give you the largest total income for the two-week period?

6. Parents commonly use savings accounts to accumulate funds for the college education of their children. Frequently the funds are deposited over a period of several years, and the funds together with accumulated interest are used to pay the costs of education.

Suppose that $1000 is to be deposited each year into an account that pays interest at a rate of 8% per year compounded annually. If the first payment is made when a child is 2 years old and the last is made when the child is 17 (16 payments), how much is available when the child begins college at the age of 18?

Hint: For any real number x and any positive integer n, the following relation holds:

$$1 + x + x^2 + x^3 + \cdots + x^n = \frac{x^{(n+1)} - 1}{x - 1}$$

In this formula the dots . . . mean that all powers of x up to the nth (x^n) are included in the sum on the left-hand side.

7. (Continuation of 6) A recent (1986) advertisement reads as follows: "If you had put $100 per month in this fund starting in 1976, you'd have $37,747 today." How much money was deposited during this period? What rate of interest (yearly rate, compounded monthly) would lead to the result described in the advertisement?

8. In some cases it is important to look at problems that are in a sense the reverse of those discussed in this chapter. In a financial setting such a question might be: How much do you need to deposit today in an account that pays interest at a known rate in order to have a specified amount at a specified time in the future? This question is crucial in certain financial planning considerations, for instance in planning for a major purchase in the future.

The question is easy to answer. Indeed, suppose that we ask for a specified amount, say A dollars, at a specified time in the future, say n years, and we know that the account pays interest at the rate r per year. The unknown quantity — namely, the amount that must be deposited today — will be denoted by P. Using our basic formula, we know that A, P, r, and n are related by $P = A/(1 + r)^n$. The quantity P is known as the *present value* of an amount A to be paid n years in the future.

    a. Suppose that you will need $15,000 to pay for a year of college 8 years in the future, and you can buy a certificate of deposit whose interest rate of 10% compounded quarterly is guaranteed for that period. How much do you need to deposit?

    b. If the interest rate drops to 8%, how much do you need to deposit?

c. If the interest rate increases to 12%, how much do you need to deposit?

9. The situation described in Exercise 8 gives the rationale for the pricing of the so-called zero coupon bonds. These securities pay no current interest but are sold at a substantial discount from redemption value. The difference between purchase price and redemption value provides, at the time of redemption or resale, income to the bond-holder. If the interest rate in the economy is now 7%, what should be the price of a zero coupon bond that will pay $10,000 eight years from now?

10. Kenya was one of the most rapidly growing nations in the world in 1985, with an estimated growth rate of over 4% per year. The populations of Kenya for the decades 1950 through 1980 are as shown in the table below. Projected population sizes for 1990 and 2000 are also shown. What assumptions about the rate of growth of the Kenya population lead to these projected population sizes?

Year	Population size
1950	6,018,000
1960	8,115,000
1970	11,225,000
1980	15,667,000
1990	25,000,000 (projected)
2000	38,500,000 (projected)

11. Suppose that a population of size P grows by an amount $Pk(1 - P/100)$ between observations. Here it is reasonable to view the parameter k as an intrinsic growth rate (the rate of population growth without resource constraints) and the number 100 as a carrying capacity of the environment.

a. If the population is initially at size 10 and $k = 0.8$, find the size of the population at the first 10 observations.

b. Answer the question in **a** if the population is initially of size 110. What differences do you observe between the results of **a** and **b**?

c. Answer the questions in **a** and **b** with $k = 1.8$.

d. On the basis of your analysis of these situations, what can you say about the dependence of the population growth on the parameter k?

12. Suppose that a population of size P grows by the amount $Pk(1 - P/M)$ between observations, where k is an intrinsic growth rate and M is a carrying capacity of the environment. Suppose also that the carrying capacity is 100 for the first 5 observations and then drops to 70. (There is an environmental catastrophe at that time; for instance, a flood wipes out much of the food supply). If the initial population is of size 20 and $k = 0.9$, find the population sizes for the first 10 observations.

13. Suppose that a population of size P grows by the amount $Pk(1 - P/M)$ between observations, where k is an intrinsic growth rate and M is a carrying capacity of the environment. Suppose also that the carrying capacity M is increasing steadily. This might be the case if, for example, the food supply were increasing steadily. Suppose that at the nth observation the carrying capacity is $100 + 5n$.

a. If the initial population is of size 10 and $k = 0.7$, find the population sizes for the first 10 observations.

b. If the initial population is of size 10 and $k = 2$, find the population sizes for the first 10 observations.

c. What do you observe? Describe what is happening in terms of the setting of the problem.

14. Reproduction curves for two populations are shown in graphs (a) and (b) below. Determine the equilibrium population sizes and the maximum sustainable yields for each of these populations.

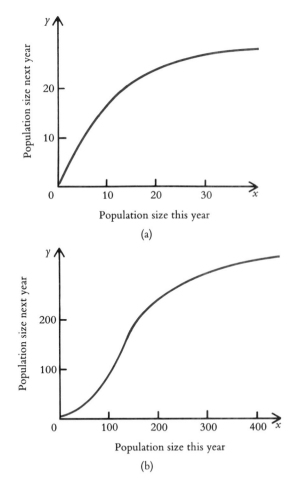

(a)

(b)

15. Using the notation and terminology developed in the discussion of the cost of harvesting, find a formula for the cost $G(x, h)$ of harvesting h individuals from a population of size x.

16. Suppose the cost of harvesting function g is $c = 200/(10 + 5x)$. That is, the cost of harvesting one individual from a population of size x is $200/(10 + 5x)$. Find the cost of harvesting 5 individuals from a population of size 20. Find the cost from a population of size 80.

17. Suppose that the curve shown below gives the cost of harvesting an individual from a population of size x. Use this graph to estimate the cost of harvesting 5 individuals from a population of size 50. Find the cost from a population of size 80.

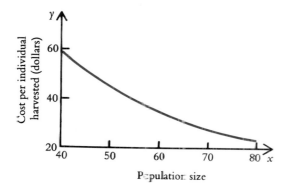

18. Suppose that the reproduction curve for a population is as shown in graph (a) below and that the cost-of-harvesting function is as shown in graph (b). The price obtained for one harvested individual is $8. Find the net yearly revenue when a sustained-yield harvesting policy is used that maintains a population of size 20.

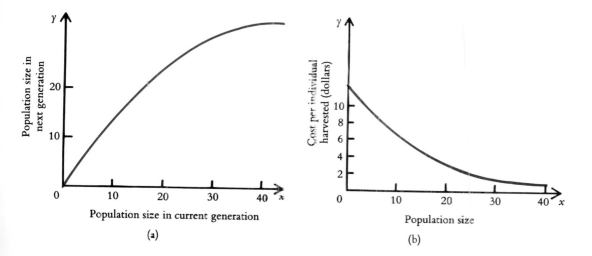

(a)   (b)

19. Solve the same problem as in Exercise 18 in the case where the price for one harvested individual is $5 and the population is maintained at size 15.

20. Discuss the question posed in Exercise 19 when the price for a single harvested individual is $4 and the population is maintained at size 15.

16

Measurement

Before 1600, people had nothing more than their eyes and sighting rods with which to see the universe. Nevertheless, they made precise measurements of the size of objects on the earth, the size of the earth itself, and the distances to the moon and sun. The mathematics used to determine these distances is the geometry Euclid set down in his *Elements,* written sometime around 300 BC and consisting of 13 chapter-long "books" (see Box 16.1). Although *The Elements* organized all the geometrical and arithmetical knowledge accumulated by the Greeks, thus incorporating earlier knowledge from Babylonia and Egypt as well, its greatest achievement was to show a natural sequence by which one result could be derived logically from another. For example, Thales (c. 600 BC) is said to have discovered that every angle inscribed in a semicircle is a right angle (see Box 16.2). We don't know how — or in fact whether — he proved that this is so, but in *The Elements* Euclid shows just how it can be derived from much simpler facts.

The basic ideas of *congruence* and *similarity* are developed in *The Elements.* Two triangles are congruent if one is a copy of the other. One congruent triangle can be made to fit exactly onto the other. Two triangles are similar if they have the same shape, but not necessarily the same size. If a triangle is enlarged or reduced by a photocopy machine, then the resulting triangle is similar to the original triangle. Thus, corresponding angles of similar triangles have the same size, and the lengths of corresponding sides are in the same ratio.

Book I of *The Elements* contains most of the standard facts about congruence for triangles. These facts are followed by the introduction of the famous *parallel postulate,* which proved to be so important in the history of ideas and paved the way for Einstein's theory of relativity and other modern theories about the struc-

BOX 16.1 Euclid's *Elements*

Among the many editions of Euclid, one of the most interesting is that written by Oliver Byrne and titled *The First Six Books of the Elements of Euclid in which Coloured Diagrams and Symbols are Used Instead of Letters for the Greater Ease of the Learner.* Byrne's Euclid was published in 1847 by William Pickering of London. Below is Byrne's colored proof of the Pythagorean theorem. Byrne was Surveyor of Her Majesty's Settlements in the Falkland Islands and author of numerous mathematical works. (Proof is from the library of Professor G. L. Alexanderson, University of Santa Clara.)

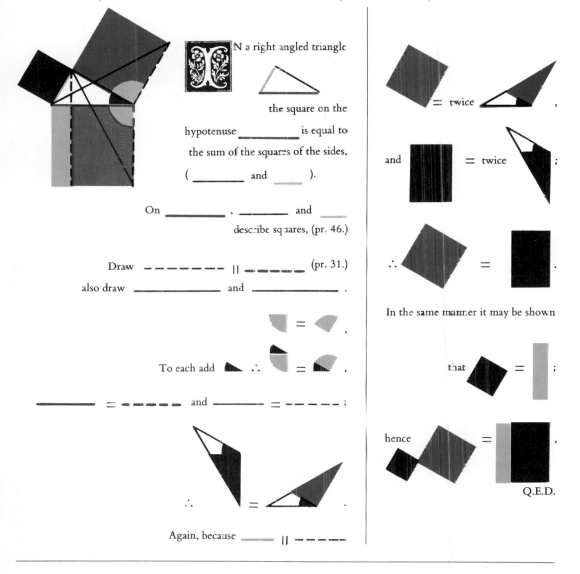

BOX 16.2 Thales (c. 600 BC)

We know very little for certain about Thales, who lived 2600 years ago. He is usually considered the founder of Greek philosophy and is one of the "seven sages" honored by the Greeks and Romans. He taught that in spite of the vast differences in the appearance of things, there is an underlying unity in the world. He also believed that everything originates in water. According to an often-repeated anecdote, Thales once fell into a well while looking at the stars, and a servant girl laughed at him, saying that he wanted to know what happens in the heavens but couldn't even keep track of his own feet.

However, if the ancient stories are true, he was also a versatile and practical man. He is said to have predicted the year in which a solar eclipse turned "day into night" during a battle of 585 BC. He also invested in olive oil at a time when he knew of a forthcoming shortage, thereby making a great deal of money.

Thales was not the first to unravel the intricacies of mathematics and astronomy. For example, his prediction of an eclipse could not have been accomplished without the knowledge, observations, and geometric facts known before him in Egypt and Babylon. He probably began a systematic organization of these teachings just as Euclid later organized knowledge in *The Elements.*

---

ture of the universe. Book I culminates with the *Pythagorean theorem,* which can be used to calculate distances along a slant when the corresponding horizontal and vertical distances are known. This theorem is the basis of the standard ways of determining distance between two points on a straight line or on a curve, and hence for all of analytic geometry. It is in turn the necessary preliminary for calculus, the tool that Newton invented for understanding the motions of the planets and of falling apples (see Box 16.3).

In later books of *The Elements,* the standard theory of similarity appears, which is the basis for all map making and most other methods of calculating and representing inaccessible distances, including the tools of trigonometry. In the remainder of this chapter we will show how these simple ancient tools, which are still routinely taught to all engineers, have shaped the way we view our universe.

Estimating Inaccessible Distances

Our story concerns four men: Thales, Euclid, Aristarchus, and Eratosthenes, each of whom developed new techniques for measuring ever more distant objects. The last three lived about 300 BC and were probably born in the order they are named. Some 300 years earlier, Thales is supposed to have made two difficult measurements: (1) the distance of ships at sea, using *congruence* of triangles, and (2) the height of the Great Pyramid in Egypt, using *similarity* of triangles.

Suppose you want to find the distance of a ship at position B straight out at sea from your position A on the shore, using the method attributed to Thales (see Figures 16.1 and 16.2). Starting from A, walk along the shore any distance in a direction perpendicular to AB. Put a marker at S, making sure it is tall enough to see from a distance. Then walk the

BOX 16.3 The Pythagorean Theorem

The Pythagorean theorem states that for any right triangle, the sum of the squares on the two short sides, or legs, is equal to the square on the long side, the hypotenuse. That is, S_1 and S_2 together have the same area as S_3.

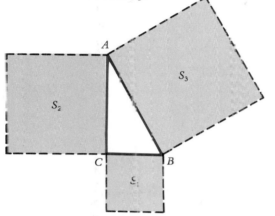

One simple way to prove this theorem is to draw the squares S_1 and S_2 in this way:

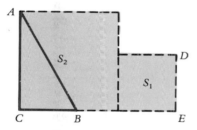

If you now cut the joined squares S_1 and S_2 into three pieces by a cut at AB and a cut at BD, you can move piece ABC (T) and piece BDE (S) to fill up square S_3 exactly:

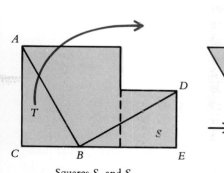

Squares S_1 and S_2
cut into three pieces

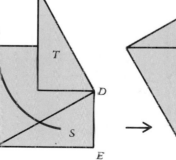

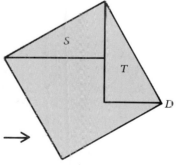

Square S_3, formed
from the same three pieces

Not only do the areas of S_1 and S_2 add up to the area of S_3, but S_1 and S_2 can actually be cut into (a very few) pieces that can be reassembled to make square S_3. You can make these pieces out of cardboard; it is surprisingly difficult to take the three pieces into which the joined S_1 and S_2 are cut and reassemble them to make a single square.

The fundamental role of the Pythagorean theorem is illustrated by the following straightforward problem: In a city laid out on a grid of east-west and north-south streets, a house B is known to be four blocks south and three blocks east of another house, A. What is the straight-line distance from A to B?

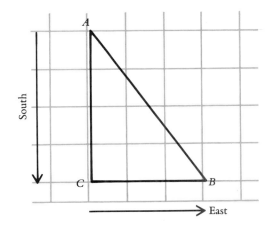

Here is the answer: because $AC = 4$ and $BC = 3$, by the Pythagorean theorem we know that $(AB)^2 = (AC)^2 + (BC)^2 = 4^2 + 3^2 = 16 + 9 = 25$. If $(AB)^2 = 25$, we know immediately that $AB = 5$. The Pythagorean theorem gives us a way to calculate the slanted distance when its horizontal and vertical components are known.

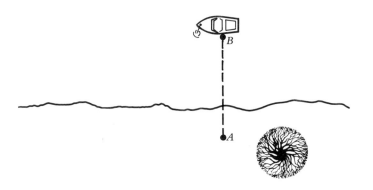

Figure 16.1 *Problem:* Find the distance of the boat from shore using elementary geometry.

same distance to point C. Now turn at a right angle and walk away from the shore until you reach a point E from which your marker S is exactly lined up with the ship B. The distance CE you walked away from the shore is exactly the same as the distance AB of the ship from shore.

Figure 16.2 shows why this works. The simple geometry that Thales used, long before Euclid, tells us that the two angles at S are equal: opposite angles are equal. (If the sides of an angle are extended through its vertex, then the other angle that is formed is called the *opposite angle.*) The angles at *A* and *C* are equal because both are right angles. The sides *AS* and *SC* are equal because they were paced off to be equal. By one of the **congruence theorems** for triangles, we know that triangle *ABS* is congruent to triangle *CES* and hence that the corresponding lengths *AB* and *CE* are equal, as we claimed. The congruence theorem that we use here says that two triangles are congruent if two angles and the adjacent side of one are equal to two angles and the adjacent side of the other.

Two figures are said to be **similar** if they have the same shape, but not necessarily the same size. For example, a photograph and its enlargement are similar to each other and to the original (see Chapter 14). In the special case of triangles, Thales may have known that if right triangles had equal angles, they would be similar, with proportional sides. With this knowledge he could calculate the height of the Great Pyramid.

To take the height of the Great Pyramid (or any other vertical object such as a tall tree), hold an upright stick on the ground at the site of the object and measure its length and the length of its shadow. The right triangle whose legs are the stick and the shadow line is similar to the right triangle whose legs are any other vertical object and its shadow at the same place and time. Hence the ratio of these two lengths, stick and shadow, is the same as the ratio of the length (height) of the Great Pyramid to the length of its shadow (see Figure 16.3). Suppose, to be specific, that in this case the length of the stick is 10 feet and it casts a 16-foot shadow. Then Thales's measurement of the shadow of the Great Pyramid would have been 770 feet. From the equality of these ratios, $\frac{10}{16} = h/770$, he could immediately calculate the height *h* of the Great Pyramid to be $(\frac{10}{16}) \times$ 770, or 481 feet.

Notice that ratios (such as $\frac{10}{16}$) of the height of the stick to the length of its shadow are the key to finding all sorts of inaccessible heights.

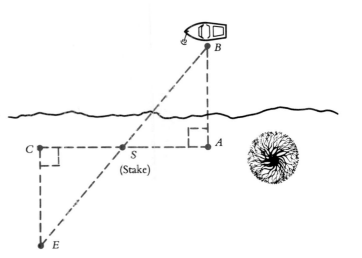

Figure 16.2 *Solution:* Thales found the distance of the boat from shore by putting a stake at *S* and then showing that triangles *SAB* and *SCE* are congruent.

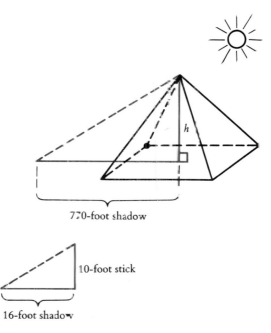

770-foot shadow

10-foot stick

16-foot shadow

Figure 16.3 Thales used similar triangles to determine the height of the Great Pyramid. Since the shadow of the Great Pyramid is 48.1 times the shadow of the stick, then its height must be 48.1 times the height of the stick, 481 feet.

In fact, if the ratio of stick to shadow is a/b and the length of shadow of any object is s, we can find the height of that object simply by multiplying the two: $h = (a/b) \times s$.

In later times, extensive tables of such ratios showed, for any given angle of inclination of the sun, what the ratio of the length of the stick to its shadow would be (see Table 16.1).

With such a table, the heights of inaccessible objects can be calculated without the use of the stick if a suitable instrument, such as a sextant, is available for measuring the angle of the sun above the horizon. Table 16.1 immediately tells us, for each angle of the sun, the ratio for

$$\frac{\text{Length of stick}}{\text{Length of its shadow}}$$

To summarize, the procedure for measuring height is:

1. Measure the length s of the shadow of the object whose height you want to determine.

2. Measure the angle of the sun above the horizon, and look up the corresponding ratio r in the table.

3. Multiply r times s to find the height of the object.

This procedure was the beginning of trigonometry (triangle measurement) as we know it. The ratio we have tabulated is now called the *tangent* of the angle of inclination. We can verify the entries in the table by using a pocket "scientific" calculator — one with buttons for the "trig" functions, sin, cos, and tan (= tangent). Try this on the calculator: hit the "tan" button for various angles and check that you get the same values given in Table 16.1. On some calculators there may be slight round-off errors, so you may get tan 45 = .99999 instead of tan 45 = 1.00000. Essentially, however, you should be able to reproduce the table and even fill in the omitted entries if you wish.

Digging Straight Tunnels

The famous Greek historian Herodotus, who lived some 100 years after Thales, described three engineering achievements that had occurred on the Greek island of Samos. One was a tunnel that brought water through Mount Castro to the capital city, Samos.

Nearly 2500 years later, in 1882, archeologists rediscovered the tunnel, exactly as Herodotus had described it. It was 1 kilometer (about 0.6 mile) in length and more than 2 meters (about 6 feet) high and wide. A deep ditch in its floor contained pipes, and the tunnel had vertical vents for changing the air and cleaning away rubble, and niches where workers placed their lamps. The ditch had a depth of some 2 meters at the upper end and 8 meters at the lower end, and was probably dug

TABLE 16.1 Table of tangents

Angle of the sun above the horizon	Tangent of the angle $= \dfrac{\text{length of stick}}{\text{length of its shadow}}$
5° (Sun nearly on the horizon)	.08749
10° (Sun somewhat higher)	.17633
20°	.36397
30°	.57735
40°	.83910
45° (Sun exactly halfway between horizon and directly overhead)	1.00000
50°	1.19175
60°	1.73205
70°	2.74748
80°	5.67128
87°	19.08114
89° (Sun almost directly overhead; higher than it ever gets in the United States)	57.29000

because the drop that had originally been planned turned out to be too small.

The remarkable thing about this tunnel was that the digging teams, proceeding from each end, met at the center with an error of only 10 meters (33 feet) horizontally and 3 meters (10 feet) vertically. We know this because at the center of the tunnel there is a jog of that size to make the two ends meet.

King Hezekiah of Judea was less successful. When he had a similar aqueduct constructed through the rocks near Jerusalem around 700 BC, his workers had to check the direction of digging in a very primitive way, by means of vertical shafts from the top. The result was a zigzag tunnel twice as long as the distance between its ends.

How was the Samos tunnel dug without the benefit of guiding shafts? We do not know for sure, but a later writer, Heron, described a likely method, which we modify slightly here to bring out the essentials. In his view, the method used similar triangles in a considerably more complicated way than Thales had used them. We will describe this method in detail, following Figure 16.4.

Suppose the tunnel entrances are to be at A and B, on opposite sides of Mount Castro in Figure 16.4. Begin by marking off some convenient distance BE on any line at B. Following the figure, at E make a right turn and go to F. At F turn again and go to G, then turn again and go to H, which is chosen so that a right turn takes you straight to A. Suppose, to be specific, that the distances in this detour around the mountain are $BE = 750$ meters, $EF = 1000$ meters, $FG = 2000$ meters, $GH = 800$ meters, and $HA = 250$ meters. For the right triangle ABC, which is underneath the mountain and therefore not directly accessible, we know, by subtraction, that $AC = 200$ meters and $BC = 1000$ meters. Any right triangle with short sides in this same ratio (5:1) will be *similar* to ABC, thus, its angles will be equal to those of ABC.

To find the direction to dig, we construct triangles BOP (say, $BO = 50$ meters, $OP = 10$ meters) and AQR (say, $AQ = 50$ meters, $QR = 10$ meters) outside the proposed entrances A and B to the tunnel. (These are not drawn to scale in Figure 16.4.) The angle OBP is the same as the angle CBA, which tells us that direction PB is the direction to dig. Similarly, at the other end, RA is the direction to dig. Thus, the clever use of similar triangles over 2600 years ago helped to solve a major problem of civil engineering.

Measuring the Earth

The modern use of similar triangles in engineering projects also depends on similarity principles. Our next example makes a big jump in the gradually increasing scale of dis-

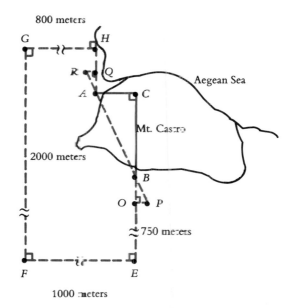

Figure 16.4   The plan for digging a tunnel through Mount Castro on the island of Samos. Using this plan, two digging teams starting at opposite ends A and B met at the center with only a small error.

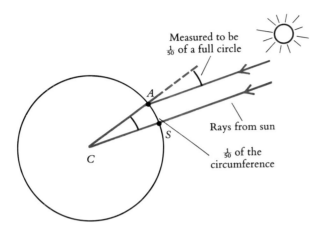

Figure 16.5 Eratosthenes noticed that the sun is about 7° (one-fiftieth of a circle) south of the zenith at Alexandria when it is directly overhead at Syene. Thus the distance between Alexandria and Syene must be about one-fiftieth of the earth's circumference.

tances: here we learn how to find the size of the earth itself. One of the truly spectacular achievements of ancient mathematical science was the determination, by a very simple method, of the circumference of the earth. The most accurate of these calculations was that of Eratosthenes in about 200 BC. His method is described in Figure 16.5.

It was known that at a certain time the sun was directly overhead at a place called Syene, point S, in Egypt. At exactly the same time in Alexandria, lying straight north of Syene at point A, the position of the sun was measured to be $\frac{1}{50}$ of a full circle (that is, 7.2°) away from directly overhead. Because the sun is so far away from the earth, the two arrows in the figure that point to the sun are essentially parallel lines. Hence, the angle at C, the center of the earth, is also $\frac{1}{50}$ of a full circle because it is the *alternate angle* when the two parallel lines are cut by the transversal AC.

Then, because the angle at C is $\frac{1}{50}$ of 360 , the full circle, the distance AS, from Alexandria to Syene, is also $\frac{1}{50}$ of the complete circumference of the earth. It is only necessary to measure the distance from Alexandria to Syene

(not a triviality in those days!) to have all the information needed. When the distance from Alexandria to Syene was found to be 5000 *stades,* this yielded $5000 \times 50 = 250,000$ *stades* for the circumference of the earth.

Although we are not sure how large the *stade* unit was, one estimate from Pliny is that a *stade* was 157.5 meters. Using this value, we get $157.5 \times 250,000$ meters, or 157.5×250 kilometers, for the circumference of the earth. This number is 39,375 kilometers. The kilometer was originally defined as $\frac{1}{10,000}$ of the distance from the north pole to the equator— one-quarter of the earth's circumference—so that its total circumference is 40,000 kilometers or 24,800 miles. We see that Eratosthenes' result is nearly on the mark. We can be excused for thinking that some of Eratosthenes' numbers seem to be rounded off, and hence only *accidentally* accurate. Even so, we must admire his achievement. After all, in later years there was even some doubt that the earth was round!

Measuring Astronomical Distances

Knowing the circumference of the earth (and hence its radius), the astronomer Aristarchus could consider even greater distances. His measurements of the distances of the moon and the sun from the earth were not as accurate as Eratosthenes' determination of the size of the earth, but his ingenious method is worth looking at. It shows how even a very simple understanding of triangle and circle geometry yielded information that completely revised his contemporaries' picture of the universe; they had imagined celestial distances to be much smaller than he showed them to be.

Using simple geometry, Aristarchus first determined how much farther the sun is than the moon from the earth. He noticed that when the moon is exactly half full, that is,

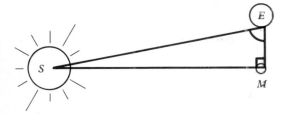

Figure 16.6 Aristarchus' method for estimating the ratio of the distances of the sun and moon from the earth.

when we see exactly half the moon in shadow and half in the sun's light, the triangle *MES* formed by the moon, earth, and sun is a right triangle, with its 90° angle at the moon, *M* (see Figure 16.6). (The two days in each month when the moon is exactly half full are marked on many modern calendars.)

By measuring the angle at *E*, we would be able to read off the ratio *MS/EM* from the tangents in Table 16.1. (We actually want the ratio *ES/EM*, which for such large distances is only slightly different. Another table, the "cosine" table, would give the ratio *EM/ES* if we wanted it.) Aristarchus tells us that angle *E* is 3° less than a right angle, that is, $E = 87°$. From the table we see that $\tan 87° = 19$; the sun is therefore 19 times as far from the earth as the moon is.

Of course, neither Thales nor Aristarchus actually had a table of tangents, but they were able to find these strictly geometric ratios in other, more complicated ways. Moreover, it is likely that Aristarchus realized that he had measured the very difficult, crucial angle *E* inaccurately. Perhaps he only made an intelligent estimate. A glance at the table shows that for large angles *E*, a small error makes a very large difference in the resulting ratio. The true value of *E* differs from 90° by less than one-sixth of a degree, so that *E* is more than $89\frac{5}{6}°$, and tan *E* is about 390, the true ratio of the distance of the sun to the distance of the moon from the earth. Even though Aristarchus' calculation was off by a factor of 20, his method

was sound, and his results revised upward the Greek estimates of the size of the universe by an enormous amount.

You will have noticed that Aristarchus' simple method gave only a *ratio*. To determine the actual distance to the sun, he needed to know the actual distance to the moon. By observing the time it takes for the shadow of the earth to cross the moon during a total eclipse we can estimate this distance very accurately. Although Hipparchus, some 100 years after Aristarchus, used this method to come within 1% of the value we know today, the rougher estimates already available to Aristarchus were sufficiently accurate for his purposes.

Before describing this method, we need to remind ourselves of two simple facts from Euclidean geometry. In Euclid's treatise, facts about *congruence* appear at the very beginning. It is only after Euclid introduces properties of parallel lines that he is able to prove the fundamental result that the sum of the angles of a triangle is a *straight angle,* or 180° (half of a complete circle).

This property of triangles—that all of them have the same angle sum, which is 180°—is one that dramatically distinguishes Euclid's geometry from **non-Euclidean geometries:** the *spherical geometry* of the surface of the earth and *hyperbolic geometry.* (Chapter 17 treats non-Euclidean geometries in greater detail.) In spherical geometry, the angle sum in a triangle is always greater than 180°, and it is not the same for all triangles. We find triangles whose angle sum is 190° as well as triangles whose angle sum is 250°. However, any two triangles having the same area do have the same angle sum (see Box 16.4). In hyperbolic geometry (the geometry used by the artist M. C. Escher in his "Circle Limit" prints), the angle sum in a triangle is always less than 180° but is not the same for all triangles (see Box 16.5).

A simple property of circles is also used in the derivation, namely, that the arc of a circle subtended by, or opposite, an angle at its center

BOX 16.4 Angle Sums in a Triangle

The following simple proof shows that the angle sum in a triangle is equal to a straight angle. We will use triangle *ABC* as an example. Our goal will be to prove that $A + B + C$ = a straight angle.

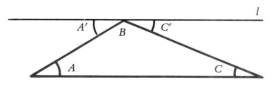

The Proof

At vertex *B*, draw a line *l* parallel to side *AC*. Then, by one of the first properties of parallels, which states that when parallel lines are cut by a transversal, the alternate interior angles are equal, angle A = angle A'. Likewise, angle C = angle C'.

Now clearly $A' + B + C'$ = a straight angle. Hence, by substituting A for A' and C for C',

$$A + B + C = \text{a straight angle}$$

which is what we set out to prove.

We can't use this same proof on the sphere because it has no parallels at all. On the sphere every two straight-line paths, or great circle routes, eventually cross. In fact, on the sphere the sum of the angles of a spherical triangle is always *greater* than a straight angle.

On the other hand, in the plane of hyperbolic geometry there are "too many" parallels, and when the line *l* is drawn at *B* so that angle A' = angle A, then angle C is always *less* than angle C'. Hence, the angle sum $A + B + C$ in triangle *ABC* is always less than the straight angle $A' + B + C'$.

BOX 16.5 Angels and Devils

The Dutch artist M. C. Escher (1898–1972) was particularly interested in the *metamorphoses* of figures that change almost imperceptibly into other figures or into larger or smaller versions of themselves. He was able to discover a way to draw inside a circle or square figures that gradually get larger as they approach the outside of the enclosure. But it wasn't until he was shown a mathematician's representation of hyperbolic geometry (in which the sum of the angles of a triangle is always *less* than 180°) that he discovered how to make figures gradually get *smaller* toward the outside of a circle.

Recently, mathematician Douglas Dunham devised a computer program based on hyperbolic geometry that can produce an infinite variety of the type of drawings Escher so ingeniously drew. One of these, "Angels and Devils," is shown here.

The "straight lines" of this geometry are arcs of circles that are perpendicular to the outside circle. (The "straight lines" of spherical geometry are great-circle routes on the sphere.) This particular print is based on a regular tiling of the hyperbolic plane. The tiles shown here are regular quadrilaterals — their vertices are the points where the feet of three angels meet the feet of three devils. Six of these tiles meet at each vertex. Thus, the angle of each is 60° instead of 90°, which it would be for the corresponding tiling of the Euclidean plane, where four squares meet at each vertex.

The edges of some of the tiles (of the underlying tiling) have been drawn in so that they can be seen, and two are shaded. Although the tiles get smaller (in the Euclidean sense), toward the edge of the outside circle, all the tiles, including the two that are shaded, are *congruent* within the hyperbolic geometry itself.

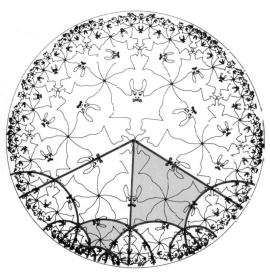

Douglas Dunham's "Circle Limit IV" plot. This computer-generated tiling in the hyperbolic plane creates an image very similar to that of M. C. Escher's "Angels and Devils." Note how the positions of feet and heads are related by radii and arcs that define the underlying tiling pattern. [Courtesy of Douglas Dunham.]

M. C. Escher's "Angels and Devils" is based on a tiling of the hyperbolic plane. [Escher Foundation, Haags Gemeentenmuseum, The Hague. © M. C. Escher Heirs, c/o Cordon Arts — Baarn, Holland.]

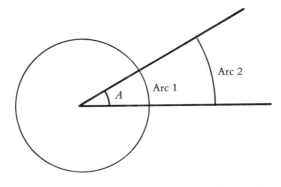

Figure 16.7   Arc 2 is twice as long as arc 1 because its circle has twice the radius of the other circle.

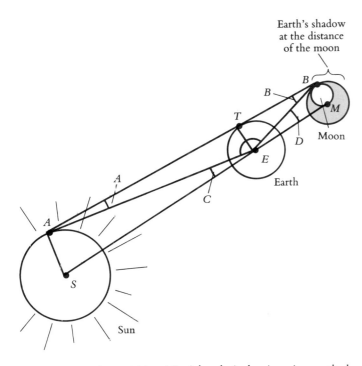

Figure 16.8   Aristarchus devised an ingenious method for determining the ratio between the moon's distance and the radius of the earth from the average duration of a lunar eclipse and the length of the month. His method resulted in an estimate for the moon's distance of 80 earth radii.

is proportional to the radius of the circle. In Figure 16.7, the arc subtended by angle A on the large circle is twice as long as the arc on the small circle because the radius of the large circle is twice that of the small one. When we measure the angle in degrees, the factor we have to multiply by to get the arc length is $\pi/180$ (about $\frac{3}{180}$, or $\frac{1}{60}$). Thus, for example, if a circle has a 10-foot radius, an angle of $90°$ at its center subtends an arc of $(\pi/180) \times 90 \times 10$, or approximately 15 feet.

Let's return to our main story: Aristarchus' measurement of the distance of the sun and the moon from the earth. Using the fact that the angle sum in a triangle is equal to a straight angle and that the arc subtended by a central angle of a circle is $\pi/180$ times the radius times the angle, we can understand the following calculation of the distance of the moon from the earth. We here combine Aristarchus' own method, a similar method used by Hipparchus a century later, and modern notation.

By observing the amount of time it takes the earth's shadow to cross its moon, Aristarchus knew that the diameter of this shadow was about two times the diameter of the moon, as shown in Figure 16.8.

Because the moon's and sun's discs are both about the same size in the sky and because both are about $\frac{1}{720}$ of the whole circumference of the circle they trace through the sky, the angles at C and D are easily found: $C = \frac{1}{1440}$ and $D = \frac{1}{720}$ of a complete circle. Hence $C + D = \frac{1}{1440} + \frac{1}{720} = \frac{1}{480}$ of a complete circle $= 0.75°$. Now, $A + B + E = C + D + E$ because $A + B + E$ is the angle sum in triangle AEB and hence is equal to the straight angle $C + D + E$. Because $A + B$ is equal to $C + D$, we know it is also $0.75°$. Moreover, angle A is very small compared to B (since the sun is much farther away than the moon), so angle B itself is approximately $0.75°$.

For angles as small as B, segment TE is essentially equal to the arc TE of the circle with center at B having radius BT. That is, because the arc TE corresponding to angle B is proportional to the radius BT, we can write

$$TE = \frac{\pi}{180} \times BT \times B$$

Solving for angle B, we have

$$B = \frac{180}{\pi} \times \frac{TE}{BT}$$

Substituting $B \approx 0.75$ and rearranging, we get

$$BT \approx \frac{180}{\pi} \times \frac{TE}{0.75}$$

$$\approx 80 \times TE$$

That is, the distance from earth to moon, by this simple calculation, is about 80 earth radii. In fact, the distance is about 60 earth radii, a result that we get from only slightly more refined versions of this same calculation.

Once the distance to the moon has been calculated, the ratio of the sun's distance to the moon's distance (which Aristarchus thought to be about 19, but which is in fact nearly 400) yields the distance to the sun in earth radii.

Combined with the still earlier calculation of the earth's radius, this gives the distance to the sun, a value which we now know to be 93,000,000 miles. Using the same elementary ideas, we can calculate the distances to and between other planets in the solar system.

In summary, a few simple tools of elementary geometry, combined with imaginative questions and insights about the possible structure of the universe, led to the calculation of inaccessible distances long before the development of sophisticated equipment. In more modern times, mathematicians have tried to understand the relation of Euclid's geometry to the physical world of light rays and the pull of gravity. These attempts led directly to the theory of relativity, which is based on a geometric model of the universe that includes time as one of the coordinates.

We know that at the microcosmic level, the other extreme of size, the structure of certain viruses is based on the icosahedron, described long ago by Euclid. The same geometric principle supports Buckminster Fuller's geodesic domes. In the future, geometry, in both its simple and sophisticated forms, will remain essential to our understanding of both the large and small aspects of our world.

REVIEW VOCABULARY

Congruent Two geometric figures are congruent if they have the same shape and the same size. In effect, they are the same figure, just moved to a different position.

The Elements Euclid's compilation and organization of the geometric and arithmetic knowledge of his time. Most high school geometry texts are strongly influenced by this work.

Proof A sequence of statements, each logically derived from earlier ones, leading to some conclusion.

Pythagorean theorem "The sum of the squares on the two short sides of a right triangle is equal to the square on the long side." A fundamental tool for calculating distances, as in surveying.

Similar Two geometric figures are similar if they have the same shape, but not necessarily the same size — like a photograph and an enlargement of the same photograph.

EXERCISES

1. What is the main practical difficulty in the way Thales measured the height of the Great Pyramid (short of getting a passport or the expense of traveling to Egypt)?

2. The Great Pyramid is no longer as high as it was in Thales's time. A modern measurement using the same 10-foot stick with a 16-foot shadow would find a pyramid shadow of only 720 feet. How high is the pyramid now?

3. Two smaller pyramids near the Great Pyramid were, in Thales's time, 471 feet and 215 feet tall, respectively. What lengths of shadows would they have cast when Thales's 10-foot stick was casting its 16-foot shadow?

4. Suppose a 4-foot stick casts a 5-foot shadow at the same time that a pine tree casts a 50-foot shadow. How tall is the pine tree?

5. Suppose a woman notices that when her shadow is exactly the same length she is, the shadow cast by a neighboring building is 100 feet long. How tall is the building? (The earliest commentators say that this was the problem actually solved by Thales, not the slightly more complicated case requiring ratios.)

6. From Table 16.1, what is the angle of inclination of the sun to the nearest 10° when a 10-foot stick has a 16-foot shadow? Use a calculator to refine your answer by calculating several nearby values to find the angle of inclination to the nearest 1°.

7. An earlier calculation of the circumference of the earth was made using the same technique Eratosthenes used. The only difference was that the angle was measured at Lysimachia (now near Gallipoli, Turkey) instead of at Alexandria. This angle was found to be $\frac{1}{15}$ of a complete circle, or 24°. It was thought that Lysimachia was 20,000 *stades* straight north of Syene. Using these figures, what would be the circumference of the earth?

8. Assuming that the necessary measurements could actually be carried out, the following figure suggests a conceptually simple way of finding the distance of the moon from the earth.

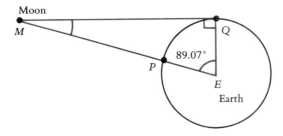

An observer at P sees the moon M directly overhead at exactly the same time as an observer at Q sees the moon right on the horizon. The two observers will be nearly a quarter of the way around the earth from each other. In fact, the central angle E is 89.07°. From a more detailed tangent table, such as Table 16.1, it can be found that tan 89.07° = 61.60295. Question: Assuming the earth's radius is 4000 miles, what is the distance PM (from the earth to moon) in miles? (Note that $PM = EM - 4000$ and that EM can be assumed to be equal to QM.)

9. The following diagram shows a simple and practical method for finding the radius RM of the moon once the distance to the moon ER is known:

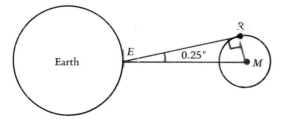

Assume that $ER = 240,000$ miles. The angle subtended by the whole moon, as seen by an observer at E on the earth, is $0.5°$; hence the angle E subtended by half the moon is $0.25°$. Given that the tangent table says $\tan 0.25° = .00436$, find RM.

10. The angle subtended by the sun is also $0.5°$. Using the fact that the sun is 93,000,000 miles from the earth, find the radius of the sun in miles.

11. If we want to find the distance of the sun from Venus by the methods used to find the distance from Earth to the moon or Earth to the sun we would have to be on Venus. Because this is not possible, we can use a more subtle and practical method. We can think of Earth and Venus as points E and V moving in circular orbits around the sun. Notice that the angle at E varies as the two planets travel around the sun and that it reaches its maximum value when the angle at V is a right angle.

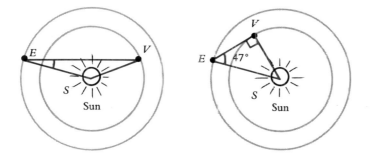

Suppose we observe this angular separation of Venus and the sun throughout the year and find that the maximum value of angle E is $47°$. In right triangle EVS, we know $\tan 47° = VS/EV$. However, because we don't know either VS or EV, the tangent table is of no help. But another table, the "sine table," would give the ratio VS/ES for various values of E. Using the value $\sin 47° = VS/ES = .73135$, find the distance of Venus from the sun to the nearest million miles.

17

Measuring the Universe with Telescopes

Modern science arose in the seventeenth century with the work of Galileo Galilei (1564–1642), Johannes Kepler (1571–1630), and Isaac Newton (1642–1727). The first distinguishing characteristic of the new science was its experimental method. The second was its quantitative character. The physics of Aristotle (384–322 BC), which still held sway in the intellectual world of Galileo's time, gave only qualitative explanations for physical phenomena; for example, the old science tried to explain *why* an apple falls downward from a tree. In contrast, Galileo was more interested in *how;* he dismissed bare qualitative explanations as "fantasies" that are "not really worthwhile." He sought instead a mathematical description of such events as the motion of a freely falling body.

The third and crowning characteristic of modern science was its striving for a mathematical theory, which was highlighted in the work of Isaac Newton. A mathematical theory enables us to make predictions and—often incidentally—explains a wide variety of related phenomena. For example, Newton's work coordinated Kepler's observed laws of planetary motion with the laws of mechanics that appeared to govern terrestrial phenomena. His mechanics explained much about gravitation and ocean tides as well as planetary motion.

From Greece to Galileo

Galileo began his assault on astronomy in 1609, after he learned that a Dutch lens maker had discovered how to achieve great magnification by arranging two lenses in a special way in a long tube. This, of course, was the invention of the telescope. Galileo then proceeded to build his own telescopes. He first achieved a

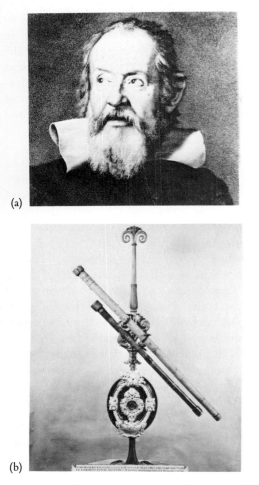

(a)

(b)

Figure 17.1 (a) After inventing the telescope, Galileo made some astonishing astronomical discoveries. (b) Two of Galileo's telescopes and the lens from another. [Photo (b): Scala/Art Resources.]

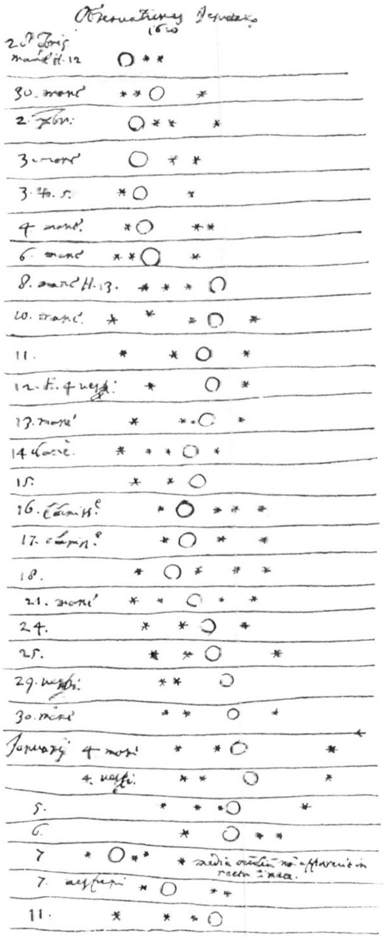

Figure 17.2 Galileo's drawings of Jupiter and its moons. He discovered the moons when noticing that four shining objects moved back and forth across Jupiter from one night to the next. [Yerkes Observatory.]

threefold magnification. Then, after mastering the problems of grinding and polishing lenses and experimenting with the arrangement of the lenses in the tube, he was able to construct a telescope that magnified approximately 33 times. These instruments, although modest by today's standards, revealed some astonishing astronomical sights to Galileo.

Turning the telescope to the moon, he saw immediately that the surface of the moon had mountains and valleys and was not the "perfect" sphere of accepted Aristotelian theory. Later, Galileo discovered sunspots, showing that the sun, too, was not "perfect." These observations shocked his contemporaries by contradicting long-held beliefs about the nature of the universe.

The prevailing conception of the universe in Galileo's time derived primarily from the Greek philosopher Aristotle and the Alexandrian astronomer and geographer Claudius Ptolemy (second century AD). Briefly, the Aristotelian and Ptolemaic view maintained that the earth is the immovable center of the universe, which is a large celestial sphere that rotates about the earth and on which all the stars are fixed. Referred to as a **geocentric,** or earth-centered, theory, this view had prevailed for well over a thousand years; it had the support of almost all academicians and the official support of the Catholic Church, of Martin Luther, and of Jewish leaders. In casting doubt on the geocentric theory, Galileo's observations were most certainly heretical.

An alternative to the geocentric theory had already been proposed by the ancient Greek astronomer Aristarchus of Samos in the third century BC, but his work was largely ignored. Aristarchus held to a **heliocentric** theory, placing the sun at the center of the universe. This theory was revived in a modified form approximately 1800 years later by a young Polish student, Nicolaus Copernicus (1473–1543). Copernicus argued that all the planets, including the earth, moved on concentric spheres, with only slight modification, about the sun.

Perhaps most devastating to proponents of the geocentric theory was Galileo's discovery of four moons revolving about the planet Jupiter. If Jupiter, a planet, possesses moons, then Earth, too, might also be a planet. Moreover, these newly discovered moons of Jupiter were not circling the Earth, the presumed center of the universe around which all bodies should revolve.

Improving the Telescope

A half-century after Galileo built his first telescope, Sir Isaac Newton turned his genius to improving the instrument. Galileo's was a refractor telescope, one that bent light rays by means of lenses. Such instruments have two

shortcomings: (1) the glass used for the lenses must be of high quality and free of flaws in order to minimize distortions, and (2) the bending of the light rays separates the colors contained in white light, introducing a distortion called chromatic aberration. The first of these problems is eliminated, and the second reduced, by using a mirror instead of a lens for the light-gathering work of the telescope. It was this idea that Newton exploited when he constructed the first reflector telescope using a mirror to replace the light-gathering lens of the refractor telescope.

As a close student, and great admirer, of Greek geometry, and himself one of the most profound geometers in all history, Newton (and others, as well) certainly knew that the best shape for a light-gathering mirror would be parabolic, a shape related to the well-known curve discovered 2000 years earlier. Parabolas possess a remarkable feature, illustrated in Figure 17.3, that makes them especially suitable. At the point V, where the parabola crosses its axis of symmetry, the curve is "sharpest." Point V is called the **vertex** of the

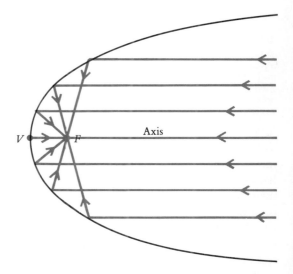

Figure 17.3   Parallel light rays reflect off a parabola and meet at its focus (F).

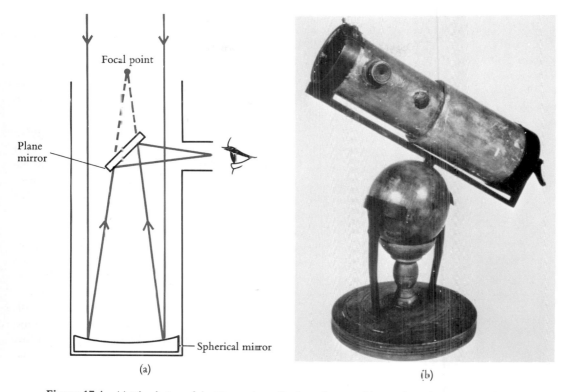

Figure 17.4   (a) The design of the Newtonian reflecting telescope (b) A replica of Newton's reflecting telescope. [The Granger Collection.]

parabola and the axis of symmetry simply the **axis.** Lines parallel to the axis that come in from afar meet the parabola (on its concave side) at some acute angle and then "bounce off" the curve at the same angle. The parabola's remarkable feature is that all the bouncing-off lines pass through a single point F called the **focus** of the parabola. If the parabola is a reflecting surface, then the lines parallel to its axis can be regarded as light rays radiating from a distant heavenly body; these light rays reflect off the surface and accumulate, or focus, at the focus of the parabola, as in Figure 17.3.

Because of the difficulties of grinding a parabolic mirror, Newton compromised and constructed a spherical mirror, instead. Such a mirror, whose surface is a portion of a sphere,

gathers light from afar and tends to accumulate it at the center, or focus, of the sphere. But such a light-gathering mirror presented another problem: the observer would have to be placed at the center of the sphere — directly in front of the mirror — thus blocking all the incoming light.

To make his telescope, Newton placed the spherical mirror at the bottom of a cylindrical tube so that the mirror would reflect the incoming rays of light onto one image point, the focus. To view the image, or focused light, from outside the tube, Newton placed a small plane mirror close to the focus in order to reflect the image to the side of the telescope, where he made a small hole in the cylindrical housing (see Figure 17.4).

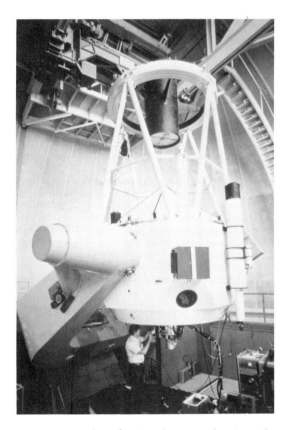

Figure 17.5   This reflecting telescope at the Kitt Peak National Observatory near Tucson, Arizona, uses a mirror measuring 2.1 meters across. [The Association of Universities for Research in Astronomy, Inc.]

Less than four years after Newton built his first telescope, a report came to the French Academy that someone else, Cassegrain, had invented still another reflecting telescope. Cassegrain had succeeded in grinding and polishing a parabolic mirror that would gather light in the new telescope.

Although the focal property of parabolas had been described by Apollonius (c. 260 – 190 BC), the reflecting telescope appears to be its first technological application. The key ideas upon which the usefulness of the parabolic mirror rests are the focal property of the parabola and the fact that when light rays reflect off a smooth surface, the angle of incidence equals the angle of reflection (see Figure 17.6). The conjunction of these two ideas have found a number of applications, including the automobile headlight, which reverses the job of the telescope. Instead of gathering incoming light and bringing it into focus at a point, the automobile headlight has the light source (a bulb) at the focus of the parabolic surface, so that the light is reflected outward along rays that are approximately parallel to the axis (see Figure 17.7a). Another major application is the dishlike antenna used for radio telescopes and radar. Like faint light rays, faint or individually weak signals are caught up by the an-

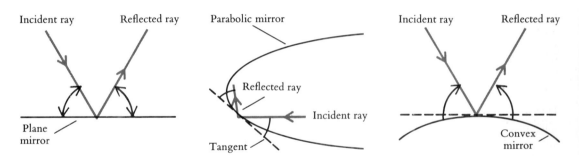

Figure 17.6   When light rays bounce off a smooth surface, the angle of incidence equals the angle of reflection.

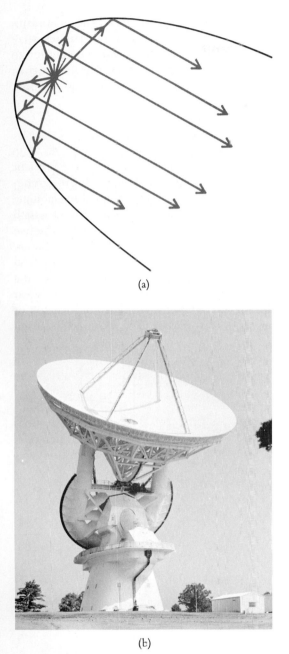

(a)

(b)

Figure 17.7 (a) In an automobile headlight, a parabola directs rays of light outward in parallel lines. (b) A 140-foot-wide radio telescope at Green Bank, West Virginia. [The National Radio Astronomy Observatory, operated by Associated Universities, Inc. under contract with the National Science Foundation.]

tenna and magnified into a strong signal when they accumulate at the focus (see Figure 17.7b).

Conic Sections

Imagine a circle drawn on a flat surface, such as a table top, with a line through the center of the circle perpendicular to the surface. Choose a point V sticking up above the table on this line. The surface consisting of all the lines that simultaneously pass through both V and the circle is called a **cone** with vertex V. (Occasionally, such a surface is called a **right circular cone**.) The vertex separates the surface into two **nappes** (see Figure 17.8).

If we slice this cone with a plane, we get a curve called a **conic section**. In more formal language, we say that the intersection of this surface with a plane is a conic section. By changing the angle of the slice, we can see the variety of possibilities for conic sections. The

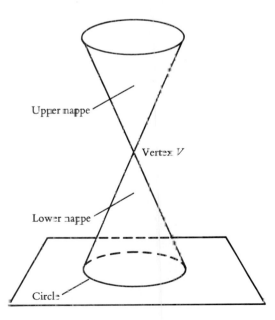

Upper nappe

Vertex V

Lower nappe

Circle

Figure 17.8 A cone.

plane we started with — the table top — obviously intersects the cone in a *circle*. When we tilt the plane a bit, the intersection becomes an **ellipse.** As we continue tilting the plane, the intersection remains an ellipse as long as the plane cuts the one nappe of the cone in a closed curve. However, there comes a point when the cutting plane, while still intersecting only one nappe, no longer intersects the nappe in a closed curve. At this point, the intersection is a **parabola.** Finally, by tilting the plane still further, it will cut both nappes of the cone, in which case we call the intersection curve a **hyperbola** (see Figure 17.9). These four curves, the circle, ellipse, parabola, and hyperbola, constitute the class of conic sections.

The conic sections were probably discovered first by the Greek geometer Menaechmus in the fourth century BC, and they were appar-ently studied by other Greek mathematicians, particularly Apollonius, whose studies were remarkably complete. It is no exaggeration to say that we now know only slightly more than he did about the properties of conic sections.

Johannes Kepler

Johannes Kepler (1571–1620) was a brilliant mathematician with a keen interest in geometry (see Figure 17.11). He devised an elaborate mystical theory of the solar system, in which the six known planets were related to the five Platonic solids (see Box 17.1 and Figure 17.12). In attempting to establish his mystical theory of celestial harmony, he had to use the ambiguous astronomical data available at the time. He realized that the construction of any theory would require more precise data.

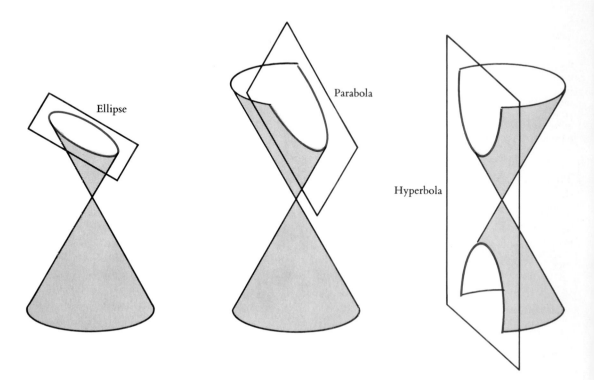

Figure 17.9   The conic sections. The circle is a special case of the ellipse.

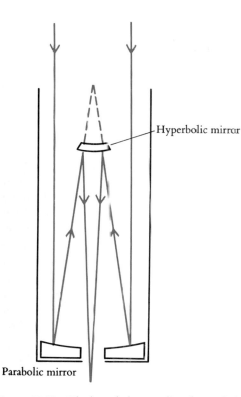

Parabolic mirror

Hyperbolic mirror

Figure 17.10   The hyperbola, as well as the parabola, is applied in the Cassegrain telescope. Whereas Newton's telescope used a plane mirror to transmit the focused image out of the cylindrical tube and to the viewer, the Cassegrain telescope uses a hyperbolic mirror for this purpose, again taking advantage of the focal properties of conic sections.

That data, he knew, was in the possession of the Danish astronomer Tycho Brahe (1546–1601), who had spent 20 years making extremely accurate recordings of the planetary positions and the positions of 1000 stars.

Figure 17.11   Johannes Kepler.

Kepler became Tycho Brahe's mathematical assistant in February of 1600 and was assigned a specific problem: to calculate an orbit that would describe the position of Mars at any time to within the accuracy allowed by the method of observation, which was at that time 4 seconds of arc, or $\frac{1}{15}$ of a degree. Kepler boasted that he would have the solution in eight days. Both the Copernican and the Ptolemaic theories held that the orbit should be circular, perhaps with slight modification. Thus, Kepler sought the appropriate circular orbits for Earth and Mars. (The orbit for Earth, from which all the observations were made, had to be determined before one could satisfactorily use the data for the positions of the

Tetrahedron   Cube   Octahedron   Dodecahedron   Icosahedron

Figure 17.12   The five Platonic solids.

BOX 17.1 Kepler's Model of the Solar System

Kepler, in the *Mysterium Cosmographicum (The Cosmographic Mystery),* published in 1596 a cosmological interpretation of the Platonic solids (here as translated by Koyré, *The Astronomical Revolution,* p. 146):

The Earth [the sphere of the Earth] is the measure for all the other spheres. Circumscribe a Dodecahedron about it, then the surrounding sphere will be that of Mars; circumscribe a Tetrahedron about the sphere of Mars, then the surrounding sphere will be that of Jupiter; circumscribe a Cube about the sphere of Jupiter, then the surrounding sphere will be that of Saturn. Now place an Icosahedron within the sphere of the Earth, then the sphere which is inscribed is that of Venus; place an Octahedron within the sphere of Venus, and the sphere which is inscribed is that of Mercury.

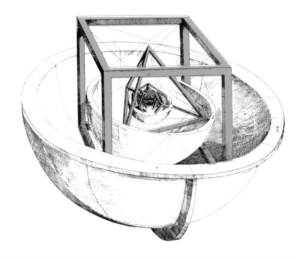

planets.) After four years, Kepler found a solution that seemed to fit Brahe's observations. However, on checking his orbits — by predicting the position of Mars and comparing it with more of Brahe's data — he found that one of his predictions was off by at least 8 minutes of arc!

This shocking failure led to two more years of struggle, in which Kepler finally took the revolutionary step of discarding the long-held conviction that all heavenly bodies move in circular paths (or circular paths modified in

some way by the imposition of smaller circles). This decision permitted him to find an accurate solution to the Mars problem and to put forth a new theory of planetary motion. The results of Kepler's six years of research were published in 1609 in his *Astronomia Nova,* in which he announced two of his three remarkable laws (see Figure 17.13 and Table 17.1):

1. *Law of elliptical paths.* The orbit of each planet is an ellipse with the sun at one focus.

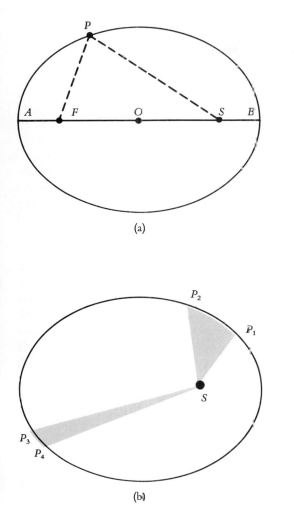

(a)

(b)

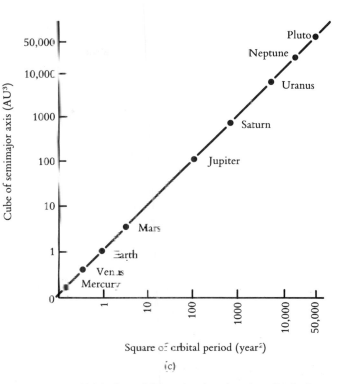

Square of orbital period (year²)

(c)

Figure 17.13 (a) The law of elliptical paths. The orbit of each planet is an ellipse with the sun at one focus. The sum ($PF + PS$) of the distances from any point P of an ellipse to the two foci F, S is equal to the major diameter AB. (b) The law of areas. The shaded parts are of equal area; thus, the sun takes an equal amount of time to move from P_1 to P_2 as to move from P_3 to P_4. (c) The law of times. The points in the graph fall along a straight line, verifying Kepler's discovery that the square of the orbital period equals the cube of the semimajor axis (the planet's average distance from the sun). (1 AU = 93,000,000 miles.) (c) is from William J. Kaufmann, III, *Universe*, W. H. Freeman, 1985.]

2. *Law of areas.* During each time interval, the line segment joining the sun and planet sweeps out an equal area anywhere on its elliptical orbit. (A common brief version of this law is: equal areas are swept out in equal times.)

Kepler's third law was published later and helped Isaac Newton formulate his law of gravity:

3. *Law of times.* The square of the time of revolution of a planet about the sun is proportional

TABLE 17.1 A demonstration of Kepler's third law

Planet	Sidereal period P (in years)	Semimajor axis a (in AU)	P^2	a^3
Mercury	0.24	0.39	0.06	0.06
Venus	0.61	0.72	0.37	0.37
Earth	1.00	1.00	1.00	1.00
Mars	1.88	1.52	3.53	3.51
Jupiter	11.86	5.20	140.7	140.6
Saturn	29.46	9.54	867.9	868.3

to the cube of that planet's average distance from the sun.

Once again, conic sections played a crucial role in the development of science. We now recognize, more than 2000 years later, the incredible genius of the ancient Greeks in identifying and exploring this and other fundamental areas of knowledge. Their work with conic sections developed a subject that is now known to be fundamental to the study of physics, astronomy, architecture, and engineering.

The ellipse, for example, has many applications beyond the magnificent ones in Kepler's work. The focal property of an ellipse is used by acoustical engineers in designing whispering galleries, such as the Mormon Tabernacle in Salt Lake City and the Capitol building in Washington, D.C. If the shape of the cupola of a gallery or auditorium is ellipsoidal, a weak whisper at one focus may be barely audible — even inaudible — in most of the room, except at the other focus, where the reflections of the whisper are brought together again. A visual illustration is provided by an elliptic pool table with a single pocket at one focus (see Figure 17.14): any shot without spin that passes over

Figure 17.14   An elliptical pool table. [From *Inventing, Discovery, and Creativity,* by A. D. Moore. Copyright © 1969 by Doubleday & Company, Inc. Reproduced by permission of the publisher.]

one focus will bounce off the cushion directly into the pocket.

Newton's Great Unification

About 50 years after the death of Galileo, Sir Isaac Newton turned his attention to some of the same problems that had engaged Galileo and Kepler, particularly to the problems of terrestrial and celestial mechanics. In his famous *Principia,* whose full title is *Philosophiae Naturalis Principia Mathematica,* he unified terrestrial and celestial mechanics into one deductive mathematical science.

Writing in the spirit of Euclid, Newton began his *Principia* with definitions of terms such as *mass, force, inertia,* and *momentum.* He then presented three laws of motion, which were direct analogues of Euclid's axioms for geometry. That is, the laws of motion were assumptions that constituted the starting point for his deductive system:

1. *Law 1.* A body continues in a state of rest or in a state of constant unaccelerated motion in a straight line unless it is acted upon by an external force.

2. *Law 2.* At any instant of time, the force acting on a body is equal to the product of its mass and acceleration.

3. *Law 3.* To every action there is always opposed an equal reaction.

Using Kepler's third law, Newton was led to the formulation of his universal law of gravitation: between any two bodies is a gravitational force of attraction that is proportional to the mass of each and inversely proportional to the square of the distance between them.

Firmly convinced of the validity of the universal law of gravitation, Newton used it as an assumption. Together with his three laws of motion, the law of gravitation enabled Newton to erect a masterpiece of mathematics in which he deduced the dynamics of Galileo, the statics that was developed by Archimedes and Galileo, the planetary laws of Kepler, and much more. Imagine — all this in one mathematical system that simultaneously vindicated the "heresies" of Copernicus, Kepler, and Galileo.

The *Principia* was published in 1687, but Newton had discovered many of its great ideas at a much earlier date. In fact, his law of gravitation must certainly have been known to him a decade earlier, for in 1679 he verified the law by calculations based on a new measurement of the earth's radius, together with observations of the moon's position. His friend Sir Edmund Halley persuaded Newton to publish his discoveries and financed the publication of his *Principia*. It contained a wealth of mathematical and physical discoveries even beyond those already mentioned. It explained the perturbations in the path of the moon, the motion of comets, the flattened shape of planets, and the phenomenon of tides. On the basis of Newton's work, later astronomers were able to predict the existence and the positions of the planets Neptune and Pluto.

New Geometries to Measure a New Universe

Another sort of mathematics was conceived during the eighteenth century but was not to be born until the nineteenth century, when it became the basis for the next major revolution in physics and cosmology — the theory of relativity. We refer to **non-Euclidean geometry** — any set of postulates, theorems, and corollaries that differs from Euclid's.

Ordinary geometry consists of statements, called theorems and corollaries, that are logical deductions from assumptions (other statements) called postulates. Euclid presented five postulates from which he developed a large

body of theorems. We call this system **Euclidean geometry.**

From the fifth century BC until late in the nineteenth century, this geometry — including its extension to three dimensions — was thought to be the only science of space. Its theorems were thought to be statements of truth about the world we live in. No one could imagine a different geometry.

Euclid's five postulates, paraphrased somewhat, are as follows:

1. Two points determine a line.

2. A line segment can always be extended.

3. A circle can be drawn with any center and any radius.

4. All right angles are equal.

5. If l is any line and P any point not on l, then there exists exactly one line m through P that does not meet l.

These five statements were supposed to be absolute, self-evident truths. The first four are rather simple statements and are sufficiently unrelated that we may consider them **logically independent;** that is, none of them can be derived from the others by deduction. However, postulate 5, known as the **parallel postulate** because the lines m and l are parallel (i.e., m and l do not intersect), is another matter. We know from historical writings that from the earliest times, some geometers thought this postulate was a logical consequence of the first four. In fact, Euclid himself may have thought his parallel postulate to be an unnecessary assumption for his geometry, for he derived nearly thirty theorems before using it.

There followed a long series of attempts to prove postulate 5 the consequence of the first four postulates. The history of these attempts makes a fascinating story of failures, for whenever someone discovered a supposed proof, it

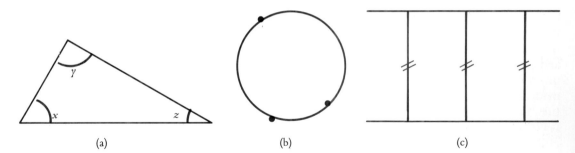

Figure 17.15 Equivalent statements: (a) The sum of the angles of any triangle is equal to 180°. (b) There is exactly one circle through any three points that are not on a line. (c) Parallel lines are equidistant. These three statements are all equivalent to the parallel postulate.

was always found to be tacitly based on some assumption not contained in postulates 1 through 4. Attempts to prove the parallel postulate always failed because they assumed something that was logically equivalent to the parallel postulate; the reasoning was circular, hence invalid.

Among the hidden assumptions that were used in these purported proofs and that are logically equivalent to the parallel postulate are:

1. The sum of the angles of a triangle equals 180°.

2. There is exactly one circle through any three points that are not on one line.

3. Parallel lines are equidistant.

Hyperbolic Geometry

Nicolai Ivanovich Lobachevsky (1793 – 1856) and Janos Bolyai (1802 – 1860) are credited with independently inventing non-Euclidean geometry (see Box 17.2). Both men considered Euclid's parallel postulate to be logically independent of the first four postulates and therefore reasoned that Euclid's postulate 5 could be replaced by a contradictory assumption. They both chose the following:

• *Postulate H.* If *l* is any line and *P* is any point not on the line, then there exists more than one line through *P* not meeting *l*.

The use of postulate H led to an entirely new system of theorems and corollaries, which we now call **hyperbolic geometry.** Some of the theorems in this system were exactly the same as those of the old, for those theorems that are derived only from postulates 1 through 4 must be valid in both systems. However, hyperbolic geometry provided some new and very surprising theorems.

Figure 17.16 shows a line *l* and a point *P* not on *l*. We drop a perpendicular from *P* to *l*, calling *A* the foot of the perpendicular. Now, consider the line *PE*, which is perpendicular to *PA*. According to postulate 5, *PE* is parallel to *l* and is the only parallel to *l* through *P*. How–

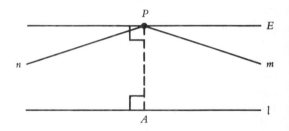

Figure 17.16 In hyperbolic geometry there is more than one parallel through a point *P* not on a given line *l*.

BOX 17.2 Nikolai Lobachevsky, 1793 – 1856, and Janos Bolyai, 1802 – 1860

Nikolai Ivanovich Lobachevsky.
[Novasti Press Agency (A.P.N.).]

Nikolai Ivanovich Lobachevsky (1793 – 1856) and Janos Bolyai (1802 – 1860) independently discovered non-Euclidean geometry. Lobachevsky was the first to publish an account of non-Euclidean geometry (1829), which he first called "imaginary geometry" and later "pangeometry." His work attracted little attention when it appeared, largely because it was written in Russian and the Russians who read it were very critical.

Bolyai published his work on non-Euclidean geometry as a 26-page appendix to a book (the *Tentamen,* 1831) by his mathematician father Wolfgang, who proudly sent the work by his son to Carl Friedrich Gauss, the leading mathematician of his day. Gauss replied to Wolfgang that he had earlier discovered non-Euclidean geometry!

---

ever, if we are using postulate H, then there is another line *m* that passes through *P* and is parallel to *l*. Let us assume that *m* makes an acute angle with *PA*, as shown in Figure 17.16. Then there must be another line *n* that makes the same acute angle with *PA* on the other side of *PA* and that is therefore also parallel to *l*. From this construction, it is easy to see the first astonishing conclusion: *through P there are infinitely many parallels to line l.* This is clear as soon as one considers the set of all lines through *P*, which are separated into two classes by *m* and *n*; one class contains *PA* and the other contains *PE*. The lines in the second class lie between *n* and *m*. All the lines in the second class are parallel to *l* (see Figure 17.17).

Using similar, albeit more complex, reasoning, Bolyai and Lobachevsky discovered many unusual theorems, of which we will list only three (Figure 17.18 illustrates theorems 1 and 3):

1. The sum of the angles in any triangle is less than 180°.

2. Any two similar triangles are congruent — that is, triangles having the same shape also have the same size.

3. Given two parallel lines, there exists a third line perpendicular to one and parallel to the second.

Similar results were obtained by Carl Friedrich Gauss (see Box 17.3), who also investigated these unusual geometries.

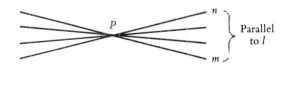

Figure 17.17 In hyperbolic geometry there are infinitely many parallels through a point *P* not on a given line *l*.

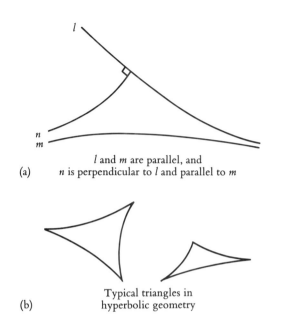

(a)   *l* and *m* are parallel, and
n is perpendicular to *l* and parallel to *m*

(b)   Typical triangles in
hyperbolic geometry

Figure 17.18 In hyperbolic geometry, (a) the sum of the angles of any triangle is less than 180° and (b) given two parallel lines *l* and *m*, there exists a third line *n* perpendicular to one and parallel to the second.

Elliptic Geometry

A generation after the discovery of hyperbolic geometry, G. F. Bernhard Riemann (1826–1866), a young German mathematician and disciple of Gauss, further scrutinized the basic assumptions of Euclidean geometry. He analyzed postulate 2 and observed that "A line segment can always be extended" should be distinguished from "A line is infinite." That is, *unboundedness* does not imply *infinite length*. Think of the geometry on the surface of the earth. Going along what you would imagine was a line around the earth, you can travel another mile and another mile, and so on, and you would eventually return to your starting point. You have traveled on an unbounded but finite path. When Riemann investigated the consequences of a line coming back on itself, he came to the conclusion that his understanding of Euclid's postulate 2 permitted him to abandon Euclid's postulate 5. Riemann re-

---

BOX 17.3 Carl Friedrich Gauss, 1777–1855

Carl Friedrich Gauss.
[Photo Deutsches Museum München.]

Nicolai Ivanovich Lobachevsky and Janos Bolyai are justly given the credit for the invention of non-Euclidean geometry because they had the courage to publish their revolutionary work. However, Carl Friedrich Gauss (1777–1855) is also given credit as a coinventor. It appears from correspondence and private papers that became available after his death that Gauss, too, long believed that Euclid's parallel postulate could not be proved from the first four postulates.

Like Lobachevsky and Bolyai, Gauss derived many theorems based on postulate H and in fact produced some of the most beautiful proofs in hyperbolic geometry. The one most familiar to all students of the subject is Gauss' proof that triangles cannot be arbitrarily large; the maximum area that a triangle may possess is the area contained by the *trebly asymptotic triangle*, the figure consisting of three lines, each of which is simultaneously parallel to the other two.

Figure 17.19 Georg Riemann. [Photo Deutsches Museum München.]

placed postulate 5 with **postulate E:** *Every two lines intersect.*

We can see the plausibility of postulate E when we consider the geometry of the earth's surface, the **spherical geometry** that is necessary for the science of navigation. What is a line in this geometry? Although the shortest distance between two points may be a chord of the sphere, the interior of the sphere is out of bounds in this "plane" geometry; we must stay on the surface of the sphere.

If we intersect the sphere in Figure 17.20 with a plane through A and B, the section is a circle passing through the given points, and the shortest arc, from A to B, of this circle is a candidate for the shortest path from A to B. Every plane section of the sphere is a circle, each with different curvature. *The larger the circle, the smaller the curvature; the smaller the curvature, the shorter the path between A and B.* Thus, the largest circle obtained as a section of the sphere gives rise to the shortest path. The largest circle, of course, is the **great circle,** obtained by having the plane cut through A and B and the center of the sphere.

This result of spherical geometry is familiar to anyone who flies. If a pilot is on the equator and wishes to fly to another point on the equator, he or she simply flies along the equator, a great circle. However, if the pilot were at 10° latitude north and wished to fly to a destination also at that latitude, then the shortest distance would require going further north than the starting and finishing latitude. The pilot flying from New York to Naples, both at the same latitude, would actually have to travel quite far north in the Atlantic Ocean; the shortest flight path from Washington, D.C. to

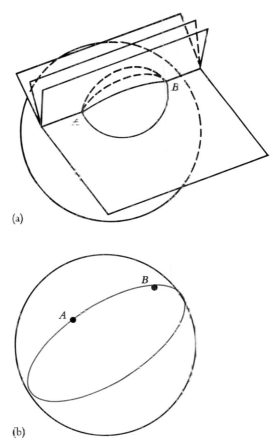

(a)

(b)

Figure 17.20 (a) The planes through A and B intersect the sphere in circles, the largest of which has the smallest curvature. (b) The great circle through A and B.

Figure 17.21 The shortest path between two cities is an arc of a great circle.

Ho Chi Minh City, Vietnam, is over the North Pole (see Figure 17.21).

Returning to the idea of parallelism, we see that this concept simply doesn't exist in spherical geometry! Every pair of lines — that is, every pair of great circles — intersect. On the surface of the globe and other spheres, we see that Riemann's postulate E is quite reasonable. Postulate E leads to **elliptical geometry.** The example of spherical geometry is in fact the best-known example of elliptic geometry. On the surface of the globe we note that triangles can have two or even three right angles: just put two vertices on the equator and one at a pole. In fact, a general theorem of elliptic geometry states that *the sum of the angles of any triangle is greater than 180°*. Hence, in each of the three geometries we have a different theorem about the sum of the angles in a triangle:

- <180° in hyperbolic geometry

- =180° in Euclidean geometry

- >180° in elliptic geometry

As we have seen, any geometry that differs from Euclidean geometry is a non-Euclidean geometry. The first departures from Euclid's postulates involved denying, in some way, his parallel postulate. This led to the development of hyperbolic geometry and elliptic geometry, which are now considered the *classical* non-Euclidean geometries. However, we now have many more geometries that differ from Euclid's in a variety of ways.

The Theory of Relativity

In 1905, Albert Einstein (Box 17.4) put forth his *special theory of relativity,* a complicated theory that constituted the first step in the greatest revolution in physics since Newton's *Principia.* Einstein proposed a new way of thinking about events in the history of the universe. An event takes place in our three-dimensional space at a specific time in history. Thus, an event is located in **space-time** by four coordinates: three determine its position in space, and the fourth determines its position in time. Of course, these coordinates locate the event relative to a specific coordinate system. Einstein observed that the location of an event in space-time therefore depends on the position of the observer — that is, on the origin and orientation of the coordinate system being used. Different observers may obtain very different views of events, especially if one observer is traveling very fast with respect to the other.

Let's consider these ideas geometrically. The *distance* between two events, usually called an **interval,** is split into two parts: a *space-part* and a *time-part.* The space-part will be the part of the interval that comes from the position of the events in three-dimensional space, and the time-part will be the length of time that separates the events. This splitting up depends on the coordinate system and its orientation, so different results may be obtained by different observers (see Figure 17.22). However, the interval, being a line segment joining the two events in four-di-

BOX 17.4 Albert Einstein, 1879–1955

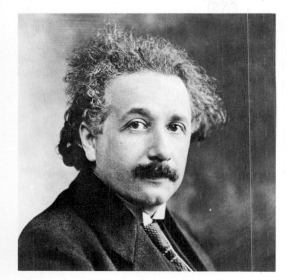

Albert Einstein. [Yerkes Observatory.]

Einstein was born in Ulm in a German Jewish family with liberal ideas. Although he did show early signs of brilliance, he did not do well in school. He especially disliked German teaching methods. In the mid-1890s, he went to study in Switzerland, a country much more to his liking. Einstein burst upon the scientific scene in 1905 with his theory of special relativity. In 1916 he published his theory of general relativity. General relativity was successfully tested in 1919, and his fame grew enormously. Nazism forced Einstein to leave Europe. He settled at the Institute for Advanced Study at Princeton, where he remained until his death at age 76.

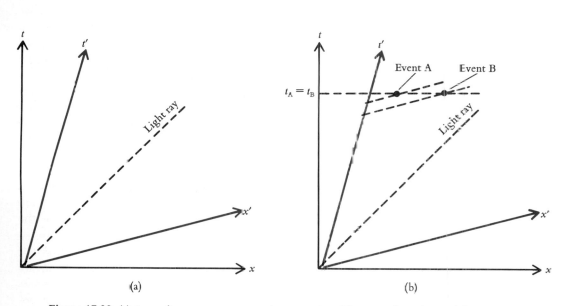

(a)   (b)

Figure 17.22 (a) A coordinate system representing space-time. The t axes show time and the x axes space. The black axes are a system at rest. Note that the red, moving system tilts toward the 45° light ray line. (b) Observers in the system at rest (black axes) will say the events A and B occur at the same time. Observers in the moving red system will say that event B occurs before event A.

mensional space-time, is absolute—in the sense that it is the same for an observer at rest and for all other observers who are traveling at a constant velocity with respect to the one considered at rest.

For example, let's imagine that the eruption of Mount St. Helens took place at the very same time that someone on Mount Palomar observed an astronomical phenomenon 100 light-years away. (A light-year is the *distance* that light travels in a year.) For those of us on earth—at rest relative to the earth—the two events took place one century apart on our time scale; the interval between the two events has a space-part of 100 light-years and a time-part of 100 years. For an observer traveling at constant velocity with respect to the earth, say at 50 light-years away from earth, the space-part and time-part of the interval will be very different. There will be observers who determine that the two events took place 200 years apart, and somewhere there is a point where an observer would see the two events taking place simultaneously. Their splitting of the interval into space-parts and time-parts will be very different from ours.

The geometry of space-time is indeed strange: in its four-dimensional space, the distance between two points—now the interval between two events—remains invariant (in the sense we have described), but its respective parts vary. Three years after Einstein published his first paper on the subject, the mathematician Hermann Minkowski (1864–1909) gave Einstein's work a geometric interpretation that accepted Einstein's strange calculation of intervals and greatly simplified the theory. The geometry that was used, justifiably called Minkowskian geometry, is certainly non-Euclidean. Further, it makes use of one of Riemann's far-reaching ideas—that the nature of a mathematical space is determined by the way distance is measured; the distance formula therefore determines the nature of the geometry.

If the coordinate representations of two events are given by

$$(x_1, y_1, z_1, t_1) \qquad \text{and} \qquad (x_2, y_2, z_2, t_2)$$

then the interval I between them, in Minkowskian space, is calculated by the formula

$$I = [(t_2 - t_1)^2 - (x_2 - x_1)^2 - (y_2 - y_1)^2 - (z_2 - z_1)^2]^{1/2}$$

whereas the distance d between them, if the points are in Euclidean four-dimensional space, would be computed by the formula

$$d = [(t_2 - t_1)^2 + (x_2 - x_1)^2 + (y_2 - y_1)^2 + (z_2 - z_1)^2]^{1/2}$$

The second formula is a direct generalization of the Pythagorean theorem from Euclidean plane geometry.

Little more than a decade after introducing his special theory of relativity, Einstein came forth with his *general theory of relativity*. This work, too, astonished the scientific world. Among other revolutionary ideas was his contention that space was "curved." By this, he meant that light rays, which are considered to travel in straight lines, actually bend. They even bend to different degrees, depending on where in the universe they are; if they pass through a strong gravitational field, then they bend considerably.

A test of this contention was made in 1919 during a total eclipse of the sun, when the light rays from a distant star passed close to the sun and could be studied. Einstein was right; the rays did bend—and in an amount very close to his predictions. This observation showed that lines in the geometry of general relativity are not of the same character as Euclidean lines.

What sort of geometry was Einstein using? There are several answers to the question. First, the idea of "curved" space smacks of elliptic geometry, in the sense that a line through the universe comes back on itself. **Riemannian geometry** also applies because the formula for distance is primary: Einstein noticed that the distance formula that is most appropriate to the needs of physics varies from place to place in the universe, depending on the strength of the gravitational field. Thus, Einstein was using Riemannian geometry along with some considerably modified ideas of elliptic geometry. An appreciation of these non-Euclidean geometries very likely motivated his remark about the postulates of geometry in a famous 1921 lecture: "[they] are voluntary creations of the human mind. . . . To this interpretation of geometry I attach great importance, for should I not have been acquainted with it, I would never have been able to develop the theory of relativity."

REVIEW VOCABULARY

Axis of symmetry (of a parabola) The line dividing a parabola into two identical portions.

Circle The conic section formed when the cutting plane is parallel to the plane P (see **Cone**).

Cone If C is a circle in a plane P and V is a point not in the plane, then the set of all lines simultaneously passing through V and the circle is called a cone.

Conic section The curve formed when a plane intersects a cone.

Ellipse A conic section formed when a plane intersects one nappe of a cone in a closed curve.

Elliptic geometry A system of geometry in which there are *no* parallel lines.

Euclidean geometry The "ordinary" system of geometry based on the parallel postulate.

Focus (of a parabola) The single point at which rays parallel to the axis of a parabola come together after "bouncing off" a parabola. (An analogous definition can be given for the foci of ellipses and hyperbolas.)

Geocentric (earth-centered) Refers to theories in which the celestial sphere rotates about the earth.

Great circle The set of points that is the intersection of the sphere and a plane containing its center.

Heliocentric (sun-centered) Refers to theories in which the sun is at the center of the universe.

Hyperbola A conic section formed when a cutting plane intersects both nappes of a cone.

Hyperbolic geometry A system of geometry in which there exists more than one parallel line through a point P not on a given line l.

Logically independent Refers to a set of statements, no one of which can be logically deduced from the others.

Nappes The two surfaces of a cone separated by its vertex.

Non-Euclidean geometry Any geometry that differs from Euclidean geometry.

Parabola A conic section formed when a cutting plane is parallel to a generating line of the cone, thus cutting only one nappe to result in a curve that is not closed.

Parallel postulate A basic assumption of geometry that states whether through a point P, not on a given line l, there exists none, one, or more than one line parallel to given line l.

Riemannian geometry A system of geometry in which there are *no* parallel lines.

Right circular cone A cone in which the line joining the vertex V to the center of the circle C is perpendicular to the plane P.

Space-time Refers to events that occur in three-dimensional space at a specific time. Thus, events are located by four coordinates in space-time.

Spherical geometry The geometry of a sphere or the earth's surface.

Vertex (of a parabola) The point where a parabola crosses its axis of symmetry. (Also see **cone**.)

EXERCISES

1. When Galileo observed the moon he noted "lofty mountains and deep valleys." He proceeded to measure the shadows cast by the mountains in order to compute the approximate height of the mountains. Galileo concluded that the moon's mountains were 4 miles high (which he thought to be higher than any mountain on the earth).

a. Galileo determined that the ratio of the diameter of the earth to that of the moon was 7 to 2, and he believed that the earth's diameter was 7000 miles. (Today, we know that the earth's diameter is closer to 7900 miles.) Using his data, compute the diameter of the moon, the circumference of the earth, and the circumference of the moon.

b. The following figure represents a view of the moon that shows one-quarter of its surface illuminated. Point T represents the top of a specific lunar mountain whose height Galileo was calculating. Line TB, which is tangent to the moon at B, represents a ray of sunlight. Thus, the surface of the moon immediately below segment TB was in the shadow of the mountain.

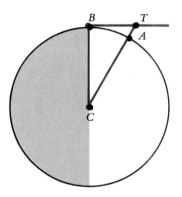

Draw segment CT, calling A the intersection of CT with the surface of the moon. Galileo measured the small arc AB to be $\frac{1}{20}$ of the diameter of the moon. He reasoned that this length was approximately equal to the length TB. What figure did Galileo use for the length TB?

c. What theorem about circles enabled Galileo to conclude that triangle BCT was a right triangle? You can now do as he did and calculate the length CT, namely, the distance from the center of the moon to the top of the lunar mountain.

d. Finally, you can determine the length of segment AT, the height of the lunar mountain. How close have you come to Galileo's calculation for the height of the mountain?

2. We saw earlier how Eratosthenes was able to compute the circumference and hence the radius of the earth. From this fact, we can easily determine the distance from the earth to the moon:

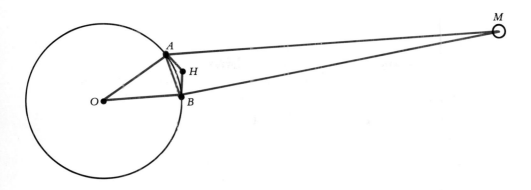

a. In the figure above, *A* and *B* denote two points that measure 500 miles apart on the earth's surface. Compute the measure of angle *AOB*.

b. Compute the measure of angles *OAB* and *OBA*.

c. *AH* and *BH* represent the earth's horizontals at *A* and *B*; they are therefore tangent to the earth at these points. Use this fact to explain how to calculate the measures of angles *ABH* and *BAH*.

d. Simultaneous observations of the moon are made from points *A* and *B* so that angles *MAH* and *MBH* are determined. Given these angles, you can easily determine the measures of angles *MAB* and *MBA*. Now, explain how to calculate the measure of angle *AMB*.

e. Finally, explain how to calculate the distances from points *A* and *B* to the moon.

3. Since the earliest time that a celestial sphere was conceived, it has been known by observers in the northern hemisphere that the North Star was directly — or very close to directly — over the north pole; hence, the name Polaris for the North Star. In the accompanying diagram, an observer at *O* is sighting Polaris in the direction of *P'*.

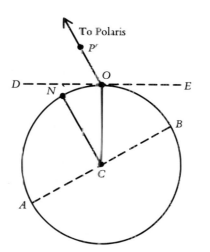

a. Which line in the diagram represents the equator and which line represents the horizon line for the observer?

b. The angle between the horizontal and the sighting of a star is called the altitude of the star. What angle is the altitude of Polaris for the observer at O?

c. What angle in the diagram represents the latitude of the observer?

d. What is the relationship between the latitude of the observer and the altitude of Polaris?

4. The following alternative definition of *parabola* does not depend on three-dimensional considerations: a parabola is the set of points in a plane equidistant from a fixed point (called the *focus*) and a fixed line (called the *directrix*) in that plane. Carry out the following steps in drawing a parabola according to this new definition.

a. On an ordinary-size sheet of paper, draw a line d and mark a point F one or two inches away from d.

b. Locate a point V that is half way between F and d. Why is V a point of the parabola?

c. Using a set of compasses or by trial and error, locate a point P (different from V) whose distance from F is equal to the perpendicular distance from P to d.

d. Repeat step **c** several times, locating five or six such points P on the parabola.

e. How does symmetry help you locate still more points on the parabola?

f. Now connect the points of the parabola to obtain a smooth curve.

5. The point V constructed in the previous exercise is the *vertex* of the parabola, and line VF is the *axis* of the parabola.

a. Explain why the axis of the parabola is the axis of symmetry of the parabola.

b. On a fairly accurate drawing of a parabola, draw its axis.

c. Draw FP, where P is a point of the parabola, and then draw a line through P that is parallel to the axis.

d. Draw a tangent (by the eyeball method) to the parabola at P, and measure the angles that this tangent makes with FP and with the line through P parallel to the axis. Does this verify the focal property of the parabola?

e. Repeat steps **c** and **d** with another point of the parabola that is not the reflection of P in the axis.

f. What are the measures of the angles of incidence and reflection if $P = V$?

6. Here's a similar two-dimensional definition of an ellipse: an *ellipse* is the set of all points P having the property that the sum of the distances from P to two fixed points F_1 and F_2 (called the *foci*) is constant.

a. Draw a horizontal line on a sheet of paper and mark two points F_1 and F_2 approximately 4 inches apart. Construct the perpendicular bisector of segment F_1F_2.

b. On the perpendicular bisector you constructed in **a**, mark a point Q approximately 2 inches from the line F_1F_2. Off to the side, draw a segment whose length is the sum of the lengths of segments QF_1 and QF_2. Point Q will be a point of the ellipse you are constructing, so the length $QF_1 + QF_2$ will serve as the constant referred to in the definition.

c. Using a set of compasses, determine several other points X, such that $XF_1 + XF_2 = QF_1 + QF_2$, thus giving you more points on the ellipse.

d. Explain why, in general, determining one point of the ellipse gives you three others almost immediately.

e. Now connect the points of the ellipse that you've constructed to form a smooth curve.

7. Pick a general point P on an ellipse and draw a tangent at P. Construct line segments from the two foci to P and measure the angles of incidence and reflection in a manner analogous to the one you used with the parabola. Do you have an analogous result?

8. Here is a two-dimensional definition of a hyperbola: the set of all points P having the property that the *difference* of the distances from P to two fixed points F_1 and F_2 (called the *foci*) is constant. Using this definition, construct a hyperbola in the same way you constructed the parabola and ellipse.

9. A hyperbola (with two branches) divides the plane into three mutually exclusive regions: one containing focus F_1, one containing the other focus, F_2, and the other containing no focus.

a. Let P be a general point on that branch of the hyperbola that isolates F_1. Draw a tangent to the hyperbola at this point. Now, draw a segment starting at F_1 and meeting the hyperbola at P. If you imagine F_1 to be a source of light and the hyperbola to be a mirror, then you have an angle of incidence. Finally, draw the line r that represents the ray of light reflected off the hyperbolic mirror at P, according to the usual law: the angle of incidence equals the angle of reflection.

b. Extend line r "backwards" and see how close it comes to passing through F_2. If your construction of the hyperbola is fairly accurate and if you have guessed right in constructing the tangent line at P, then your line r should pass very close — if not right through — F_2. (It can be proved, theoretically, that line r does pass through F_2.)

c. Let Q be a point outside the region containing F_1. The line segment joining G to F_1 meets the hyperbola at a point that we call P. As before, draw a tangent at P and imagine the hyperbola to be mirrored on its convex side. A ray of light emanates from Q, hits the hyperbolic mirror at F, and is then reflected. Draw a line that represents the reflected ray and examine how close this line comes to passing through F_2.

d. Summarize the discoveries you made in parts **b** and **c** to give a full and clear statement of the focal properties of the hyperbola.

10. The French astronomer N. Cassegrain, a contemporary of Newton, proposed a design for a reflector telescope that simultaneously makes use of the focal properties of the parabola and hyperbola. Several different Cassegrain mountings are employed with the 200-inch telescope on Mount Palomar. Examine the diagram of the Cassegrain telescope in Figure 17.10 and explain why it works, being careful to first note which point in the diagram represents the focus for the parabolic mirror.

11. Conic sections that we haven't mentioned in this chapter occur when the plane that sections the cone passes through the vertex of the cone. These sections are called *degenerate.* Can you find three distinct types of degenerate conic sections?

12. Most of the modern applications of conic sections depend on the use of analytic geometry, an alliance of algebra and geometry. When Descartes first developed analytic geometry, he discovered the remarkable fact that the graphs of all second-degree equations are conic sections. Verify Descartes' discovery by graphing the following equations:

    a. $y = x^2$

    b. $y = 2x^2 - 4$

    c. $x^2 + 4y^2 = 1$

    d. $x^2 + y^2 = 4$

    e. $x^2 - 4y^2 = 1$

13. A stone is thrown upward with motion described by the equation

$$y = -16t^2 + 48t + 32$$

where y is the height (in feet) of the stone above the ground at time t (seconds).

    a. The person throwing the stone at time $t = 0$ is atop a building. How high above the ground is his hand when he lets go of the stone?

    b. When will the stone hit the ground?

    c. How high does the stone go? (Hint: What kind of curve is the graph of the equation of motion, and what kind of symmetry does it have?)

    d. Galileo knew the equation of general projectile motion, which is described by the formula

$$y = \tfrac{1}{2}gt^2 + at + b$$

in which g is the acceleration owing to gravity and a is the velocity of the body at $t = 0$. What is the physical meaning of the remaining letter b?

14. The surface of a solid figure whose faces are polygons is called a *polyhedron* (meaning "many planes"). The polygonal faces have vertices and edges that are the vertices and edges of the polyhedron. We want to explore the numerical relationship between the vertices, edges, and faces of an ordinary polyhedron (or solid) in which there is nothing peculiar, such as a hole.

    a. If V is the number of vertices, E the number of edges, and F the number of faces, find V, E, and F for the cube and the triangular prism.

    b. Notice that $V + F$ is larger than E. How much larger in the case of the cube? And how much larger in the case of the cube with a square pyramid pasted on a face? Write your discovery as a formula: $V - E + F = \underline{\quad}$.

    c. Check this formula on all polyhedrons shown at the top of the next page.

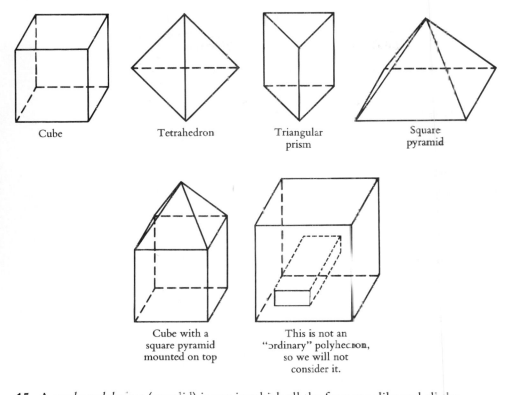

| Cube | Tetrahedron | Triangular prism | Square pyramid |

Cube with a square pyramid mounted on top

This is not an "ordinary" polyhedron, so we will not consider it.

15. A *regular polyhedron* (or solid) is one in which all the faces are alike and all the vertices are alike. To be more precise, we require that all faces be congruent regular polygons and that each vertex be surrounded by the same number of faces. Let p denote the number of edges on each face and q denote the number of faces that meet at a vertex. (A cube, for example, yields $p = 4$ and $q = 3$.) The climaxing theorem in Euclid's work on geometry is a proof that there are only five regular solids, now known as the Platonic solids (see Figure 17.12).

a. Determine p and q for the tetrahedron and the octahedron.

b. Compare the values of pF with qV for the cube, tetrahedron, and octahedron.

c. What is the meaning of pF?

d. Show that $qV = 2E$.

e. The prefix *dodeca-* derives from the Greek word for "twelve"; hence, the regular solid with 12 faces is called a dodecahedron. Determine the number of vertices and edges of a regular dodecahedron.

f. The prefix *icosa-* derives from the Greek word for "twenty." Determine the number of vertices and the number of edges of a regular icosahedron.

16. Newton's proof of Kepler's second law: Newton knew, from the work of Galileo and the Dutch scientist Simon Stevin (1548–1620), that the combined effect, or *resultant,* of two given forces acting on a body is found by the so-called parallelogram of forces. In the following figure, the two forces F_1 and F_2 are represented by arrows whose directions represent the direction of the forces and whose lengths represent the magnitude of the forces. The resultant force is then represented by the arrow along the diagonal of the parallelogram, as indicated in the figure.

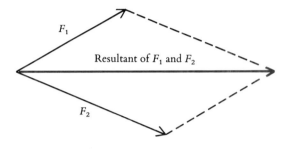

a. Imagine a heavenly body moving at a rate of v feet per second, with no external force acting on it. Let S be any point in space not on the line of motion of the planet and let the planet be at P_0 at the outset, at P_1 after 1 second, at P_2 after 2 seconds and at P_3 after 3 seconds (see diagram (a) below). Compare the areas of triangles SP_0P_1, SP_1P_2, and SP_2P_3.

b. Now, imagine the sun at S exerting a gravitational force of attraction G. At time $t = 1$, for example, G pulls the planet toward S along the line P_1S while the inertial force is pulling the planet from P_1 to P_2. In diagram (b) below, which directed segment represents the resultant of the graviational and inertial forces on the planet in the time interval $t = 1$ to $t = 2$?

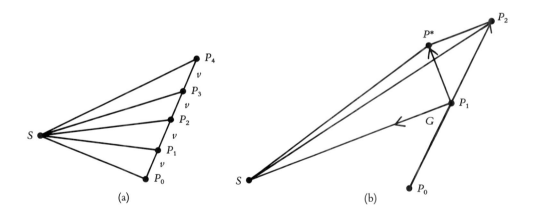

(a)   (b)

c. Compare the area of triangle SP_1P^* with that of SP_1P_2. (Hint: the two triangles have the same base.) Then explain why Kepler's second law actually follows from this result.

17. Let *ABC* be any triangle. Euclid's fifth postulate implies that there is exactly one line *l* through *C* that is parallel to *AB*. Draw line *l* and use it to prove that the sum of the angle measures of *ABC* is 180°.

18. Prove the converse of the previous exercise: if the sum of the angles of a triangle is 180°, then Euclid's parallel postulate holds. (Considerably harder than Exercise 17.)

19. The quadrilateral *ABCD* in the following figure is drawn to suggest the situation in hyperbolic geometry. The angles at *A* and *B* are right angles, whereas the angles at *C* and *D* are acute.

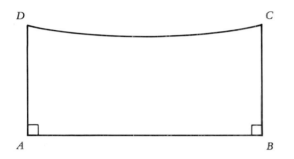

a. Prove that if *AD* = *BC*, then angles *C* and *D* are of equal measure. (You are permitted to use the Euclidean theorems on the congruence of triangles, for they do not depend on the parallel postulate.)

b. Prove that the sum of the angles of triangle *ABD* or of triangle *ABC* is less than 180°. (In fact, the sum of the angles of each of the triangles is less than 180°, but that is harder to prove.)

20. By producing a specific example, show that there is a triangle in elliptic geometry in which all three angles are right angles, so that the sum of the angles of the triangle is 270°. (Hint: spherical geometry is an elliptic geometry.)

COMPUTERS

In Part V, we examine the role of mathematics in the development and workings of modern computational machines. To many people computers and mathematics appear to be one subject, but the relationship is more subtle than that. To get a handle on that relationship, we need to look back and ask a fundamental question: What is mathematics? What lies at the foundation of the subject?

At heart, mathematics works with the notions of proof and truth, and we use proof to discover truth. We manipulate symbols, without regard to their meaning, according to fixed logical rules. The conclusions we draw reveal true statements in the real world, where our assumptions and symbols make sense. We begin our investigation with this relationship between mathematics and the world we live in — between proof and truth.

This study became truly focused in 1900 when David Hilbert, the most influential mathematician of his time, set forth his famous

Enlargement of the Mandelbrot set known as "Tail of the Seahorse." [From H.-O. Peitgen and P. H. Richter, *The Beauty of Fractals*, Heidelberg, Springer-Verlag, 1986.]

research program in a speech to the International Congress of Mathematics. Two major goals of the Hilbert program were to prove that mathematics is consistent (gives rise to no contradictions) and complete (what is true can be proven). In the 1930s, after years of intense work on Hilbert's program, Austrian mathematician Kurt Gödel proved a remarkable result: he showed that mathematical theories powerful enough to do arithmetic contain true statements that cannot be proven—so-called undecidable statements.

This result shocked the mathematical world. It meant that mathematics has inherent limitations. British mathematician Alan Turing tried another tack. Gödel's result guaranteed the existence of "undecidable" propositions. Could we perhaps decide in advance which ones they were? To look for some automatic procedure to detect the undecidable statements, Turing first had to formalize the notion of procedure. He came up with the concept of a computing machine. Now, this idea of a machine was simply a mental construct that enabled Turing to describe what he meant by a computable procedure, or function; in fact, he discovered that it is not possible to determine in advance which statements are undecidable. But the door was opened. What began as an idea for defining a computational procedure quickly became a practical reality.

The story involves many characters, but perhaps the most important is John von Neumann, one of the finest mathematicians of this century. A pure mathematician, von Neumann was fascinated by practical applications. One of von Neumann's many interests was fluid dynamics, a branch of physics with applications from weather forecasting to wing design. But the equations that represent moving fluids are so complicated that even a single problem could take a roomful of clerks with desk calculators weeks to solve. Although this was acceptable in peacetime, World War II brought urgent demands for faster results. Enormous fluid-dynamics calculations were needed for technological developments, including the atom bomb.

This type of time constraint led von Neumann to begin work on computer development. He not only theoretically designed a machine but actually directed the construction of a computer embodying his ideas. This machine was built in the late 1940s at Princeton University and became the prototype of the modern computer.

Since that time, improvements and innovations have come at an incredible pace. New developments in computer graphics have led to new mathematical discoveries. Computer verifications have even been incorporated into proofs of major new results. So, in a sense, we've come full circle. The computer began as an idea to help us understand the meaning of mathematical proof. But mathematics is always growing and expanding. Today, the computer not only helps us to do our calculations and draw our pictures, but it is even changing our notion of proof and the very image of mathematics itself.

CHAPTER 18

Computer Algorithms

If we define "computer" as a nonhuman device that computes, we include as examples both early devices, such as the sun dial and the Chinese abacus, and modern ones, such as the Apple Macintosh and the IBM-PC personal computers and the CRAY-XMP and NASA's MPP supercomputers. Contemporary computing devices have had a significant effect on our lives for two reasons. First, the modern computer is significantly faster than any of its predecessors. Personal computers are capable of performing several hundred thousand computations per second, and the fastest supercomputers exceed a billion computations per second. This awesome speed, in itself, isn't enough. Equally important is the ability of modern computers to store programs and data.

To make a computer do what we want it to do, we need to provide it with a **program,** or detailed sequence of instructions, describing the task to be performed. By controlling a computer's behavior through programs, we can vary the task of the computer without changing the equipment itself, making the computer an immensely versatile device. In this sense, computers and programs can be likened to record players and records, respectively: the output from a computer is determined by the program it is executing in much the same way as the output of a record player is determined by the particular record being played. To see the advantages of storing programs in computers, consider the following problem:

Mary intends to open a bank account on the first day of the month with an initial deposit of $100. She intends to deposit an additional $100 into this account on the first day of each of the next 19 months, for a total of 20 deposits (including the initial deposit). The account pays interest at the rate of 5% per annum, compounded monthly. Mary would like to know what the balance in her account will be at the end of each of the 20 months in which she will be making a deposit.

In order to solve this problem, we need to know how much interest is earned each month. Because the annual interest rate is 5%, the monthly interest rate is $\frac{5}{12}$%. Consequently, the balance at the end of a month is

$100.00 + interest on $100.00

$$= \$100.00 + \frac{5}{12}\% \text{ of } \$100.000$$

$$= \$100.00 + \left(\frac{5}{1200} \times \$100.00\right)$$

$$= \$100.00 \times \left(1 + \frac{5}{1200}\right)$$

$$= \$100.00 \times \frac{241}{240}$$

More generally, we can compute the balance at the end of the month from the initial balance by

$$\text{End of month balance} = \text{initial balance} \times \frac{241}{240}$$

With this analysis out of the way, we can use the following steps to compute the balance at the end of each month:

STEP 1. Let balance denote the current balance. The starting balance is $100, so we set balance = 100.

STEP 2. The balance at the end of the month is $\frac{241}{240}$ × balance at the beginning of the month.

STEP 3. If 20 months have not elapsed, then add 100 to balance to reflect the deposit for the next month and go to step 2; otherwise, we are done.

Suppose that we have to compute the monthly balances using a computing device that cannot store the computational steps and associated data. A nonprogrammable calculator is one

such device. The above steps translate into the following process:

STEP 1. Turn the calculator on.

STEP 2. Enter the initial balance as the number 100.

STEP 3. Multiply by 241 and then divide by 240.

STEP 4. Record the result as a monthly balance.

STEP 5. If the number of monthly balances recorded is 20, then stop.

STEP 6. Otherwise, add 100 to the previous result.

STEP 7. Go to step 3.

If you try this process out on any electronic calculator, you will notice that the total time you spend is not determined by the speed of the calculator but by how fast you can enter the required numbers and operators (add, multiply, and so forth) and how fast you can record the monthly balances. Even if your calculator could perform a billion computations per second, you would not be able to solve Mary's problem any faster.

When we use a stored-program computing device, we need to enter the instructions into the computer just once. The computer can then execute these instructions at its own speed. Because the instructions are entered just once (rather than 20 times), we get almost a twentyfold speedup in the computation. If the balance for 1000 months is required, the speedup increases almost by a factor of 1000. We have achieved this speedup without making our computing device any faster. We have simply cut down on the input work required by the slow human.

Instruction sequences are provided to a computer through a programming language. Over a thousand programming languages are

```
10   balance = 100
20   month = 1
30   balance = 241 * balance/240
40   print month, "$", balance
50   if month = 20 then stop
60   month = month + 1
70   balance = balance + 100
80   goto 30
```

Figure 18.1 A BASIC program for Mary's problem.

```
line   program account (input, output);
 1     {compute the account balance at the end of each month}
 2     const  InitialBalance = 100;
 3            MonthlyDeposit = 100; {additional deposit per month}
 4            TotalMonths = 20;
 5            AnnualInterestRate = 5; {percent rate}
 6     var balance, interest, MonthlyRate : real; month : integer;
 7     begin
 8            MonthlyRate := AnnualInterestRate /1200; {rate per $}
 9            balance := InitialBalance ;
10            writeln('    Month      Balance');
11            for month := 1 to TotalMonths do
12            begin
13                interest := balance* MonthlyRate ;
14                balance := balance+interest ;
15                writeln(month :10,    ', balance :10:2);
16                balance := balance+MonthlyDeposit ;
17            end;
18            writeln;
19            writeln('Balance is balance at end of month');
20     end
```

Figure 18.2 A Pascal program for Mary's problem.

in use today. Some of the more popular ones are BASIC, COBOL, FORTRAN, and Pascal. Our seven-step computational process translates into the BASIC program shown in Figure 18.1. In Pascal, it takes the form shown in Figure 18.2. These two programs are not only written in different languages but represent different programming styles. The Pascal program has been written to permit us to make changes with ease. The number of months, interest rate, initial balance, and monthly additions are more easily changed in the Pascal program.

Algorithms

Tasks such as the computation of monthly balances are called **computational procedures.** In most cases, we want these procedures to conform to a more-stringent requirement. A computational procedure consists of a finite number of steps, each of which is definite and effective. By **definite,** we mean that the outcome of each is well defined. The step "set x to 1/0" is not definite because 1/0 is not a well-defined quantity. Each of the steps in our seven-step calculator example and in the corresponding BASIC and Pascal programs is well defined. By **effective,** we mean that the

step can be completed in a finite amount of time, using a finite amount of computational resource. It is easy to see that each of the steps in our calculator, BASIC, and Pascal examples is effective. The three forms of our solution to Mary's problem are therefore all computational procedures.

A computational procedure that terminates on every input after executing a finite number of steps and produces some output is called an **algorithm.** To decide if an algorithm terminates, we need to determine if the number of steps executed by the algorithm is finite. In obtaining the execution step count, each repetition counts as an additional step; a single step executed seven times is therefore equivalent to seven different instructions executed once each. The computational procedures we have seen so far are all algorithms. Each terminates after executing a finite number of steps. The following sequence of BASIC steps is a computational procedure that is not an algorithm:

$$10 \quad x = 1$$

$$20 \quad \text{goto } 10$$

Generally, we want our computational procedures to be algorithms. However, in certain real situations, such as in nuclear-reactor monitoring, the procedure should never terminate. In practice, we want an algorithm to terminate in a reasonable period of time; after all, an algorithm that does not produce an answer in time for us to use it is not worth much. Take chess, for example, in which a kind of brute-force algorithm will determine whether white can always win. We have 20 possible opening moves for white (the 16 pawn moves shown in Figure 18.3 and 2 moves for each of the two knights) and 20 responses for black. An algorithm to analyze chess could examine each of

the 20×20, or 400, board configurations that are possible after the first pair of moves. The algorithm would figure out all the moves white could make next, then what black could do against all those configurations, and so on. In a typical game of chess, 40 pairs of moves add up roughly to 10^{95} board configurations. Even with a superfast computer, this algorithm would require fantastic amounts of time. Suppose we examine 1 billion configurations per second: it would still take 10^{77} centuries to determine all the possible game outcomes. By comparison, the age of the universe is only 10^9 centuries. Even though the algorithm terminates in theory, it takes so long that it is useless in practice. Actual computer chess systems use a completely different method, modeled more on the way human chess masters play the game.

It should be noted that an algorithm is a method for obtaining a solution to a problem, not the solution itself; an algorithm for baking a cake is a recipe, not the cake. An algorithm for computing the taxes we owe is embedded in the IRS 1040 form shown in Figure 18.4. This form should lead us through all the steps necessary to calculate the tax we owe. Although some people might quibble about how well defined some of the instructions are, let's agree that if we follow them precisely, we will get an answer. This algorithm tells us the method, but the actual solution only comes by performing the computations according to the algorithm.

It is a common misconception that in order to solve a difficult problem, you have to understand it thoroughly. In truth, we can solve problems without understanding them if we have methods that yield solutions to such problems. For example, we don't need to learn the entire U.S. tax code to figure out our tax obligation. As long as we have a good algorithm available, we don't have to understand the subject itself at all. We just follow the steps

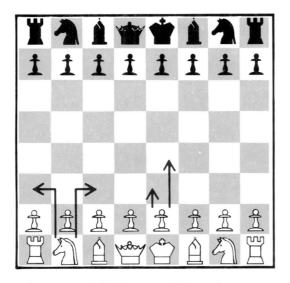

Figure 18.3   White has 20 possible opening moves.

Form 1040A Department of the Treasury—Internal Revenue Service

U.S. Individual Income Tax Return (5)   **1986**

OMB No. 1545-0085

Step 1 Name and address

Use the IRS mailing label. If you don't have one, print or type.

Your first name and initial (if joint return, also give spouse's name and initial)   Last name

Your social security no.

Present home address (number and street). (If you have a P.O. Box, see page 7 of the instructions.)

Spouse's social security no.

City, town or post office, state, and ZIP code

If this address is different from the one shown on your 1985 return, check here

Presidential Election Campaign Fund
Do you want $1 to go to this fund? . . . . ☐ Yes ☐ No
If joint return, does your spouse want $1 to go to this fund? ☐ Yes ☐ No

Step 2 Check your filing status (Check only one)

1 ☐ Single (See if you can use Form 1040EZ.)
2 ☐ Married filing joint return (even if only one had income)
3 ☐ Married filing separate return. Write spouse's social security number above and spouse's full name here.
4 ☐ Head of household (with qualifying person). If the qualifying person is your unmarried child but not your dependent, write this child's name here.

Step 3 Figure your exemptions

Attach Copy B of Forms W-2 here

Always check the exemption box labeled Yourself. Check other boxes if they apply.

5a ☐ Yourself   ☐ 65 or over   ☐ Blind
 b ☐ Spouse   ☐ 65 or over   ☐ Blind
 c First names of your dependent children who lived with you _____
 d First names of your dependent children who did not live with you (see page 1). (If pre-1985 agreement, check here ☐ .)
 e Other dependents:

1. Name	3. Relationship	2. Number of months lived in your home.	4. Did dependent have income of $1,080 or more?	5. Did you provide more than one-half of dependent's support?

Write number of boxes checked on 5a and b

Write number of children listed on 5c

Write number of children listed on 5d

Write number of other dependents listed on 5e

 f Total number of exemptions claimed. (Also complete line 18.)

Add numbers entered on lines above

Step 4 Figure your total income

Attach check or money order here

6 Total wages, salaries, tips, etc. This should be shown in Box 10 of your W-2 form(s). (Attach Form(s) W-2.)   6
7 Interest income. (If the total is over $400, also attach Schedule 1, Part III.)   7
8a Dividends (If the total is over $400, also attach Schedule 1 Part IV.) Total.   8a   8b Exclusion (see page 16).   8b
 c Subtract line 8b from line 8a. Write the result on line 8c.   8c
9a Unemployment compensation (insurance), from Form(s) 1099-G. Total received.   9a
 b Taxable amount, if any, from the worksheet on page 17 of the instructions.   9b
10 Add lines 6, 7, 8c, and 9b. Write the total. This is your total income. ▶   10

Step 5 Figure your adjusted gross income

11 Individual retirement arrangement (IRA) deduction, from the worksheet on page 19.   11
12 Deduction for a married couple when both work. Complete and attach Schedule 1, Part I.   12
13 Add lines 11 and 12. Write the total. These are your total adjustments.   13
14 Subtract line 13 from line 10. Write the result. This is your adjusted gross income. ▶   14

For Privacy Act and Paperwork Reduction Act Notice, see page 41.   Form **1040A** (1986)

1986   Form 1040A   Page 2

Step 6 Figure your taxable income

15 Write the amount from line 14.   15
16a If you made charitable contributions, write your cash contributions. (If $3,000 or more to any one organization, see page 21.)   16a
 b Write your noncash contributions. If over $500, you must attach Form 8283.   16b
 c Add lines 16a and 16b. Write the total.   16c
17 Subtract line 16c from line 15. Write the result.   17
18 Multiply $1,080 by the total number of exemptions claimed on line 5f. See the chart on page 22 of the instructions.   18
19 Subtract line 18 from line 17. Write the result. This is your taxable income. ▶   19

Step 7 Figure your tax, credits, and payments (including advance EIC payments)

If You Want IRS to Figure Your Tax, See Page 22 of the Instructions

20 Find the tax on the amount on line 19. Use the tax table, pages 31–36.   20
21a Credit for child and dependent care expenses. Complete and attach Schedule 1, Part II.   21a
 b Partial credit for political contributions for which you have receipts. See page 24 of the instructions.   21b
22 Add lines 21a and 21b. Write the total.   22
23 Subtract line 22 from line 20. Write the result. (If line 22 is more than line 20, write -0- on line 23.) This is your total tax. ▶   23
24a Total Federal income tax withheld. This should be shown in Box 9 of your W-2 form(s). (If line 6 is more than $42,000, see page 25 of the instructions.)   24a
 b Earned income credit, from the worksheet on page 27 of the instructions. See page 26 of the instructions.   24b
25 Add lines 24a and 24b. Write the total. These are your total payments. ▶   25

Step 8 Figure your refund or amount you owe

26 If line 25 is larger than line 23, subtract line 23 from line 25. Write the result. This is the amount of your refund.   26
27 If line 23 is larger than line 25, subtract line 25 from line 23. Write the result. This is the amount you owe. Attach check or money order for full amount payable to "Internal Revenue Service." Write your social security number, daytime phone number, and "1986 Form 1040A" on it.   27

Step 9 Sign your return

Under penalties of perjury, I declare that I have examined this return and accompanying schedules and statements, and to the best of my knowledge and belief, they are true, correct, and complete. Declaration of preparer (other than the taxpayer) is based on all information of which the preparer has any knowledge.

Your signature   Date   Your occupation
X
Spouse's signature (if joint return, both must sign)   Date   Spouse's occupation
X

Paid preparer's use only

Preparer's signature   Date   Preparer's social security no.
X
Firm's name (or yours, if self-employed)   Employer identification no.
address and ZIP code   Check if self-employed

Figure 18.4 The IRS 1040 form contains a familiar algorithm.

---

of the algorithm and arrive at the answer. Computers do the same thing; they do not "understand" problems but are able to follow algorithms through to a solution. Of course, it's not quite that easy; the algorithm has to be written so the computer can interpret it, which means specifying all the steps in a language the computer can understand. And that's what most computer programs actually are: algorithms translated into a precisely defined language the computer can interpret. We can program computers to handle tasks ranging from the lofty to the mundane, from space exploration to printing junk mail. But underlying that amazing task variety are some fundamental procedures that computers do over

and over. For example, computers do a lot of **sorting,** putting in order items such as numbers, names and addresses, or pictures from Saturn. So computer scientists have given a lot of attention to developing algorithms that tell a computer how to sort things.

Without getting too technical, let's say that sorting means putting a list of similar things into a sequence determined by some specified attribute. Numbers might be sorted according to size, for example, from smallest to largest, whereas names and addresses would probably be sorted alphabetically. Suppose you want to sort the following 10 numbers according to their size, from smallest to largest, on a computer that can do essentially two things: it can compare two numbers, and it can move a number from one position to another.

$$42 \quad 12 \quad 17 \quad 98 \quad 56 \quad 63 \quad 34 \quad 72 \quad 25 \quad 83$$

Here is an algorithm for sorting these randomly arranged numbers, using only comparisons and movements.

INSERTION SORT ALGORITHM

Start with the second number and move it to a temporary holding place. Compare it with the first number. If the first number is larger, move it to the right and insert the one from the holding position in its place. (This sequence of data moves is shown in Figure 18.5.) This guarantees that the first two numbers will be in increasing order. Next, move the third number into the holding place.

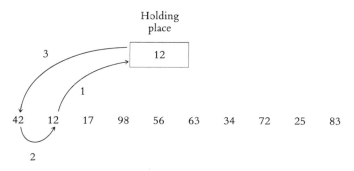

Figure 18.5 The initial data movements for insertion sort.

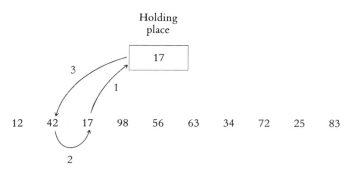

Figure 18.6 A second set of data movements for insertion sort.

Move those numbers to the left of the third number (the first and second) to the right if they happen to be larger than the number in the holding place. (In our particular example, the second number, 42, will move to the right, but the first number will remain in its current location.) Now insert the number from the holding position into the vacant spot. As shown in Figure 18.6, these moves guarantee that the first three numbers on our list are in ascending order.

Continuing this way, we can sort all 10 numbers. Notice that at each step, we move one of the numbers to the holding place. As we do that, we know that all the numbers to its left are already sorted. Now we move these sorted numbers to the right, one at a time, until we find the correct spot, and we insert the number from the holding place at that spot. Subject to our initial constraints, we have limited ourselves to the movement and comparison of numbers.

Figure 18.7 shows a Pascal program segment for this algorithm. Before we compare algorithms that perform the same task, let's look at merge sort, another sorting algorithm.

MERGE SORT ALGORITHM

Start with a list of numbers, but this time imagine that each number constitutes a separate sublist. Because each sublist contains only one number, all initial sublists are already sorted. Now we merge each adjacent pair of sorted sublists, producing half as many new sorted lists of two numbers each. The algorithm simply tells us to take successive passes in which adjacent sublists that are already sorted are merged to form fewer sublists with a larger number of entries, until all the numbers appear in a single sorted list. Figure 18.8 shows the structure of the sublists in successive passes of merge sort over the 10 numbers we previously used in the insertion sort algorithm.

When we have two or more items in each list, we need to be more explicit about how to

```
line  procedure insertsort        (var k              : datarray; {array to be sorted}
  1                                N                   : integer); {# of items in the array}
  2
  3    {this procedure sorts the array k (of size N) using the insertion sort
  4    method. insertsort only traverses the array; it calls the subprocedure
  5    insert to insert specific items into the array.}
  6
  7                  var      ctr  : integer;           {used to step through the array}
  8    {. . . . . . . . . . . . . . . . . . . . . . . . . . . . . . . . . . . . . . . . . . . .}
  9
 10                  procedure insert(var sortarray     : datarray;
 11                               loc                   : integer); {of next item in file}
 12
 13                  {this procedure is a subprocedure to insertsort for inserting the
 14                  locth entry of the array sortarray into its proper position among
 15                  sortarray [1], sortarray [2], ... , sortarray [loc − 1].}
 16
 17                       var  cnt   : integer;
 18                            record: integer;         {variable for array entry}
 19
 20                  begin {of insert}
 21
 22                       record := sortarray [loc];    {hold the value at the loc}
 23                       cnt := loc − 1;               {start at previous item}
 24
 25                       while record < sortarray [cnt] do    {compare to find place}
 26                            begin {of while}
 27                                 sortarray [cnt + 1] := sortarray [cnt]; {move next item up}
 28                                 cnt := cnt − 1                 {so to previous item}
 29                            end; {of while}
 30
 31                       sortarray [cnt + 1] := record          {put item in correct place}
 32
 33                  end; {of insert}
 34    {. . . . . . . . . . . . . . . . . . . . . . . . . . . . . . . . . . . . . . . . . . . .}
 35
 36
 37    begin {of insertsort}
 38
 39                  k [0] := − maxint;                 {initialize to small value}
 40
 41                  for ctr := 2 to N do               {call on insert to locate}
 42                       insert (k, ctr);              {k[2], . . . , k[N] into k}
 43
 44    end; {of insertsort}
```

Figure 18.7 A Pascal program for insertion sort.

Initial	{42}	{12}	{17}	{98}	{56}	{63}	{34}	{72}	{25}	{83}
Pass 1	{12	42}	{17	98}	{56	63}	{34	72}	{25	83}
Pass 2	{12	17	42	98}	{34	56	63	72}	{25	83}
Pass 3	{12	17	34	42	56	63	72	98}	{25	83}
Pass 4	{12	17	25	34	42	56	63	72	83	98}

Figure 18.8 Successive passes of merge sort.

merge adjacent lists. We start with markers positioned on the first number in each list, compare these numbers, and determine which number is smaller. Move that number to the first place in a new list, and advance the marker to the next item in the old list. Now repeat the process, comparing the two numbers now opposite the markers.

When the number of lists is uneven—five, for example—pairing leaves out one list. In this example, the right-hand list of two numbers doesn't get paired until the fourth round, when we have to merge one list of eight with this list of two. At the in-between stages, we keep copying that pair intact. On the other hand, a shortcut will speed up the process: when one list is empty, the algorithm instructs the computer just to move down the remaining numbers in the other list.

Comparison of Algorithms

Because we can have more than one algorithm for a single problem, our obvious question is, Which is the best algorithm? Before we confront this issue seriously, we need to clarify what we mean by "best": we could mean the most simple and direct, the algorithm that uses the least amount of computer memory, or the fastest one. To compare the two sorting algorithms, insertion sort and merge sort, let's agree that by better we mean faster.

One way to determine which algorithm is faster is to take performance measurements, wherein we monitor the actual time taken by a computer to execute the algorithms. To do this, we need to refine our algorithms into computer programs written in a specific programming language. When this is done, the two programs can be given worst-case data (if worst-case times are desired) or average-case data (if average times are desired) and the actual time taken to sort is measured for lists of different sizes. The generation of worst- and average-case data is itself a challenge. When it becomes difficult to generate the worst-case or average data, we resort to simulations.

Suppose we wish to measure the average performance of our two sorting algorithms, using the programming language Pascal on a VAX-11/780, a minicomputer widely used in universities and in business. For this experiment, 2000 random numbers are generated in groups of 100. Table 18.1 gives the comparative sorting times for insertion sort and merge sort, and Figure 18.9 makes a graphic comparison of these sorting times. In both figures, N represents the length of the list being sorted.

A more rigorous way of analyzing algorithms is to consider how much work they do. In the case of our two sorting algorithms, this means counting the number of comparisons and data movements required by the algorithms. For the insertion sort, we start by making one comparison and at most three data movements (see Figure 18.5) for a total of four operations when we position the second number properly relative to the first one. At the next step we need at most two comparisons and four data movements; in Figure 18.6 actually two comparisons and three data movements were required. In general, when the kth element of the list is being positioned relative to its predecessors, at most $k - 1$ comparisons and $k + 1$ data movements, or a total of $2k$ operations, will be needed. Thus, the proper

TABLE 18.1 **Time (in seconds) of insertion and merge sorts on a VAX-11/780**

N	Insertion sort	Merge sort
100	0.03	0.02
200	0.14	0.04
300	0.29	0.07
400	0.57	0.09
500	0.84	0.11
600	1.19	0.15
700	1.68	0.16
800	2.15	0.20
900	2.73	0.22
1000	3.38	0.25
1100	4.12	0.29
1200	4.82	0.33
1300	5.77	0.35
1400	6.63	0.39
1500	7.59	0.41
1600	8.71	0.43
1700	9.84	0.46
1800	11.19	0.50
1900	12.19	0.52
2000	13.48	0.57

positioning of the second through Nth data items will require at most

$$4 + 6 + \cdots + 2N = N^2 + N - 2$$

operations. The worst-case computation time for sorting N numbers by insertion sort will therefore be proportional to $N^2 + N - 2$, or simply proportional to N^2.

By contrast, merge sort is more efficient. Without showing the derivation here, we assert that the number of operations needed to sort N numbers using merge sort (and hence the time required by merge sort) is proportional to $N \log N$.

The distinction of computation times proportional to N^2 and $N \log N$ is a major one. Figure 18.10 shows that as N gets large, the

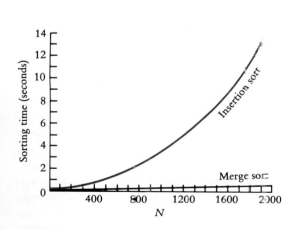

Figure 18.9 As N gets large, merge sort becomes more efficient compared to insertion sort.

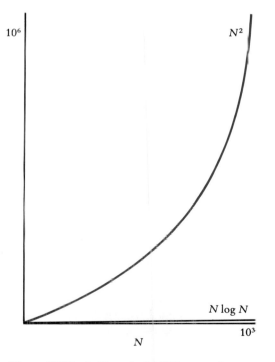

Figure 18.10 As N gets large, N^2 increases far more rapidly than $N \log N$.

TABLE 18.2 Time (in seconds) of insertion and merge sorts

N	Insertion sort VAX-11/780	Merge sort AT&T PC-6300+
200	0.14	0.96
400	0.57	2.36
600	1.19	3.27
800	2.15	4.83
1000	3.38	6.74
1200	4.82	7.60
1400	6.63	9.96
1600	8.71	10.57
1800	11.10	12.89
2000	13.48	13.53
2200	14.82	15.50
2400	18.81	16.98
2600	19.48	18.87
2800	22.59	20.08
3000	29.50	21.51
3200	33.45	23.24
3400	35.40	24.99
3600	37.35	26.14
3800	43.22	28.31
4000	49.03	29.98

TABLE 18.3 Comparative growth of some functions

N	$\log N$	$N \log N$	N^2	N^3	2^N
5	1.6	8.0	25	125	32
10	2.3	23.0	100	1,000	1,024
15	2.7	40.6	225	3.375	32,768
20	3.0	60.0	400	8,000	1,048,576
25	3.2	80.4	625	15,625	33,554,432
30	3.4	102.0	900	27,000	1,073,741,824
35	3.6	124.4	1225	42,875	34,359,738,368

difference between N^2 and $N \log N$ becomes more pronounced. The poor performance of insertion sort cannot simply be overcome by using a faster computer. We get experimental evidence for this by contrasting the times required for sorting random numbers by insertion sort on a VAX-11/780 computer and the time required for sorting the same numbers by merge sort on the much slower AT&T PC-6300 Plus microcomputer. Table 18.2 illustrates that for large N, the choice of algorithm is more critical than computer speed.

N^2 and $N \log N$, respectively, are the **time complexities** of the insertion sort and merge sort algorithms. They represent the behavior of these algorithms when N is suitably large. More precisely, an algorithm of time complexity $f(N)$ will have execution time proportional to $f(N)$ for sufficiently large N. Even if insertion sort is faster than merge sort for small values of N, when N gets sufficiently large, merge sort will be faster.

To develop insight about the execution times of algorithms with different time complexities, consider Table 18.3, which gives values of $\log N$, N^2, N^3, and 2^N for $N = 5$, 10, . . . , 35. The function 2^N grows very rapidly with N. In fact, if a program needs 2^N steps for execution, when $N = 40$ the number of steps needed is approximately 1.1×10^{12}. On a computer executing 1 billion instructions per second, this would require 18.3 minutes. If $N = 50$, the same program would run for about 13 days on this computer; when $N = 60$, about 311 years will be required; and when $N = 100$, more than 10^{11} centuries will be needed. The utility of programs with exponential time complexity is limited to small N (typically $N \leq 40$.)

The usefulness of programs with time complexities that are polynomials of high degree is also limited. For example, if a program needs N^{10} steps, then using our 1 billion instructions per second computer, we will need 10 seconds when $N = 10$; 3171 years when $N = 100$ and 3.17×10^{13} years when $N = 1000$. Had the program's time complexity been N^3 instead, then we would have needed 1 second of computation time for $N = 1000$, 111 minutes when $N = 10,000$, and 11.57 days when $N = 100,000$.

Table 18.4 gives the time needed by a 1 billion instructions per second computer to execute a program of time complexity $f(N)$ instructions. We should note that currently only the fastest computers can execute 1 billion instructions per second. From a practical standpoint, for reasonably large N (say $N > 100$) only programs of small time complexity (such as N, $N \log N$, N^2, N^3, and so forth) are feasible. Furthermore, this would be the case even if we could build computers capable of

executing 10^{12} instructions per second, because this would only decrease the computing times of Table 18.4 by a factor of 1000. The economic implications of more modest improvements of linear-programming algorithms are discussed in Box 18.1.

Although we have clear evidence of the superiority of merge sort over insertion sort, that does not necessarily mean we should always use the merge sort algorithm. We have other factors to consider. For instance, sorting is a data-sensitive task: which method is most efficient depends to some degree on the original arrangement of the input data. Imagine that your boss gives you a randomly arranged set of 500 index cards containing names and addresses and tells you to sort them alphabetically. You're probably going to start making individual piles for each letter of the alphabet,

TABLE 18.4 Time to compute $f(N)$ instructions on a 1 billion instructions per second computer

N	$f(N) = N$	$f(N) = N \log N$	$f(N) = N^2$	$f(N) = N^3$	$f(N) = N^4$	$f(N) = N^{10}$	$f(N) = 2^N$
10	0.01 μs	0.03 μs	0.1 μs	1 μs	10 μs	10 s	1 μs
20	0.02 μs	0.09 μs	0.4 μs	8 μs	160 μs	2.84 hr	1 ms
30	0.03 μs	0.15 μs	0.9 μs	27 μs	810 μs	6.83 d	1 s
40	0.04 μs	0.21 μs	1.6 μs	64 μs	2.56 ms	121.36 d	18.3 min
50	0.05 μs	0.28 μs	2.5 μs	125 μs	6.25 μs	3.1 yr	13 d
100	0.10 μs	0.66 μs	10 μs	1 ms	100 ms	3171 yr	4×10^{13} yr
1,000	1.00 μs	9.96 μs	1 ms	1 s	16.67 min	3.17×10^{13} yr	32×10^{283} yr
10,000	10.00 μs	130.3 μs	100 ms	16.67 min	115.7 d	3.17×10^{23} yr	
100,000	100.00 μs	1.66 ms	10 s	11.57 d	3171 yr	3.17×10^{33} yr	
1,000,000	1.00 ms	19.92 ms	16.67 min	31.71 yr	3.17×10^7 yr	3.17×10^{43} yr	

μs = microsecond = 10^{-6} seconds
ms = millisecond = 10^{-3} seconds
s = seconds
min = minutes
hr = hours
d = days
yr = years

Source: Sartaj Sahni, *Software Development in Pascal*, The Camelot Publishing Company, Fridley, Minnesota, 1985, p. 415.

BOX 18.1   The Economic Value of Algorithms

Algorithms that we can implement on computers have significant economic value. We can see a good example of their potential value in the case of Karmarkar's algorithm. In 1984, Narendra Karmarkar, a researcher at AT&T Bell Laboratories, announced his discovery of a revolutionary new linear-programming algorithm. As discussed in Chapter 4, linear-programming methods are used to solve complex real-world problems — especially in the transportation and communications industries — of scheduling, routing, and planning.

Following the announcement of its discovery, Karmarkar claimed that his algorithm would solve problems many times faster than the currently used simplex method, the standard tool for solving linear-programming problems for over 40 years. Because breakthroughs such as Karmarkar's are rare, his claims of superior performance in solving complicated problems met some initial skepticism. By 1986, however, growing experimental evidence showed that implementations of Karmarkar's algorithm in specialized computer programs did solve certain types of linear-programming problems between 1.7 and 4.5 times faster than the standard computer implementation of the simplex method. It is particularly noteworthy that the computer program employing Karmarkar's algorithm improved the relative speed of solving larger problems (those with 1000 constraints and between 2000 and 3000 variables) more than it improved the speed of solving problems with fewer variables. These tests were performed on the same large mainframe computer so that comparisons could be made.

Although current programs are experimental, the potential cost-saving implications for Karmarkar's algorithm are enormous. When successfully implemented in commercial software programs, Karmarkar's method could save airlines, railroads, trucking firms, federal and state agencies (such as NASA), communications companies, and others millions of dollars annually by solving complex problems more quickly and efficiently. Because the flow of goods, services, and information through intricate networks in the United States and around the world is essential to global economic health and growth, the application of Karmarkar's algorithm would affect everyone. Furthermore, such a fundamental discovery in applied mathematics can have extraordinary effects on companies that develop products based on this new mathematics and on companies that use those products to serve large populatons.

---

or maybe for groups of letters. When you have sorted all the cards into piles, you alphabetize each pile and then stack up the piles in alphabetical order.

On the other hand, imagine that the telephone company in a small town has hired you to put together next year's directory. You take the current directory and have to incorporate 20 new names at the proper alphabetic places. Are you going to put the new names at the end and then merge sort the whole list? Of course not. You're going to find the right place to insert each new name. The point is that for a common task like sorting, you have many algorithms to choose from. Which one you should pick depends on your particular prob-

BOX 18.2 How the Weather Is Forecasted

The starting point for local weather forecasters is information transmitted by the National Weather Service (NWS) in Washington, D.C. The NWS collects observations all over the world — from human observers, weather balloons, ships and planes, and satellites. Twice a day computers turn all that data into national and regional forecasts. The NWS stays ahead of nature partly by running its numerical weather prediction algorithm on one of the world's fastest supercomputers. This supercomputer is designed to repeat a sequence of calculations at maximum speed, exactly what the weather prediction algorithm calls for.

The algorithm divides the atmosphere into a grid of 256,000 points in the three-dimensional space. Observations give four basic values for each of these points: temperature, pressure, moisture content, and wind velocity. Taken together, these values yield a model of what the atmosphere looked like when the observations were taken. The next step is to apply a set of equations based on physics to the values at each point. The equations make use of what we know about surrounding points to predict new values for a given point 10 minutes in the future. This same process is applied to all points of the grid; the process is then repeated for successive 10-minute intervals until we can view a model of what the atmosphere will be like in 24 hours.

The NWS research and development team is always trying to make the forecast more accurate. But accuracy is not allowed to interfere with deadlines. Changes in the forecasting program are not adopted until they have been proven not to slow the process down.

---

lem, the nature of the data you are working with, and your goals in terms of time, money, and simplicity.

Complex Algorithms

For more-complex problems, the search for better algorithms is bound up with the development of faster computers. Everyone relies on weather forecasts, and everyone would like to have reliable predictions. At the National Weather Service in Maryland, mathematicians and computer scientists are hard at work on improving forecasting techniques (see Box 18.2). Even for complex problems such as weather forecasting, more-efficient algo-

rithms and faster computers are making it practical to obtain better results.

Some Unsolved Problems

Because algorithms are so useful, it seems reasonable to ask if there is an algorithm to solve a given problem. Unfortunately, the answer to this question is "no." Consider the equations

$$x^2 + y^2 - 2 = 0$$

and

$$x^3 + y^3 = z^3$$

Do these equations have solutions that are whole numbers? With a few guesses, we can see that $x = 1$ and $y = 1$ are whole-number solutions for the first equation. But no matter how long we keep making guesses for the second equation, all we can say is that we haven't found a solution yet. For centuries, mathematicians looked for an algorithm that would determine whether any such equation has a whole-number solution. Finally, in 1970 Yuri Matijasevic, a 22-year-old Russian mathematician, proved that such an algorithm does not exist.

For many practical problems, we simply do not know whether an algorithm exists. Suppose we wish to make leather goods, for example, coin purses and shoes. The parts for these products are leather shapes like the ones you see in Figure 18.11. To cut each of these parts from large pieces of leather, without any waste, we would like to fit the shapes together in a pattern with no gaps or overlaps. Mathematicians call this type of pattern a tiling (see Chapter 13).

Although it is easy enough to discover a tiling for rectangular pieces, it is not so clear whether we can arrange shoe shapes in a pattern without any wasted leather. It would be nice to have an algorithm that would tell us whether a particular shape will form a tiling. So far, such an algorithm has not been discovered, and we do not know if one exists; yet mathematicians are also unable to prove that it does not exist.

Concluding Comments

As we have seen, algorithms are useful in solving a variety of problems. Because many problems have multiple algorithmic solutions, we often need to choose a specific algorithm from among those available to us. In such instances, a comparative analysis of algorithms can be very beneficial. On pp. 356–361 we contrasted algorithms (insertion sort and merge sort) that executed in times proportional to N^2 and $N \log N$. We were led to the conclusion that for large N, the $N \log N$ algorithm was so superior that it outperformed the N^2 algorithm even when handicapped by a slow computer. Whenever available, the time-complexity analysis of an algorithm gives us insight into the performance quality of the algorithm.

We have four possibilities for the existence of algorithms for a given problem.

1. For problems such as sorting, many good algorithms are known.

2. For problems such as weather forecasting, better algorithms and better computer techniques are leading to better solutions.

3. For problems such as whole-number solutions to equations, we can prove there are no algorithms.

4. For problems such as the tiling of shapes, we do not know if an algorithm exists.

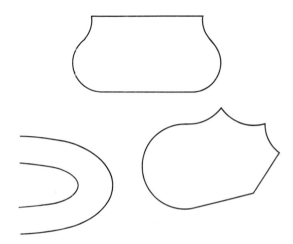

Figure 18.11 We do not know if an algorithm exists that will tell us if these leather shapes can be fit together to form a tiling.

What should keep mathematicians and computer scientists properly humble is that there is no algorithm that can determine which category a given unsolved problem belongs to. In other words, there is not, and cannot be, a comprehensive algorithm that proves or disproves the existence of an algorithmic solution for a particular problem. For this reason, the application of mathematics to computers will continue to be an experimental science.

REVIEW VOCABULARY

Algorithm A computational procedure that terminates on every input and produces some output.

Computational procedure A finite number of definite and effective steps.

Definite (step in a computational procedure) A step whose outcome is well defined.

Effective (step in a computational procedure) A step that can be computed in a finite amount of time.

Program A sequence of instructions in a language that a computer can interpret.

Sorting The arrangement of data into a prespecified order.

Time complexity (of an algorithm) The execution time (up to a proportionality constant) of an algorithm.

EXERCISES

1. Give an algorithm which for a specified integer, N, inputs N numbers and computes and prints their sum.

2. Modify the algorithm in Exercise 1 to obtain an algorithm for computing the average of the N integers.

3. Figures 18.5 and 18.6, respectively, illustrate the data movements necessary to insert the second and third data elements into their correct positions, relative to their predecessors. With similar figures, illustrate the data movements required for the insertion of the fourth through the tenth numbers.

4. By reviewing the figures from Exercise 3, determine the exact number of data movements and comparisons used by insertion sort when sorting the 10 numbers that were considered in Exercise 3.

5. In the section on comparison of algorithms, we derived $N^2 + N - 2$ as the worse-case total of comparisons and data movements in insertion sort. Give a similar derivation for comparisons and data movements for the average case.

6. In selection sort, the largest number in a list is determined in the first pass and moved to position N. Then the pass is repeated on the remaining $N - 1$ numbers, moving the second-largest number to position $N - 1$. Additional passes (a total of $N - 1$ successive passes) eventually sort the entire list. Determine the number of comparisons and data movements required by the selection sort algorithm. How do the average and worst cases compare?

7. Algorithms A and B perform a certain task on a set of N records. For the execution times of the two algorithms given below, which algorithm is the faster one for large values of N? For what values of N (if any) should we prefer the other algorithm?

A	B
$2N^2 + 250N$	$5N^2 + 4N$
$2N^3 + 3N$	$10{,}000N$
$1000N^3$	$\frac{1}{5} \times 2^N$

8. (For students with programming experience.) Write and test programs in BASIC or Pascal (or another programming language of your choice) to implement the algorithms that were developed in Exercises 1 and 2.

9. (For students with programming experience.) Write a BASIC or Pascal program to implement the selection sort described in Exercise 6.

Codes

We can view computing systems as systems that process information. Using such a system, we can enter data into computers and extract data from computers; we can store information and manage it to produce new results organized in useful ways. Computers can process many different kinds of information, and different computers often deal with a given type of data in different ways. In this chapter, we look at some of the coding schemes associated with the storage of certain types of information in modern computing systems.

Thirty years ago, during the computer's early stages of development, computers were used almost exclusively to do numerical calculations. Although we continue to use computers this way, most computers today are also used to perform manipulations on nonnumeric data. If we are to view computers as information processors, we must interpret *data* somewhat more broadly than simple numeric information. Here are some examples of data commonly managed by computers:

1. Numeric data such as integers and real numbers.

2. Textual data consisting of the letters of the alphabet, punctuation marks, and special symbols.

3. Information that computers use to generate visual patterns on display units.

4. Information that can be used to generate sounds through such devices as music synthesizers.

5. Complex information structures such as records of employees of a corporation. Such records might contain subsections that could be numeric or textual.

6. Information that identifies where other data is stored. This type of data is called address, or pointer, data.

7. Information that tells the computer what it should do, for example, programs in any programming language.

In order to store, reproduce, and manipulate such diverse types of information, computers use **codes** to represent the specific type of information being processed; codes are mechanisms for representing information. For example, our alphabet is a code of 26 symbols that we use to represent our speech in written form; a biological coding scheme, based on chemical patterns that make up DNA molecules, is used to transmit biological information from one generation to the next.

One powerful insight from mathematics is that codes can be used to represent information. In a sense, all mathematics exploits this idea. For example, we can represent things in the real world with equations—a code of numbers and symbols—and then manipulate the equations to find results that we can apply to the real world. In mathematics, a code consists of two elements: a group of symbols and a set of rules for interpreting those symbols. In music, those same elements make up a code called a score. The musical score helps us to understand the codes that are used by computers.

The basic musical symbol is the note

Of course, there are other symbols, such as

which signifies a longer duration of the sound. But what about interpretation? How does the musician know which notes to play? In other words, how can a code with just one symbol encode so many different sounds? The answer is that the information encoded by the note varies according to its location on the staff. Mathematicians refer to such encoding systems as **place-value systems** because the value of the symbol is determined in part by its position. Notes at different locations on the staff represent different sounds.

We all use a place-value system everytime we look at numbers. There's no mystery to the difference between 20 and 200. We recognize without thinking about it that the symbol "2" in 20 represents two tens and that the same symbol in 200 represents two hundreds. In music, the same principle allows a trained performer to quickly recognize the values of the notes.

Binary Codes

Because computers are built from two-state electronic components, it is natural to use two-state codes to represent information in computing systems. For example, floppy disks and magnetic tapes, the common storage media of computers, are covered with a thin layer of ferromagnetic material that can be magnetized into one of two states: positive or negative. (If you have used a computer, you are probably familiar with floppy disks. The drives into which these disks fit can read data stored on the disk by sensing magnetic states, or they can write data onto the disk by altering existing magnetic states. The reading and writing are done through the oval window in the disk, which exposes the disk to the disk drive.) Each cell on a floppy disk can be polarized by electric current into positive or negative states. The magnetic patterns created by these two possible orientations determine the information contained on the disk. The key to this system is a scheme for representing data with only two symbols: the **binary code.** In binary, we customarily use the symbols 0 and 1 to represent the two states.

As so often happens, the mathematics needed to exploit these characteristics of elec-

BOX 19.1 George Boole, 1815–1864

George Boole. [The Granger Collection.]

George Boole (1815–1864) was an English mathematician and logician. According to no less an authority than Bertrand Russell, "Pure mathematics was discovered by Boole, in a work which he called *The Laws of Thought.*" Boole's singular contribution was the development of a two-valued algebra, which serves as a workable calculus for interpreting logical truth and falsity. He set up a system of logical interpretation that could be treated as a numerical algebra in which the symbols are restricted to 0 and 1. As Boole said, "We may in fact lay aside the logical interpretation of the symbols in the given equation; convert them into quantitative symbols, susceptible only of the values 0 and 1; perform upon them as such all the requisite processes of solution; and finally restore them to their logical interpretation." Although Russell's statement is an exaggeration, it is nevertheless fair to say that Boole's work was essential to the development of modern logical thought.

tricity and magnetism was developed long before modern computers were a practical possibility. In the 1850s the English logician George Boole (see Box 19.1) had an ambitious goal: to codify the laws of thought by studying the ways we draw logical conclusions from different combinations of statements. Suppose that statement A is "The balloon is red," and statement B is "The balloon has polka dots." A is true for the balloon in Figure 19.1a but is false for the one in Figure 19.1b. Boole's innovation was to create a mathematical approach to the logical relationships of statements. He assigned 0 to represent false and 1 to represent true. Thus, statement A, "The balloon has polka dots," would have value 0 in Figure

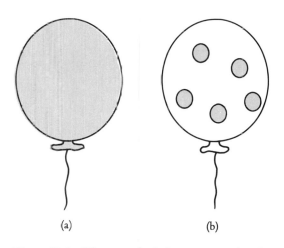

(a) (b)

Figure 19.1 We can use logical statements and truth values to describe these two balloons.

19.1a and value 1 for the one in Figure 19.1b. The two statements together can be described by four possible combinations.

	Statements	
	A	B
Truth values	0	0
	0	1
	1	0
	1	1

We can now form a compound statement: "*A* and *B*." In terms of our example, this says, "The balloon is red and has polka dots." We want "*A* and *B*" to be true (to have the value 1) only when both *A* is true and *B* is true. We can summarize the information about this compound statement by expanding our original table into the following **truth table**:

	Statements		
	A	B	A and B
Truth values	0	0	0
	0	1	0
	1	0	0
	1	1	1

This truth table transforms our intuitive notion of "and" into a mathematically precise definition "*A* and *B*" is true if and only if the balloon is red and has dots.

More than 80 years after Boole's time, in the late 1930s, a similar inspiration occurred to a number of persons working toward automated calculating machines. In the United States, that insight struck both George Stibitz (see Box 19.2), a mathematician at Bell Laboratories, and Claude Shannon, a master's degree candidate at the Massachusetts Institute of Technology. These men and others realized that Boole's algebra of logic could be physically embodied in electrical circuits. The circuits that engineers used in the 1930s were relay switches, which basically resemble light switches. They have two positions, on or off, which correspond to current flowing or not flowing. We can represent on or off with the values 1 and 0. Stibitz and Shannon's breakthrough was the recognition that circuits could represent relationships between statements.

Returning to our statements *A* and *B* (which are either true or false, that is, have truth value 1 or 0) we can think of *A* and *B* as representing relay switches in a circuit. Then we can represent the logical function "*A* and *B*" by a circuit. If current is flowing in both *A* and *B*, then the "and" circuit permits current to flow; otherwise, it does not. You can see that this circuit is like two switches wired in series, the way Christmas tree lights often used to be. In a modern computer, this circuit, called an AND gate, is represented by a tiny intersection on a microchip and is designated in circuit diagrams by the symbol given in Figure 19.2a.

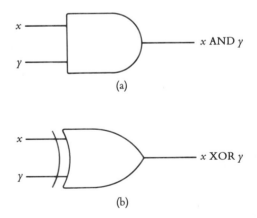

Figure 19.2 (a) The circuit diagram for an AND gate. (b) The circuit diagram for an XOR gate.

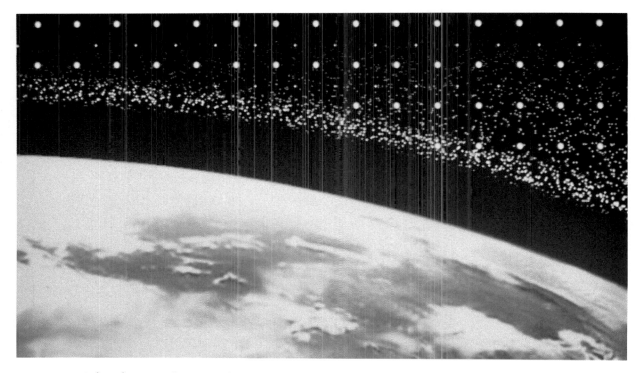

Color Plate 28   The continual development of better algorithms and faster computers is allowing scientists to devise more realistic models of complex systems such as the earth's atmosphere. (See Box 18.2.)

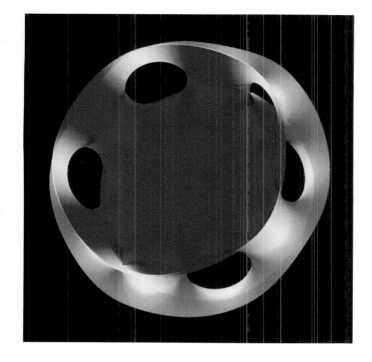

Color Plate 29   This surface represents an example in the infinite sequence of embedded minimal surfaces discovered by Hoffman and Meeks. A surface is called "minimal" when each small piece of it is curved like a soap film, which seeks to minimize its area. The example has genus 4. Genus is related to the number of holes in the surface. [Mathematics by David Hoffman and William H. Meeks III. Computer graphics by Jim Hoffman. Image produced at the University of Massachusetts using a Ridge computer and a Raster Technologies graphics controller.] (See Chapters 18 and 21.)

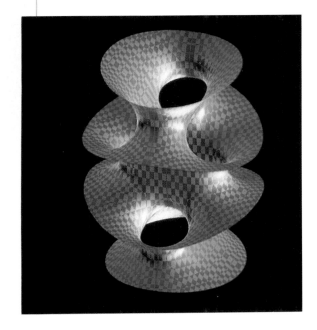

Color Plate 30 This minimal surface was discovered in March 1986 at the University of Massachusetts and is rendered here by a computer graphics system designed by Jim Hoffman. This is in fact a "complete embedded minimal surface with four ends," meaning that it can be extended forever, without intersecting itself, in four flat sheets or "ends." [Mathematics by Michael Callahan, David Hoffman, and William Meeks III. Computer graphics by Jim Hoffman. Image produced at the University of Massachusetts using a Ridge computer and a Raster Technologies graphics controller.] (See Chapters 18 and 21.)

(a)

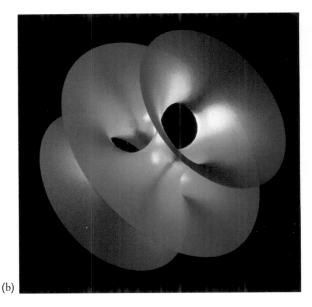

(b)

Color Plate 31 (a) This is a rendering of the Costa-Hoffman-Meeks surface (side view), which was shown in 1984 to be the first new "complete embedded minimal surface of finite total curvature" to be discovered in two centuries. That is, this is a minimal surface that goes on forever without intersecting itself, becoming flat as it extends outward. (b) The Costa-Hoffman-Meeks surface rotated and viewed from the top. [Mathematics by Celso Costa, David Hoffman, and William H. Meeks III. Computer graphics by Jim Hoffman. Image produced at the University of Massachusetts using a Ridge computer and a Raster Technologies graphics controller.] (See Chapters 18 and 21.)

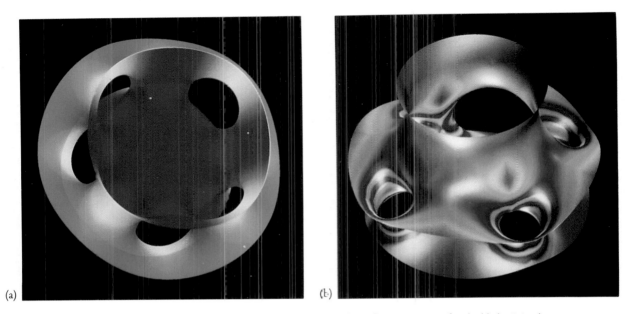

(a)   (b)

Color Plate 32   (a) This surface represents an example in the infinite sequence of embedded minimal surfaces discovered by Hoffman and Meeks. The example has genus 3. Genus is related to the number of holes in the surface. (b) This surface is a deformation of the minimal surface in part a. Hoffman and Meeks proved that this minimal surface could be deformed in a continuous way to yield a one-parameter family of complete minimal surfaces. During this deformation the surface changes by expanding on its right side and contracting on its left. It retains two reflective symmetries in planes but loses the straight lines and the rotational symmetries about these lines. [Mathematics by David Hoffman and William H. Meeks III. Computer graphics by Jim Hoffman. Image produced at the University of Massachusetts using a Ridge computer and a Raster Technologies graphics controller.] (See Chapters 18 and 21.)

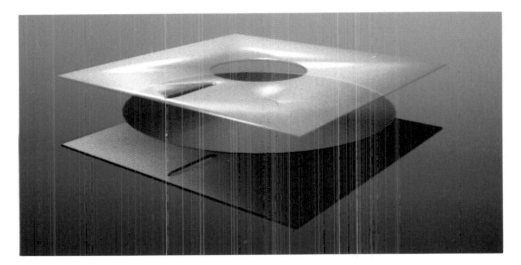

Color Plate 33   An exploded view of a floppy diskette in which information is stored on the circular magnetic media. When the diskette is placed in a disk drive, a magnetic head (similar in principle to that found in a tape recorder) can "read" the information stored on the spinning disk by sensing magnetic states (either positive or negative) and sending the patterns electronically to the computer. Data can be written to the disk by altering existing magnetic states. (See Chapter 19.)

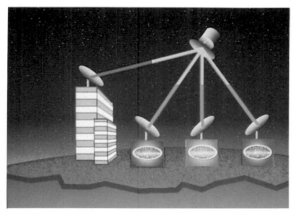

(a)

(b)

Color Plate 34 (a) Television signals are transmitted from the studio to a satellite and then to viewers with receivers on the ground. Viewers also receive a digital encryption signal that scrambles the audio or video portion of the programming signal. (b) The scrambled signal also contains the password encrypted as a string of binary bits. A decoder box can then read the password and unscramble the signal. (See Chapter 19.)

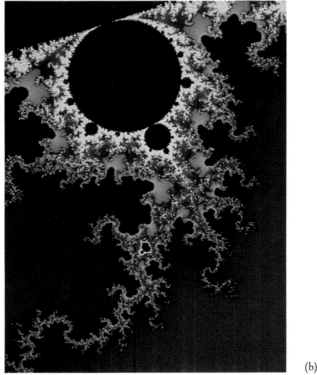

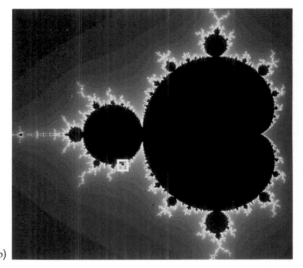

(a)

(b)

Color Plate 35 (a) Computer-generated enlargement of part of the boundary of the Mandelbrot set, whose members are complex numbers graphed as points in the complex plane. The points in the set are shown in black; the colored regions of the image represent numbers outside the set that, following a repeated operation, flee the boundary. The boundary of the Mandelbrot set is a fractal. (b) The Mandelbrot set. Note the area in the small square, which is seen enlarged in part a. [From H.-O. Peitgen and P. H. Richter, *The Beauty of Fractals,* Heidelberg, Springer-Verlag, 1986.] (See Box 21.2.)

(a)

(b)

Color Plate 36 (a) Enlargement of
the Mandelbrot set known as the "Tail
of the Seahorse." (b) Boundary region
showing area (in square) enlarged in
part a. [From H.-O. Peitgen and P. H.
Richter, *The Beauty of Fractals,*
Heidelberg, Springer-Verlag, 1986.]
(See Box 21.2.)

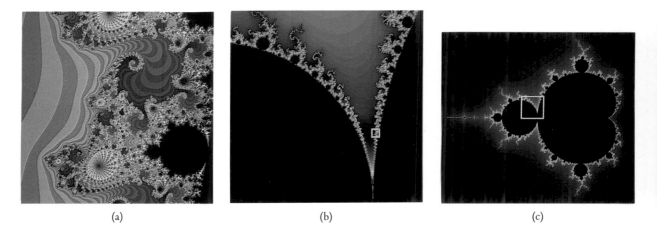

(a) (b) (c)

Color Plate 37 (a) Enlargement of a boundary portion of the Mandelbrot set known as "Seahorse Valley." (b) Small square represents region of enlargement in part a. (c) The Mandelbrot set showing the region (in the square) of enlargement for part b. [From H.-O. Peitgen and P. H. Richter, *The Beauty of Fractals,* Heidelberg, Springer-Verlag, 1986.] (See Box 21.2.)

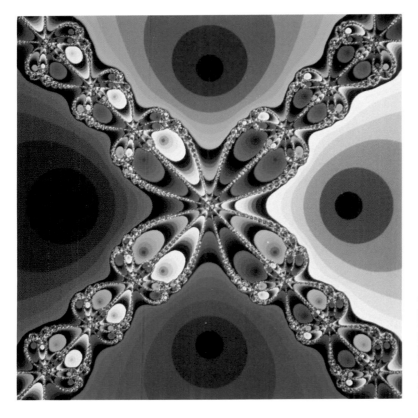

Color Plate 38 The four basins of attractions of the four solutions of $z^4 - 1 = 0$ by Newton's method. All regions of one particular color represent the set of starting points that converge to a common solution. Shades of the color indicate relative rates of convergence. [Image by H. E. Benzinger, S. A. Burns, J. Palmore/University of Illinois at Urbana-Champaign.] (See Box 21.3.)

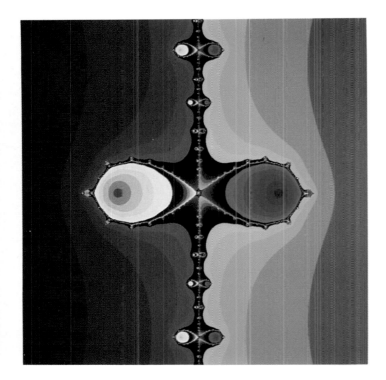

Color Plate 39 The four basins of attraction of the four solutions of $(z^2 - 1)(z^2 - 4) = 0$ by Newton's method. Red and green colored regions are the basins of attraction of the solutions $z = 1$ and $z = -1$. The complicated boundary occurs when all four basins meet at every point of the boundary of the regions. [Image by H. E. Benzinger, S. A. Burns, J. Palmore/University of Illinois at Urbana-Champaign.] (See Box 21.3.)

Color Plate 40 Each color represents a basin of attraction of one of the 17 zeros of $z^{17} - 1$. The black regions reveal a limitation of the computer: the computation needed to assign a color produces a number too small to be represented by the computer. [Image by H. E. Benzinger, S. A. Burns, J. Palmore/University of Illinois at Urbana-Champaign.] (See Box 21.3.)

Color Plate 41 Computer-generated image. [Computer graphics by Philip Zucco on equipment by Symbolics, Inc.] (See Chapter 21.)

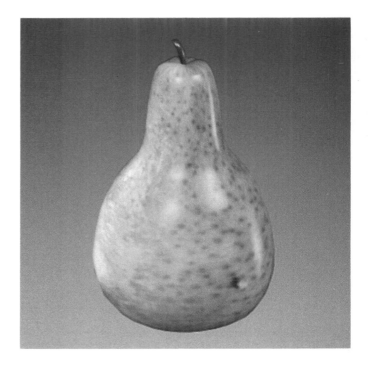

Color Plate 42 Three-dimensional computer-generated solid image of a pear. [Computer graphics by Philip Zucco on equipment by Symbolics, Inc.] (See Chapter 21.)

BOX 19.2 George Robert Stibitz, 1904–

George R. Stibitz first used telephone relays to calculate in binary notation and to control calculation steps in an early electromechanical computer. [Photo courtesy of Denison University Archives.]

George Stibitz (1904–) received a B.S. degree from Denison University in 1926 and went on to earn a Ph.D. in mathematical physics from Cornell University in 1930. From 1930 to 1941 he was a research mathematician with the Bell Telephone Laboratories. A search for mathematics that would define the state (open or closed) of switching circuits led him to use telephone relays both to calculate in binary notation and to control calculation steps in a computer that would replace a desk calculator. The Complex Calculator, first in a series of relay computers, was designed for calculation with complex numbers and time-sharing operation from remote consoles. It was placed in service in 1939 and was demonstrated in 1940, with a terminal in Hanover, New Hampshire, and the computer in New York.

Stibitz was a technical aide with the National Defense Research Committee, later reorganized as the Office of Scientific Research and Development, from 1941 to 1945. His model 2, the Relay Interpolator, used for ballistic calculations for gun directors, was placed in service in 1943. It utilized changeable programs and "self-checking." Later models in this series of relay computers included memory indexing, floating-point arithmetic, parallel operation, a logic calculator, automatic selection of programs, and jump program instructions.

From 1946 to 1954 Stibitz consulted in applied mathematics with clients in private industry and the government. Among the many subjects he investigated were propeller vibration computations, stability of discontinuous servos, an electronic computer for commercial application, small-tool design, blood flow in an elastic artery, a mechanical change computer for a machine vender, an electronic organ and tone synthesizer, a heart-beat rate alarm, an arterial pump, and a ballistocardiograph.

He became a professor in the physiology department at Dartmouth College in 1966 and was granted emeritus status in 1970. At Dartmouth he has engaged in applying mathematics to such biomedical problems as biochemical binding to plasma, radiation dosimetry, random-walk transmission of nutrients through capillary walls, and the electrical properties of cell membranes. The author of numerous scientific articles and two books, he received the Harry Goode Memorial Award in 1966, the Emanuel R. Piore Award in 1977, the Computer Pioneer Award in 1982, and was elected to the National Inventors Hall of Fame in 1983. He holds 34 patents and has received three honorary Doctor of Science degrees.

Other logical functions are fundamental both to Boolean algebra and to the design of computers. A particularly useful relationship, called the "exclusive or," abbreviated as XOR, has the following truth table:

	Statements		
	A	B	A XOR B
	0	0	0
	0	1	1
Truth values	1	0	1
	1	1	0

A XOR B is true if either A is true or B is true, but false if *both* A and B are true. Again, we can build a circuit to physically implement this function. An XOR circuit is designated by the engineering symbol given in Figure 19.2b. It is easy to confuse the XOR function with the OR function, which assigns a truth value of 1 to A or B if A, B, or *both* A and B are true.

Computer Addition

Modern computers derive their power not only from their ability to store information but also from their ability to perform operations on the stored data. To illustrate this latter capability of computers, we consider the simple problem of adding two binary integers, or integers represented in base 2. There are only four possible pairs of single-digit binary integers. These pairs of numbers and their corresponding sums are given by

$$\begin{array}{cccc} 0 & 0 & 1 & 1 \\ +0 & +1 & +0 & +1 \\ \hline 0 & 1 & 1 & 10 \end{array}$$

Note that in the last case, when we add 1 to 1, we get a sum of zero and a *carry* of 1. Re-

member that 10 is the way we represent decimal 2 in binary.

The simplest version of an adder, called the half adder, takes two single-digit inputs and produces a sum and a carry digit. The addition pattern we have just observed can be implemented by circuits that generate the truth values summarized below.

Input 1 x	Input 2 y	Output 1 Sum	Output 2 Carry
0	0	0	0
0	1	1	0
0	1	1	0
1	1	0	1

Because

$$\text{Sum} = x \text{ XOR } y$$

and

$$\text{Carry} = x \text{ AND } y$$

a circuit with the diagram given in Figure 19.3 will perform this addition.

Figure 19.4 shows the first adder, called the model K, which was built by Stibitz on his

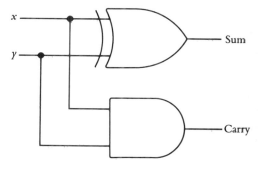

Figure 19.3 The circuit diagram for the half adder.

Figure 19.4 A replica of the model K computer built by George Stibitz. [Photo courtesy of Denison University.]

kitchen (that is where the *K* comes from) table in 1937. In the 50 years since Stibitz and Shannon, computers have evolved from automatic calculators into the most versatile machines ever designed. But they still work in essentially the same way, by acting on two-valued functions. We can now appreciate the secret of their flexibility: the 0s and 1s in these functions can represent numbers; but they can also represent other kinds of information, such as text for word processing, elements of a picture, and even musical notes.

Errors and Error Correction

Although errors by programmers and computer operators are typically called "computer errors," true computer errors occur infrequently. Errors that do occur could be caused by equipment failure or even by background radiation, which we can not control. In this section, we consider a method from Richard Hamming, a mathematician at Bell Laboratories, for detecting and correcting certain types of errors in binary data.

In 1948 Hamming proposed an efficient method of detecting and correcting errors in binary coded data. Suppose we need to store 4 bits (single binary digits) in memory in such a way that we can correct any single error in memory. We can arrange the 4 bits (0 1 1 0, in our case) in a diagram as shown in Figure 19.5a. Next, we fill the three empty spaces in the diagram with a 1 or a 0, so that each circle contains an even number of 1s, as illustrated in Figure 19.5b. We can think of these 7 bits as a representation, a redundant one, of the central 4 bits that we want to preserve in memory. The 7 bits are called an **error-correcting code** because through a simple algorithm, we can correct any single error. To understand how the algorithm works, suppose there is an error in one of the bits, the bit that is underlined in Figure 19.5c. When we count the number of 1s in each of the three circles, we find that two of them have an even number of 1s. We label these circles "good" and label the one circle

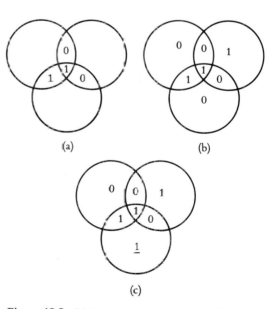

Figure 19.5 (a) We want to store the 4 bits 0, 1, 1, and 0. (b) To complete the Hamming code, we add error correction bits. (c) The underlined bit is an error.

with an odd number of 1s "bad." The bad circle tells us there is an error somewhere in the diagram. Once we have detected an error, we try to correct it.

Obviously, we should try changing the value of a bit to make all the circles good. Because changing any of the bits outside the bad circle is not going to help, we consider the 4 bits inside the bad circle. If we change the leftmost bit (from 1 to 0) or the rightmost bit (from 0 to 1), we fix the bad circle but ruin the one that was good. If we change the "innermost" 1 to 0, things get worse, because we end up fixing the bad circle but ruining both good circles. We have only one bit left to try: the one that has the error. When we change it, all the circles become good and the error is corrected.

The power of the **Hamming code** comes from the fact that there is just one way to fix a single error: the correct way. Note that in the Hamming code, each code "word" is 7 bits long (4 message bits plus 3 error-correcting bits), and every possible sequence of 7 bits is either a correct message or a message with a single correctable error. Thus, every code bit makes a contribution. Such codes are called perfect codes.

The Hamming code has its limitations, however. Suppose, owing to two errors, the code depicted in Figure 19.6a is transformed to the one in Figure 19.6b. If we count up the number of 1s, we find two good circles and one bad one. The error-correcting procedure can make all the circles good by changing one of the 1s to a 0 (Figure 19.6c). However, this makes matters worse by introducing a third error.

To appreciate the value of the Hamming code, consider a modest-size memory bank that has about a million 8-bit words. If no error-correcting mechanisms are in place, there will be a mistake due to radiation about every 43 days on the average. To use a single

error-correcting code similar to the one we just considered, we would have to enlarge the memory by a little over 20%. That should not be surprising because we cannot expect to get "error insurance" for free. This larger memory will be prone to even more frequent errors, about one every 36 days. (Each memory location has a certain chance of being struck by radiation, and the more memory we have, the greater the chance that some location will be hit.) It might seem as if we have made the problem worse. But remember that a single error can be fixed, and we have a problem only if there are two or more errors in the same code word. We can calculate that this will happen about once in 63 years. Thus, by designing a clever code, Hamming was able to reduce the frequency of errors from once every 43 days to once every 63 years.

Some computer applications, such as flight-control management, need more protection

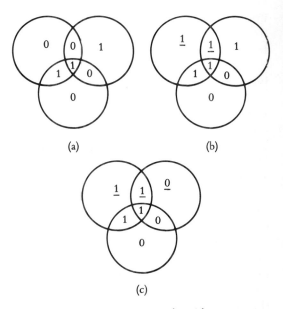

(a)   (b)

(c)

Figure 19.6 (a) A Hamming code, with error correction bits. (b) The code now has two errors (underlined). (c) A "corrected code" with three errors.

than can be provided by single-error correction. In some such cases, redundant computing units can be used to reduce the chance of error. In less-critical instances, more complex error-correcting codes can be used. A tiny flaw on a compact disk can affect a large number of bits. Therefore, a code that can correct up to 14,000 consecutive errors is used by many compact disk players. This code requires only 1 correcting bit for every 3 "sound bits," an overhead of only 33%.

Data Protection

The ability to correct errors in binary coded information protects against the loss of data from corruption owing to unpredictable circumstances. Our concern in this section is about another form of data protection, one that prevents unauthorized persons from interpreting coded data. Although the most obvious applications of data protection are in military data transmissions, we will consider a similar problem that occurs when commercial television signals are transmitted by satellite.

The new method of transmitting television pictures with the highest quality and lowest cost is by satellite transmission. The problem with this scheme is that anyone with a few thousand dollars can interpret the coded data by buying a satellite dish and bypassing the local cable company altogether, in the process avoiding payment of premiums to the local company and to national programming carriers such as HBO (Home Box Office). Because backyard dish owners could use their dishes to intercept transmissions from national program services to local cable television systems, television program services had to devise a way to protect their investment. As part of their data-protection effort, HBO, the largest of the premium television services, decided in 1984 to scramble its signals. To unscramble

the audio signal, cable television operators and authorized subscribers must have a password. Furthermore, to ensure the monthly payment of fees, HBO decided to change the password every month.

The only practical way to deliver the password every month to all paying customers is to send it by satellite right along with the scrambled signal. Thus, the password can be received by all those unauthorized backyard dish owners. To protect the secrecy of the password, HBO uses a coding scheme to alter the way the password is represented. The technical term for the process is **encryption;** encryption is just another form of **encoding** representing information by means of a set of symbols and rules for their interpretation. However, in the case of encryption, the rules for interpretation are kept secret.

HBO's monthly password is a string of 56 binary bits. The process of encrypting this password begins when a **pseudorandom-bit generator** (a computer algorithm) receives a specific 56-bit input and produces a sequence of 56 bits. The output sequence is called pseudorandom because the bits appear random, even though they are produced by a definite, repeatable procedure. "Repeatable" here means that if you restart the algorithm with the same input, you will get the identical output sequence. This pseudorandom sequence is combined with the password bit by bit to produce a **cipher,** or disguised form of the password. The cipher, not the password, is sent out by satellite, along with the scrambled TV signal. A cipher bit is computed to be 0 if the corresponding bits within the password and pseudorandom sequence are identical; otherwise, the cipher bit is a 1. You should recognize this as the "exclusive or" function that we encountered earlier.

Notice that the cipher alone contains no information whatsoever about the original password. For example, a 0 in the cipher has a 50%

BOX 19.3 Getting Your Signals Scrambled

Paul Heimbach, a vice-president at Home Box Office, tells why Home Box Office decided to scramble its signals and explains something about how the scrambling system works: HBO started using satellite distribution in 1975. In 1980 we began to see an increase in the number of unauthorized users receiving the HBO signal from the satellites. And in 1984 it's generally accepted that there were perhaps 1,000,000 backyard unauthorized dishes in operation. In 1980, as we saw the number of backyard dishes increase, we saw that it was necessary to somehow receive or scramble the HBO signal to prevent unauthorized reception of our programming.

The video is scrambled by making the picture into a negative, that is, lights become dark and darks become light. In addition, timing information in the video signal that's used by home TV sets to position the video picture on the screen is also removed. The audio signal is digitized much like a compact audio disk and encrypted. Authorized users receive a monthly key to activate their descrambler. Without this monthly key or password, the signal is totally unintelligible.

---

chance of standing for a 0 in the password and a 50% chance of standing for a 1. So even though the would-be satellite pirate can intercept the cipher, he can't deduce the password from it. Authorized subscribers, after receiving the encrypted password from HBO via satellite, use a decoder box provided by HBO to decrypt the cipher. Owing to the elegance of the binary representation, decrypting is identical to encrypting but performed in reverse. A microprocessor inside the decoder box has access to a copy of the original pseudorandom bit generator. It is given the proper input, and it produces the same sequence that it did for encryption at HBO. This sequence is then combined with the cipher to produce the password. The same procedure used to combine the password and the pseudorandom sequence for encryption is used to combine the cipher and the pseudorandom sequence for decryption. The procedure is its own inverse.

To understand how encryption and decryption work, let us suppose we are dealing with 8-bit binary sequences. If the password is 1100 1011 and the pseudorandom algorithm pro-

duces 0101 0110, then the encryption at HBO produces

$$
\begin{array}{ll}
& 1100\ 1011 \quad \text{password} \\
\text{XOR} & \underline{0101\ 0110} \quad \text{pseudorandom sequence} \\
& 1001\ 1101 \quad \text{cipher}
\end{array}
$$

The cipher 1001 1101 is transmitted via satellite to the subscriber. The HBO-supplied decoder produces the same pseudorandom sequence, 0101 0110, and combines it with the transmitted cipher to produce the password.

$$
\begin{array}{ll}
& 1001\ 1101 \quad \text{cipher} \\
\text{XOR} & \underline{0101\ 0110} \quad \text{pseudorandom sequence} \\
& 1100\ 1011 \quad \text{password}
\end{array}
$$

It might come as a surprise that the pseudorandom-number algorithm is public. You can get a copy from the National Bureau of Standards for the price of a self-addressed stamped envelope. The security comes from the initial input to the algorithm. There are many possible inputs, each of which causes the generator to produce a different pseudorandom sequence.

Each local cable company or dish owner has a personal input sequence. HBO produces a unique cipher for each subscriber, combining the monthly password with a pseudorandom sequence generated by that subscriber's unique input sequence. HBO actually transmits a stream of ciphers, only one of which can be decrypted by any particular box. If a cable operator's service contract changes or a dish owner does not pay his bills, HBO can withhold its programs from that subscriber just by erasing one input sequence.

Suppose you cannot resist the challenge to "beat the system." You can start by testing all possible input sequences. Since the input itself is a 56-bit sequence, you might as well try to guess the password directly. If you test all possible 56-bit sequences, one of them is bound to work. This approach is called an exhaustive, or more appropriately, an exhausting search. How long would it take? If passwords were only 4-bit instead of 56-bit sequences, there would only be 2^4, or 16, possibilities to try. One of these passwords will be the one that unscrambles the TV signal. Checking these would be pretty easy. But how long would it take to guess a 56-bit password? There are 2^{56} possible passwords. That's more than the number of stars in 100,000 galaxies the size of the Milky Way. If you tried a million passwords per second, it would take you on the average over 1000 years to hit on the correct one. By the time you find it, the show you want to catch will be over.

REVIEW VOCABULARY

Binary code A coding scheme that uses two symbols.

Cipher A message in disguised (encrypted) form.

Code A mechanism used to represent information. A code consists of a group of symbols and a set of rules for interpreting the symbols.

Encoding The conversion of information from one code to another.

Encryption The conversion from a standard coding scheme to a protected, or secret, coding scheme.

Error-correcting code A code in which certain, but not necessarily all, types of errors can be identified and corrected.

Hamming code: An error-correcting code, proposed by Richard Hamming, that can correct all single errors.

Place-value system A coding scheme in which the value of a symbol is determined in part by its position.

Truth table A tabular representation of the truth values of a statement.

EXERCISES

1. The negation of the statement A, NOT A, is defined to have the opposite value of A. Give a truth table for NOT A.

2. The logical inclusive or function of statements A and B, A OR B, yields a true value if and only if A is true or B is true or both A and B are true. Give a truth table for A OR B.

3. Use truth tables to show that for all possible truth values of A and B, NOT (A OR B) is identical to (NOT A) AND (NOT B).

4. Use truth tables to show that for all possible truth values of A and B, NOT (A AND B) is identical to (NOT A) OR (NOT B).

5. Use truth tables to show that for all possible truth values of A and B, A XOR B is identical to ((NOT A) AND B) OR (A AND (NOT B)).

6. Starting with 1010, construct a diagram similar to that shown in Figure 19.5b. Consider the four possible single-bit data errors that may result (0010, 1110, 1000, and 1011), and show in each case that the error can be detected and corrected.

7. Assume that the 4-bit pattern, 1010, of Exercise 6 is erroneously recorded as 1100. Can this error be detected? If so, can it also be corrected?

8. In general, is it possible that errors in two of the 7-bit (4 data bits and 3 error-correction bits) positions could go undetected? If so, illustrate by an example; otherwise give an explanation.

9. In general, is it possible that errors in three of the 7-bit (4 data bits and 3 error-correction bits) positions could go undetected? If so, illustrate by an example; otherwise give an explanation.

Computer Data Storage

We saw in the preceding chapter that modern computers can store and manipulate a variety of types of data coded, in all cases, in some suitable binary form. It is possible, for example, to devise a binary code for representing sound, and with enough detail embedded within the code, it is possible to store very faithful reproductions of musical performances.

Composers have always arranged and rearranged their music until it has just the feeling or structure they desire. Until recently, the product of their creative effort — a melody or a harmony — has been stored on record albums and tape units. However, today's musicians command a new kind of tool to record everything they have played: a music synthesizer with a built-in digital computer. Modern compact disks, digital tape devices, and music synthesizers are all equipped with microcomputers and digital storage mediums for the storage and reproduction of music.

For a specific data type (for example, visual image, sound, or numeric value), we have several ways to represent the data in binary coded form. The choice of a particular code usually depends on the storage and computational efficiency associated with that code. Integers are one of the simplest types of data manipulated by computers. We will first look at how they are stored within computers and then consider the problem of storing textual and pointer information (pointers designate the location of other data). Finally, we will examine the problem of data organization and its impact on the efficiency of computation.

Integers

Integers are the simplest type of numeric information that computers process. To appreciate how computers handle this problem, it helps to review how we humans store integers, that is, how we write them on a piece of paper. When we write 5472, we use the sequence of decimal digits 5, 4, 7, and 2 to represent the integer that is equal to

$$5 \times 10^3 + 4 \times 10^2 + 7 \times 10^1 + 2 \times 10^0$$

(Recall that in exponential notation, $10^0 = 1$.) The significance of each digit in our representation of 5472 is determined by the digit itself and by the position of that digit within the sequence that comprises 5472. The two 7s in the representation of 7174 have different interpretations: the leftmost digit has the factor of 10^3 associated with it, whereas the rightmost 7 is associated with the factor 10^1. The ordinary notation that we use to write integers is called the **decimal positional notation** because we associate powers of 10 with the positions of the digits within the sequence of digits.

If we had a computer that could store items such as digits (for example, if its storage cells could assume any one of 10 different states), we could store integers in a computer in much the same way we record them on paper. The 10 states then could be used to represent the digits 0, 1, . . . , 9; numbers such as 3438 could be represented as a sequence of states within the computer that represent the digits 3, 4, 3, and 8, in that order. However, because of the electromagnetic materials used to build them, computers are generally capable of storing things only in 1 of 2 states, not in 1 of 10 states.

Representation of Positive Integers within Computers

To store information on a medium whose storage cells can be in only one of two states, we can associate the digit 1 with one of the states, say the positive state, and the digit 0

with the negative state. In this two-state storage scheme, each of the two digits, 0 and 1, is called a *bit,* short for binary digit. By combining bits in long sequences, we can store anything that can be represented by sequences of 0s and 1s.

If we choose to write integers in base-2 positional notation (instead of in base 10), then we can represent all positive integers by a sequence of 0s and 1s. Thus, the sequence 1101101 would represent

$$1 \times 2^6 + 1 \times 2^5 + 0 \times 2^4 + 1 \times 2^3 \\ + 1 \times 2^2 + 0 \times 2^1 + 1 \times 2^0 = 109$$

To store a number such as 353, we need to reverse this calculation to find the digits of the base-2 positional representation of 353. This can be done through an algorithm that successively subtracts the largest possible powers of 2 from 353:

$$
\begin{aligned}
353 &= 256 + 97 \\
&= 2^8 + 97 \\
&= 2^8 + 64 + 33 \\
&= 2^8 + 2^6 + 33 \\
&= 2^8 + 2^6 + 32 + 1 \\
&= 2^8 + 2^6 + 2^5 + 2^0
\end{aligned}
$$

Hence, $353_{10} = 101100001_2$, where the subscripts indicate the base of the positional notation.

From these illustrations, you can see that 7 binary digits (bits) are needed to store 109, whereas 9 bits are needed to store 353. In actual computer implementations, a fixed number of contiguous, or adjacent, bits are grouped to form a **word,** and all integers are stored within the bits of a single word. In a computer with 16-bit word size, 109 and 353 would be stored as

0000   0000   0110   1101

0000   0001   0110   0001

respectively. Most small microcomputers —for example, Apple II, TRS-80, Commodore, and Atari—use 16 bits for integer storage. Some of the newer microcomputers as well as most medium-size computers use 32 bits for integer storage.

Here are two more examples that illustrate the way ordinary integers are encoded into bits and then stored in words. In a computer that uses 12-bit words to store integers, the integer represented by the word 0011 1010 1101 is

$$2^9 + 2^8 + 2^7 + 2^5 + 2^3 + 2^2 + 2^0$$
$$= 512 + 256 + 128 + 32 + 8 + 4 + 1$$
$$= 941$$

Conversely, to determine the representation of the integer 1814 in a computer that uses 12-bit words, we write 1814 as a sum of powers of 2. Because the largest power of 2 less than 1814 is $2^{10} = 1024$, the calculations re-

quired by the algorithm described earlier look like this:

$$1814 = 1024 + 790$$
$$= 2^{10} + 790$$
$$= 2^{10} + 512 + 278$$
$$= 2^{10} + 2^9 + 278$$
$$= 2^{10} + 2^9 + 256 + 22$$
$$= 2^{10} + 2^9 + 2^8 + 22$$
$$= 2^{10} + 2^9 + 2^8 + 16 + 4 + 2$$
$$= 2^{10} + 2^9 + 2^8 + 2^4 + 2^2 + 2^1$$

Hence, 1814 is represented by the sequence 0111 0001 0110, where 1s represent powers of 2 that are present (in positions 10, 9, 8, 4, 2, 1) and 0s represent powers of 2 that are absent.

Once stored in words inside the computer, the patterns of bits that represent numbers can be added in a manner very similar to the method we use with a pencil and paper to add ordinary decimal numbers. The details are explained in Box 20.1.

BOX 20.1 How Computers Add

Suppose we use a computer that stores integers in 16-bit words: 4502 and 1234 are then encoded as

$$4502 \longrightarrow 0001\ 0001\ 1001\ 0110$$

and

$$1234 \longrightarrow 0000\ 0100\ 1101\ 0010$$

(The symbol $\rightarrow$ signifies "is stored as.")

To compute the sum of 4502 and 1234, the computer can work directly on the bits representing 4502 and 1234 and perform the logical equivalent of bit-by-bit addition in the same way that we do ordinary addition: add columns from right to left, carrying when necessary:

```
         0 0000 0011 0010 110      carry bits
4502 →     0001 0001 1001 0110
1234 →     0000 0100 1101 0010
6736 →     0001 0110 0110 1000
```

Here are the steps in detail:

1. Add the bits 0 and 0 in the rightmost column, producing a 0 and a carry bit of 0.

2. Next, add 1 and 1 and the 0 carry bit from the previous addition, obtaining a sum of 10, which is recorded as 0 with a carry bit of 1.

3. Next, add the bits 1 and 0 and the 1 carry bit from the previous addition, getting a sum of 10: record the 0 and carry the 1.

4. Continue the process from right to left.

Computers contain special circuitry to do addition in binary form, just as we have illustrated here.

Representation of Negative Integers

The simplest way to handle negative integers is first to reserve one of the bits, usually the leading or leftmost bit, as a designation of the sign of the integer and then to use the remaining bits for the binary representation of the absolute value of the integer. In such a scheme, if we had a 12-bit word size, we would use 11 bits to represent the integer and 1 bit for the sign. The common agreement on how to interpret the sign bit could be to use a 0 for positive integers and a 1 for negative integers. Using this scheme, 109 would be stored as 0000 0110 1101, and −109 as 1000 0110 1101. This method of storing integers is called the **sign-magnitude** method, where the leading bit represents the sign and the remaining bits represent the magnitude or absolute value of the integer.

The main disadvantage of the sign-magnitude method is that it complicates computer implementation of arithmetic. To add integers represented in this format, different logical steps must be followed, depending on the signs of the integers. If both integers are positive, that is, if both have 0 leading bits, then the sum of the two integers would have to be computed in two steps: first, the computer-stored binary numbers are added; second, the leading bit of the sum is set to 0 to indicate that the sum is positive. Similarly, if both the integers are negative, then the leading bit of the sum must be set to 1 to indicate a negative sum. A more complicated algorithm has to be followed to determine the sum of integers of different signs, because the sign of the sum would be determined by the magnitudes of the two integers, not just by their signs. Because of these complications, the sign-magnitude method is not often used to store integers.

A better strategy for storing negative integers is to use the **one's complement** of the binary number that represents the positive integer: the sequence of bits that are the exact opposite of the original bits. In this scheme, when the 12-bit representation of 109 is 0000 0110 1101, the negative integer −109 would be coded as 1111 1001 0010.

To illustrate the use of negative integers in one's-complement notation, consider the addition of 4502 and −1234.

```
             1 1110 0110 0111 100   carry bits
  4502 →       0001 0001 1001 0110
 −1234 →       1111 1011 0010 1101
  3268         0000 1100 1100 0011
```

A quick look at the result should indicate to you that something is wrong. The computed sum, 0000 1100 1100 0011, represents 3267, not 3268. (Note the 1 in the rightmost position — sure sign of an odd number.) What has gone wrong here is that the last carry bit is not being captured. To correct this error, the last carry bit is always added to the results of the bit-by-bit addition. This additional 1 will correct the rightmost bit and force a corresponding change in the rest of the numbers:

$$
\begin{array}{r}
0000\ 1100\ 1100\ \ 0011 \\
+\ \ \ \ 1 \\
\hline
0000\ 1100\ 1100\ \ 0100
\end{array}
$$

(For sums of two positive integers, as in Box 20.1, this carry bit is zero, so it can be ignored.)

The necessity of adding the last carry bit to the bit-by-bit sum complicates the logic of integer addition. However, the ability to add integers without concern for their sign, positive or negative, makes this type of addition considerably simpler than addition in sign-magnitude format.

To further simplify integer arithmetic, computer scientists developed yet another method of storing integers. In this scheme, called **two's complement,** negative integers are coded by first complementing all the bits of the original (positive) integer, and then adding 1 to the complemented bits. For example, in a 12-bit configuration, the representation of -109 would be obtained by taking the complement of the representation of 109 and then adding 1, all in binary:

$$
\begin{array}{ll}
109 \rightarrow & 0000\ 0110\ 1101 \\
-109 \rightarrow & \\
& \text{complement } (0000\ 0110\ 1101) + 1 \\
& = 1111\ 1001\ 0010\ +1 \\
& = 1111\ 1001\ 0011
\end{array}
$$

The 16-bit two's-complement representation of -4502 and -1234 would be

$$
\begin{array}{ll}
4502 \longrightarrow & 0001\ 0001\ 1001\ 0110 \\
-4502 \longrightarrow & 1110\ 1110\ 0110\ 1001 + 1 \\
& = 1110\ 1110\ 0110\ 1010
\end{array}
$$

$$
\begin{array}{ll}
1234 \longrightarrow & 0000\ 0100\ 1101\ 0010 \\
-1234 \longrightarrow & 1111\ 1011\ 0010\ 1101 + 1 \\
& = 1111\ 1011\ 0010\ 1110
\end{array}
$$

Notice that the convention of having a leading 0 bit for positive integers and a leading 1 bit for negative integers is continued. Also — and very important — the representation of x can be obtained by taking the two's complement of the representation of $-x$:

$$
\begin{array}{rl}
-4502 \rightarrow & 1110\ 1110\ 0110\ 1010 \\
-(-4502) \rightarrow & 0001\ 0001\ 1001\ 0101 + 1 \\
\rightarrow & 0001\ 0001\ 1001\ 0110
\end{array}
$$

To find the decimal value of the integer x represented in two's-complement notation by 1110 1101 0010 0010, we note first that x must be negative because its leading bit is 1. Hence, the absolute value of x must be represented by $TC(x)$, the two's complement of x:

$$
\begin{array}{rl}
TC(x) =\ & TC(1110\ 1101\ 0010\ 0010) \\
=\ & 0001\ 0010\ 1101\ 1101 + 1 \\
=\ & 0001\ 0010\ 1101\ 1110
\end{array}
$$

Converting $TC(x)$ from binary to decimal, we get (reading from right to left),

$$
\begin{array}{r}
2 + 4 + 8 + 16 + 64 + 128 + 512 + 4096 \\
= 4830
\end{array}
$$

hence $x = -4830$.

We can still do integer addition by bit-by-bit additions, but in the two's-complement case, we can ignore the last carry bit (see Box 20.2). Although two's-complement representation is somewhat more complicated, integer addition is now simplified because we can ignore the carry bit at the end of the bit-by-bit addition without jeopardizing the result.

BOX 20.2 Two's-Complement Arithmetic

To illustrate the way two's-complement notation facilitates arithmetic in computers, we will do two arithmetic problems:

$$
\begin{array}{cc}
4502 & 1234 \\
-1234 & \text{and} \quad -4502 \\
\end{array}
$$

First, we need to find the binary representation of 4502 and 1234:

$$4502 = 4096 + 216 + 128 + 16 + 4 + 2 \longrightarrow 0001\ 0001\ 1001\ 0110$$

$$1234 = 1024 + 128 + 64 + 16 + 2 \longrightarrow 0000\ 0100\ 1101\ 0010$$

Now we can compute the two's complement of each of these binary numbers to get the representations of -4502 and -1234.

$$
\begin{aligned}
-1234 \longrightarrow \ & \mathrm{TC}(0000\ 0100\ 1101\ 0010) \\
& = 1111\ 1011\ 0010\ 1101 + 1 \\
& = 1111\ 1011\ 0010\ 1110
\end{aligned}
$$

$$
\begin{aligned}
-4502 \longrightarrow \ & \mathrm{TC}(0001\ 0001\ 1001\ 0110) \\
& = 1110\ 1110\ 0110\ 1001 + 1 \\
& = 1110\ 1110\ 0110\ 1010
\end{aligned}
$$

The bit-by-bit addition for $4502 - 1234$ will be

$$
\begin{array}{lr}
& 1\ 1110\ 0110\ 0111\ 110 \qquad \text{carry bits} \\
4502 \longrightarrow & 0001\ 0001\ 1001\ 0110 \\
-1234 \longrightarrow & \underline{1111\ 1011\ 0010\ 1110} \\
& 0000\ 1100\ 1100\ 0100
\end{array}
$$

The answer, $2048 + 1024 + 128 + 64 + 4 = 3268$, is the sum of 4502 and -1234.
We handle the other problem in much the same way:

$$
\begin{array}{lr}
& 0\ 0001\ 1001\ 1000\ 010 \qquad \text{carry bits} \\
-4502 \longrightarrow & 1110\ 1110\ 0110\ 1010 \\
1234 \longrightarrow & \underline{0000\ 0100\ 1101\ 0010} \\
& 1111\ 0011\ 0011\ 1100
\end{array}
$$

However, because the answer now begins with a 1, it must represent a negative number, which we can find by taking the negative of the integer represented by its two's complement.

$$
\begin{aligned}
\mathrm{TC}(1111\ 0011\ 0011\ 1100) & = 0000\ 1100\ 1100\ 0011 + 1 \\
& = 0000\ 1100\ 1100\ 0100
\end{aligned}
$$

This yields $4 + 64 + 128 + 1024 + 2048 = 3268$; therefore, $1111\ 0011\ 0011\ 1100$ represents -3268.

Word Size

One of the decisions that must be made in designing a computer is the number of bits to be allocated to integer representation. If relatively few bits are used, then only integers with small magnitudes can be stored and manipulated directly by the computer. On the other hand, if a large number of bits are used, fewer integers can be stored in a fixed portion of a memory or storage device. Suppose a computer uses 8 bits to store integers. Regardless of the precise storage scheme (sign-magnitude, one's complement, or two's complement), only 7 bits would be available to represent positive integers. Thus, 0111 1111, or 127, would be the largest integer that could be accommodated, and in two's-complement form, 1000 0000, or -128, would be the smallest integer. For almost any application involving integer arithmetic, these would be severe limitations.

In general, if n bits are used for integer storage, then the largest integer that could be accommodated will be 011 . . . 1 representing

$$2^{n-2} + 2^{n-3} + \cdots + 2 + 1 = 2^{n-1} - 1$$

Most microcomputers use 16 bits for integer storage. The largest integer that can be directly manipulated by such machines is therefore

$$2^{15} - 1 = 32,767$$

Most medium-size computers and some microcomputers use 32 bits to store integers. In these cases, integers up to and including

$$2^{31} - 1 = 2,147,483,647$$

can be handled.

Text Data

A large number of computing applications, such as automated letter writing, billing, and interoffice communications, deal mostly, if not exclusively, with text data. Formally, **text data** signifies sequences of characters from the character set consisting of 26 uppercase letters, 26 lowercase letters, 10 digits, about 8 punctuation marks, and a number of special characters, including "@," "%," and "$." In addition to these 90 or so printable characters representing symbols that appear in written text are a number of nonprintable characters whose effects are visible on the computer screen but that do not represent written text. These are called **control characters:** they govern how data is transmitted and the way the computer screen behaves. Characters that cause carriage returns, line feeds, and page breaks are examples of control characters.

To store text data or, more simply, a single character such as "A," "7," or "?," we need to devise a code to store these characters as sequences of binary bits. If such a code consisted of 2 binary bits, then only 4 characters, represented by 00, 01, 10, and 11, could be accommodated. If 3 bits are used, 8 different characters can be represented by 000, 001, 010, 011, 100, 101, 110, and 111. If n bits are used, then 2^n characters can be represented.

Unfortunately, no single character coding is accepted as an industry standard. Currently systems employ character sets of size 64, 128, and 256. The two most widely accepted character codes are the *Extended Binary Coded Decimal Interchange Code* (EBCDIC) and the *American Standard Code for Information Interchange* (ASCII). Because ASCII is the preferred coding scheme on microcomputers, we will confine our discussion to it.

In ASCII format, a **byte** (8 consecutive bits) is used to store each character. The full set of printable characters, their binary representations, and their decimal equivalents are shown in Table 20.1. Notice that uppercase letters start with a code of 65 for *A* and progress sequentially to 90 for *Z*. The lowercase letters have codes ranging from 97 for *a* to 122 for *z*.

TABLE 20.1 Printable characters and their ASCII codes

Character	ASCII code (decimal)	ASCII code (binary)	Character	ASCII code (decimal)	ASCII code (binary)
	32	0010 0000	8	56	0011 1000
!	33	0010 0001	9	57	0011 1001
"	34	0010 0010	:	58	0011 1010
#	35	0010 0011	;	59	0011 1011
$	36	0010 0100	<	60	0011 1100
%	37	0010 0101	=	61	0011 1101
&	38	0010 0110	>	62	0011 1110
'	39	0010 0111	?	63	0011 1111
(	40	0010 1000	@	64	0100 0000
)	41	0010 1001	A	65	0100 0001
*	42	0010 1010	B	66	0100 0010
+	43	0010 1011	C	67	0100 0011
'	44	0010 1100	D	68	0100 0100
-	45	0010 1101	E	69	0100 0101
.	46	0010 1110	F	70	0100 0110
/	47	0010 1111	G	71	0100 0111
0	48	0011 0000	H	72	0100 1000
1	49	0011 0001	I	73	0100 1001
2	50	0011 0010	J	74	0100 1010
3	51	0011 0011	K	75	0100 1011
4	52	0011 0100	L	76	0100 1100
5	53	0011 0101	M	77	0100 1101
6	54	0011 0110	N	78	0100 1110
7	55	0011 0111	O	79	0100 1111

The blank character has code 32 and the null character (it is convenient to have a character that does not produce any effect on the computer screen) has code 0.

Execution of the simple BASIC program in Figure 20.1 or the equivalent Pascal program in Figure 20.2 will produce the code table used on a particular computer. These programs will transmit the printable as well as nonprintable characters to a computer terminal or printer. You can observe the effect of these nonprint-

```
10   rem Program to Generate Character Code Table
20   print "CODE","CHARACTER"
30   for I = 0 to 127
40        print I, chr$(I)
50   next I
60   end
```

Figure 20.1   A BASIC program for generating ASCII characters and code.

TABLE 20.1 Printable characters and their ASCII codes *(continued)*

Character	ASCII code (decimal)	ASCII code (binary)	Character	ASCII code (decimal)	ASCII code (binary)
P	80	0101 0000	h	104	0110 1000
Q	81	0101 0001	i	105	0110 1001
R	82	0101 0010	j	106	0110 1010
S	83	0101 0011	k	107	0110 1011
T	84	0101 0100	l	108	0110 1100
U	85	0101 0101	m	109	0110 1101
V	86	0101 0110	n	110	0110 1110
W	87	0101 0111	o	111	0110 1111
X	88	0101 1000	p	112	0111 0000
Y	89	0101 1001	q	113	0111 0001
Z	90	0101 1010	r	114	0111 0010
[	91	0101 1011	s	115	0111 0011
\	92	0101 1100	t	116	0111 0100
]	93	0101 1101	u	117	0111 0101
^	94	0101 1110	v	118	0111 0110
_	95	0101 1111	w	119	0111 0111
`	96	0110 0000	x	120	0111 1000
a	97	0110 0001	y	121	0111 1001
b	98	0110 0010	z	122	0111 1010
c	99	0110 0011	{	123	0111 1011
d	100	0110 0100	|	124	0111 1100
e	101	0110 0101	}	125	0111 1101
f	102	0110 0110	~	126	0111 1110
g	103	0110 0111			

```
10   program info1 (output);
20   {This program produces a table of ASCII for the characters with codes from 0 through 127}
30
40   var code : integer;
50   begin
60       writeln; writeln;
70       writeln(' CODE    CHARACTER');
80       for code := 0 to 127 do
90           writeln(code:4, chr(code):12)
100  end
```

Figure 20.2 A Pascal program for generating ASCII characters and code.

able control characters without any specific character actually being exhibited. If a listing of only the printable characters is desired, the loop counters in both programs should start at 32.

Pointer Data

In many computing applications, including those found in such diverse areas as banking, inventory control and payroll management, the structure of each item of data is composed of many distinct subportions. For example, a student record may consist of the following:

Full name	Address		Credits	GPA	Mailbox
25 bytes	100 bytes		4 bytes	4 bytes	4 bytes

Figure 20.3   A sample student record.

Bates David A	. . .		. . .	. . .	. . .
Bonar Daniel D	. . .		. . .	. . .	. . .
Cameron James S	. . .		. . .	. . .	. . .
Snyder Rita E	. . .		. . .	. . .	. . .
Sterrett Andrew	. . .		. . .	. . .	. . .
Stout Elliott R	. . .		. . .	. . .	. . .
Wetzel Marion D	. . .		. . .	. . .	. . .

Figure 20.4   An array of student records.

the name of the student (up to 25 characters long); the student's address (up to 100 characters long); the number of credit hours the student has accumulated (an integer); the student's grade point average (a real number); and the student's mailbox number (an integer). Figure 20.3 shows the layout of a student record in graphic form, with an indication of the number of bytes dedicated to each subsection of the record. In this example, each record would be 137 bytes long.

One way of organizing the data for, say, 10,000 students would be to take a large enough area (1,370,000 bytes) on some storage medium and alphabetically locate the 10,000 records in contiguous positions in that area. This type of structure, called an **array** of records, is illustrated in Figure 20.4.

Although structurally simple, arrays can have serious disadvantages. Suppose that in the illustration given in Figure 20.4, the second student, Daniel Bonar, drops out of school. To delete Bonar's record, records numbered 3 through 10,000 (9998 in all) would have to be moved up one spot. This means moving $9998 \times 137 = 1,369,726$ bytes of data, which would require a significant amount of computing resources. Of course, if the record deleted comes from the lower portion of the array, fewer records would have to be moved. On the average, 5000 record or 685,000 byte movements would be required for each record deletion or addition.

A more efficient but somewhat more complicated storage organization will result if we attach to each record an additional field containing the address of the record alphabetically next in line. This type of data structure is called **address,** or **pointer,** data. Because we can now move from one record to the next by following pointers, we may store records anywhere and need not locate them contiguously. Figure 20.5 gives the layout of a single record with a 4-byte pointer subfield. Figure 20.6 illustrates the **linked list** structure for the entire data set.

To delete student Andrew Sterrett's record, we follow the pointers to Rita Snyder's record and redefine the pointer embedded there, so that it points to Elliot Stout's record (see Figure 20.7). To insert a new record for Emmett Buell, we traverse the linked list until we determine the proper location for this record (between Daniel Bonar and James Cameron, in this case) and redefine the two pointers in the manner indicated in Figure 20.8.

To manage storage efficiently when records are added and deleted, the computer must have the ability (1) to handle storage that is released once records have been deleted and (2) to locate new storage areas to accommodate records that are to be inserted. Provisions for pointer

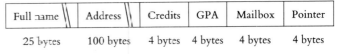

Full name		Address		Credits	GPA	Mailbox	Pointer
25 bytes		100 bytes		4 bytes	4 bytes	4 bytes	4 bytes

Figure 20.5 A single record with a 4-byte pointer subfield

data as well as mechanisms for access and release of storage areas are built into certain higher-level programming languages such as Pascal and PL/1.

Data Organization

Most computers are **byte addressable,** in the sense that a unique location number, or address, identifies each byte of memory. The

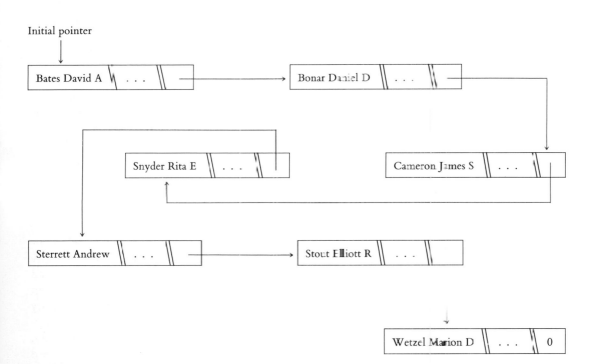

Figure 20.6 A linked list structure of student records.

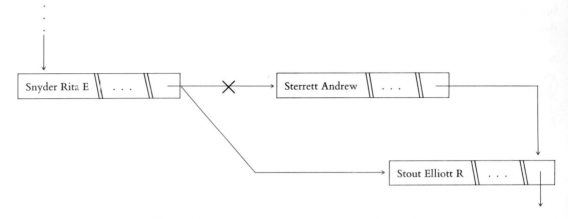

Figure 20.7 Redefining pointers for deletion of a record.

overall memory size is usually given by the total number of bytes that make up the memory. A memory size of 16K bytes (or 16 kilobytes) refers to $16 \times 2^{10} = 16 \times 1024 = 16{,}384$ bytes of memory addressed by 0, 1, . . . , 16,383.

As we have seen, the byte is a convenient unit for storing characters. Larger numbers of bits generally are required for storing integers and other, more complex types of data. This is accomplished by using 2, 3, or 4 contiguous bytes for storing 16-, 24-, or 32-bit integers. Suppose a computer uses the 8-bit ASCII code for storing characters and a 16-bit two's-complement format for storing integers. If two contiguous bytes within the memory of this

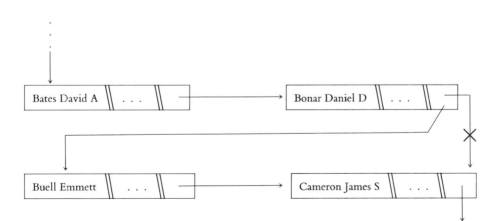

Figure 20.8 Redefining pointers for the insertion of a record.

computer contain the binary configuration

$$0100 \quad 0100 \quad 0100 \quad 0001$$

then the content of these two bytes could be the characters A and D (stored in right-to-left order), or the 16 bits could represent the integer

$$2^{14} + 2^{10} + 2^6 + 2^0$$
$$= 16,384 + 1024 + 64 + 1 = 17,473$$

Just by looking at the content of these two bytes, we cannot determine whether the data located there is the integer 17,473, the two characters A and D, or possibly some other form of encoded data.

Knowing the *context*, or how the data is to be used, removes the ambiguity in interpretation. If an integer addition is being executed and the addition instruction specifies a context for 0100 0100 0100 0001, then the data is interpreted as the integer 17,473. If, however, our task was to print out two characters and we encountered 0100 0100 0100 0001, then in this context, this bit configuration would be interpreted as the two characters A and D.

Concluding Comments

Any type of data that can be coded in binary form can be stored in modern computing systems. Specific binary coding schemes are used for representing integers (sign-magnitude, one's complement, and two's complement) and text data (ASCII code). Obviously, coding methods exist for representing other commonly used data types such as real numbers, or, as computer scientists sometimes call them, floating point numbers. It is also possible, through clever codes, to store more exotic data types such as visual images and musical scores.

As we saw in this chapter, to store a large number of records, it is better to include a pointer field within each record. The use of pointers allows records to be added and deleted more efficiently. In general, as coding schemes become more complex, the issues of storage efficiency and computational efficiency become more significant.

REVIEW VOCABULARY

Address A numeric value representing the location of data in memory.

Array A collection of similar data objects stored sequentially in memory.

ASCII (American Standard Code for Information Interchange) A binary coding scheme for representing character data.

Bit The smallest storage unit in computing systems, generally represented by a binary digit.

Byte A group of 8 consecutive bits in a computer-storage medium such as memory or floppy disk.

Byte-addressable computer A computer in which each byte of memory is uniquely addressed by successive integers starting with 0.

Control character A nonprintable character that, when transmitted to an output device such as a console or a printer, governs how the device behaves.

Decimal positional notation The usual (base 10) notation for writing integers.

EBCDIC (Extended Binary Coded Decimal Interchange Code) A binary coding scheme for representing character data.

Linked list A collection of data elements in which all elements except the last one have a pointer to their successor.

One's complement A binary representation for storing integers, in which nonnegative integers are coded by their usual binary digits. Negative integers are represented by the complements of the bits used by their positive counterparts.

Pointer data A data element whose value represents the memory address of other data.

Sign-magnitude A coding scheme for storing integers, in which the leading bit represents the sign of the integer and the remaining bits represent the magnitude (absolute value) of the integer.

Text data Data comprised of the characters consisting of the letters of the alphabet, decimal digits, punctuation marks, and special characters such as "@", "%", and "$."

Two's complement A binary representation for storing integers in which nonnegative integers are coded by their usual binary digits. Negative integers are represented by the bits obtained by complementing the bits of the corresponding positive integer and adding 1 to the result.

Word A group of contiguous bits. The number of bits within the group varies among computers.

EXERCISES

1. Assume that a computer is using 16 bits to store integers in sign-magnitude form.

   a. What bit patterns will represent the integers 3419, −3419, 17,842, and −17,842?

   b. What integers are represented by the bit patterns 0000 0011 1010 0110 and 1111 0101 1100 0010?

   c. What is the largest integer that can be stored in this system?

2. Assume that a computer is using 16 bits to store integers in one's-complement form.

   a. What bit patterns will represent the integers 3419, −3419, 17,842, and −17,842?

   b. What integers are represented by the bit patterns 0000 0011 1010 0110 and 1111 0101 1100 0010?

   c. What is the largest integer that can be stored in this system?

3. Assume that a computer is using 24 bits to store integers in two's-complement form.

   a. What bit patterns will represent the integers 3419, −3419, 17,842, and −17,842?

   b. What integers are represented by the bit patterns 0000 0000 0000 0011 1010 0110 and 1111 1111 1111 0101 1100 0010?

   c. What is the largest integer that can be stored in this system?

4. On a computer that uses a 12-bit one's-complement integer representation, if an integer has the bit configuration 1000 1001 1101, what bit configuration will this same integer have on a 12-bit two's-complement system?

5. In decimal positional notation, the presence of a 0 as the least significant digit is a simple test for divisibility by 10. Similarly, the presence of a 0 as the least significant digit is a test for divisibility by 2 when binary positional notation is used. Is the presence of a 0 in the rightmost position a test for divisibility by 2 when the sign-magnitude storage scheme is used? (Consider positive and negative integers.) What if one's-complement or two's-complement schemes are used?

6. Complete the following table, which gives the 4-bit integer representations of sign-magnitude, one's-complement, and two's-complement methods.

Bit configuration	Sign magnitude	One's complement	Two's complement
0000	0	0	0
0001	.	.	.
0010	.	.	.
0011	.	.	.
0100	.	.	.
0101	.	.	.
0110	.	.	.
0111	.	.	.
1000	.	.	.
1001	.	.	.
1010	.	.	.
1011	.	.	.
1100	.	.	.
1101	.	.	.
1110	.	.	.
1111	−1	0	−1

7. In this chapter we discussed the advantages of using pointers when storing 10,000 student records (see Figures 20.5 and 20.6). One disadvantage of using pointers is the storage space they require. How many bytes of additional storage will be necessary to store the 10,000 student records in the format described in Figure 20.5? What percentage increase of storage space does this represent over the array method of storage?

8. In the array structure, an average of $N/2$ record movements was required to insert a record in an N-record array. If we wanted initially to build the entire array by successive record insertions, N would grow as we made insertions. On the average, how many record movements will be required to build the entire array by successive insertions?

9. (For students with programming backgrounds.) Use one of the programs given in Figure 20.1 or Figure 20.2 to determine whether the computer you are using stores characters by their ASCII codes. By making a suitable adjustment to these programs, you can find out if the computer you are using has a 256 character implementation.

10. (For students with programming backgrounds.) Most BASIC compilers and interpreters allow the user to perform bit-by-bit AND operations on integers. A single-bit AND operation is given by: 1 AND 1 = 1, 0 AND 1 = 0, 1 AND 0 = 0, and 0 AND 0 = 0. The result of an AND operation on two integers is obtained by ANDing the corresponding bits of the integers. Write a BASIC program that will input an integer, A, and use the AND operation to compute and print out the bit configuration of A. Using negative values for A, find out if the system you are using stores integers in sign-magnitude, one's-complement, or two's-complement format.

CHAPTER **21**

Computer Graphics

In the last decade, innovations in computer technology have transformed the computer into an important tool for people in many areas of our society. The way in which the individual interacts with the computer has become an important concern for computer hardware and software system designers. Computer-generated graphics is an extremely important part of this computer/human interface. The more people understand about graphics, the more creatively they can use the computer, and the more effectively they can assimilate the information that the computer presents to them.

In addition to contributing to the user interface, computer graphics has become commonplace as a design and analysis tool in its own right in such diverse activities as automotive design, publishing and commercial art layout, and the crafting of architectural plans. Computer graphics is used by artists in the development of animated cartoons, the generation of geometric patterns for use in textile design,

and the production of special effects for the feature film industry.

In the scientific community, computer graphics is used by chemists to create models of complex chemical molecules for the analysis of chemical reactions. Astronomers generate graphic models of galaxies to better understand the interaction of celestial bodies. These models are created using data sets that often contain millions of values, and the graphic presentation of these data sets allows the user to more efficiently interpret these values and their interrelationships.

Computer graphics contributes to the speed, productivity, and, in some cases, the creativity with which people do their work. By becoming aware of the fundamentals of graphics algorithms and operations, the user gains a sense of the mechanics involved in presenting the computer-generated information: he or she can then use the new insight to work with the computer more easily.

Computer graphics can simply be defined as the synthesis of pictures by a computer. This synthesis may take place by the computer presenting a predefined image, such as the "icons" in the user menus of many personal computers or the "sprites" that comprise the on-screen players in a video game. Alternately, the computer may reproduce the picture on a display device by executing a program that uses an algorithm based on a mathematical formulation of an object. This algorithm is created to transform the usually three-dimensional data about the object into an internal format that the computer can process and convert into a two-dimensional picture. The picture should have enough detail so that we can easily recognize it. However, the physical constraints of display devices make it impossible to produce perfect representations of the mathematical formulation.

To reduce the difficulties involved in constructing a complex figure, we place simple objects together in a suitable combination to form the complex object. Through arrangements of line segments, for example, we can produce pictures of considerable complexity. In actual practice, special graphics packages are used to generate and manipulate a collection of primitive, or simple, objects, thereby enabling the user to design a large variety of figures made up from the primitive forms.

In the first section of this chapter, we discuss the physical characteristics of display devices and the way they produce pictures. In the remaining sections, we discuss the production and manipulation of some simple objects to give you a feel for computer graphics.

How Pictures Appear at a Display Device

A computer graphics system consists of the display device on which the picture is presented and the hardware and software necessary to formulate the data required to construct the picture. In later sections, we describe the computer and its component software; here, we consider the device used to display the picture generated by the computer. A wide range of devices are available for presenting the computer-created picture to the user. In the simplest case, devices such as line printers or plotters render the picture in a "hardcopy" format. Other devices use light-emitting diodes (LEDs) or gas-discharge technologies, such as plasma panels.

The device most commonly used—from home computer systems to the most sophisticated computer graphics environments—is the **cathode ray** tube, or CRT. We can roughly classify CRTs as belonging to the stroke, or vector family or to the raster family. Although these types function differently, they both use the same basic technology to display a picture. The front of the display surface of a vacuum tube (depicted in Figure 21.1a) is coated with a material called phosphor. If bombarded by electrons, phosphor becomes "excited" and emits a glow as it returns to its normal preexcitation state. The key to the operation of a CRT is to control this bombardment so that the phosphor glow is predictable and can represent a picture. Control is attained by using an electron "gun" in the narrow end of the tube, which emits the electrons, together with a set of deflection magnets, which control the direction of the flow of a finely focused beam of electrons to the phosphor surface.

The electron gun in Figure 21.1b consists of a heat-generating filament surrounded by a cathode. When the filament is heated, the cathode gives off electrons in all directions. A directing sheath surrounds the cathode, allowing these electrons to escape only through a hole in the sheath. Once the electrons have exited the sheath, they are focused into a precise beam by a series of electrostatic fields set

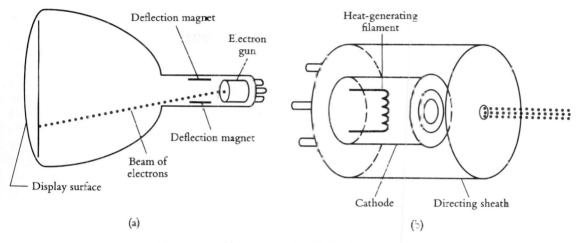

Figure 21.1 (a) A vacuum tube. (b) An electron gun.

up by several plates of different electric potential. A similar process accelerates this ray at different rates in order to control the intensity of the glowing phosphor.

Once the electron ray exits the focusing acceleration structure, the deflection system directs it to the phosphor field on the screen. Deflection is controlled by the display controller, which has its values either hardwired (built in as part of the display equipment), as in the raster system, or supplied by the computer, as in the stroke system. This difference in deflection methodology separates CRTs into the stroke and raster families.

As shown in Figure 21.2, the display controller of the stroke system receives a starting and ending value as well as an intensity value from the host computer. It then deflects the ray to the starting value and moves it in a straight line to the ending value, exciting all the phosphors it encounters on the way at the intensity level determined by the acceleration of the beam. By contrast, the display controller of the raster system (Figure 21.3) receives in-

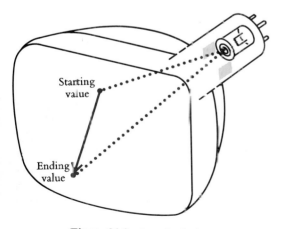

Figure 21.2 A stroke device.

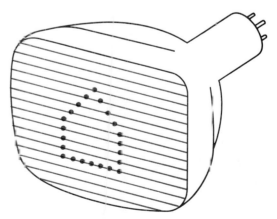

Figure 21.3 A raster device.

tensity, or color, values from the computer for the points on a horizontal line. The picture is then formed by sequentially scanning the horizontal lines of points.

In both cases, the display controller must perform its tasks in rapid succession, because the phosphor glows for only a very short time. A typical rate is once every $\frac{1}{50}$ of a second for stroke display units and once every $\frac{1}{30}$ of a second for raster display devices. The precise speed depends on the type of phosphor used. If the repetition is too slow, the picture goes away for a period, and we see a "flicker."

We get color by using phosphors with different characteristics. A CRT uses phosphors that can produce red, green, and blue. A wider spectrum of colors is achieved by exciting the different types of phosphors singly or in various combinations. The $2^3 = 8$ different colors that can be obtained from the three primary colors are shown in Figure 21.4. Yellow is derived from red and green; magenta from red and blue; cyan from green and blue; white from red, green, and blue; and black, simply from the absence of all three primary colors.

The intensity controls the amount of color, giving a smooth range of color transitions.

Simple Geometric Objects

The simplest geometric objects we study in school are points, lines, and polyhedral shapes. Although the notion of a point as a dimensionless object is appropriate in the study of geometry, a point on a display screen will necessarily have a definite size. The smallest rendition of a point on a graphics screen is called a **pixel,** and depending on the quality of the screen, it could be quite large. Objects such as straight lines or curves will be composed of a set of pixels. If pixel sizes are small and their density great, a large number of pixels could be used in the representation of a figure to make it appear smooth.

The simplest figure we can consider is a straight line segment. To display the line segment joining the two points $A = (x_1, y_1)$ and $B = (x_2, y_2)$, we need to "turn on" the pixels lying on the path from A to B. We can mathematically characterize the points on the line shown in Figure 21.5 by

1. finding the equation of the line from A to B, $y = mx + b$, in slope-intercept form

2. choosing all points (x, y) with the properties

    a. $y = mx + b$
    b. $\min(x_1, x_2) \le x \le \max(x_1, x_2)$
    c. $\min(y_1, y_2) \le y \le \max(y_1, y_2)$

A better approach would be to think of x as a combination of x_1 and x_2 in the following way

$$x = ux_2 + (1 - u)x_1$$

If $u = 0$, then $x = x_1$ and if $u = 1$, $x = x_2$. For all intermediate values of u between 0 and 1,

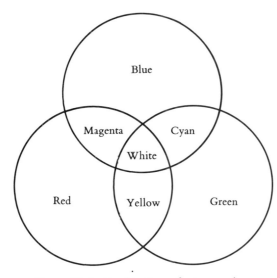

Figure 21.4 Combinations of primary colors.

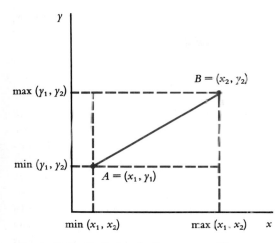

Figure 21.5 To display the line segment AB, we turn on the pixels lying in the path from A to B.

we get intermediate values of x between x_1 and x_2, as illustrated in Figure 21.6. We can now restate the formulation of the line segment:

1. Find the equation of the line from A to B, $y = mx + b$.

2. "Turn on" all pixels with coordinates (x, y) subject to

    a. $x = ux_2 + (1 - u)x_1$ $0 \le u \le 1$
    b. $y = mx + b$

We can further simplify the formulation by substituting $ux_2 + (1 - u)x_1$ for x in $y = mx + b$.

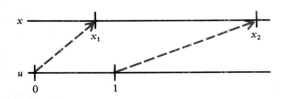

Figure 21.6 Conversion from the u scale to the x scale.

$$y = m[ux_2 + (1 - u)x_1] + b$$
$$= mux_2 + mx_1 - mux_1 + b$$
$$= mux_2 + mx_1 - mux_1 + b + ub - ub$$
$$= u(mx_2 + b) + mx_1 + b - u(mx_1 + b)$$
$$= uy_2 + y_1 - uy_1$$
$$= uy_2 + (1 - u)y_1$$

This refinement leads us to restate:

1. For each u with $0 \le u \le 1$, compute

$$x = ux_2 + (1 - u)x_1$$

and

$$y = uy_2 + (1 - u)y_1$$

2. "Turn on" the pixel at location (x, y).

The representation of a line segment with the equations

$$x = ux_2 + (1 - u)x_1$$
$$y = uy_2 + (1 - u)y_1$$

with $0 \le u \le 1$ is called the parametric form of the line segment from (x_1, y_1) to (x_2, y_2). Notice that as u moves from 0 to 1, (x, y) moves from A to B, as shown in Figure 21.7. Its value to us here is that it lends itself to computations that produce the line segment from (x_1, y_1) to (x_2, y_2). The next step is to design an algorithm that uses this mathematical formulation to display the line segment at a graphics terminal.

Before stating the line-segment-drawing algorithm, we should note that because there are infinitely many points on the line segment, infinitely many points will satisfy the condition stipulated by the parametric form of the line. We will therefore need infinitely many pixels for an "exact" rendition of the line seg-

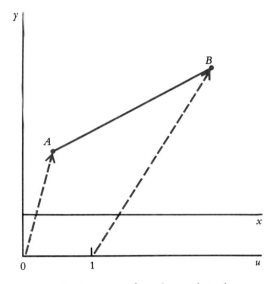

Figure 21.7 Conversion from the u scale to the segment AB.

As u is incremented 20 times, each time by $E = 0.05$, 21 pixels are "turned on" near the ideal locations of the line segment. If E were given value 0.01 or 0.005, the quality of the graphic display would significantly improve.

Once we have the ability to produce line segments, we can easily produce a polygon by specifying the coordinates of its vertices and connecting these vertices with line segments. To produce a polygonal figure that looks solid, we can fill the polygon by turning on all the pixels that lie within the polygon. The brute-force method of filling a polygon would be to test each pixel on the screen and to turn on the pixel if it lies inside the polygon. Because this approach would be extremely inefficient, we now consider an alternative method for filling polygons.

Consider the polygon shown in Figure 21.8, with vertices $(1, 5)$, $(2, 1)$, $(4, 3)$, $(7, 7)$, $(8, 0)$, and $(9, 3)$. We could fill it by scanning the figure vertically (from top to bottom) and drawing horizontal line segments that lie within the polygon; clearly, we need not consider line segments above the line $y = 7$ or

ment. A computational procedure based on this approach will not be effective, that is, will not terminate in finite time. The following algorithm, through the variable E, controls the number of pixels that will be used in the display of the line segment.

1. Input (x_1, y_1) and (x_2, y_2).

2. $u \leftarrow 0$.

3. $E \leftarrow 0.05$.

4. Repeat the following as long as $u \leq 1$:

5. $x \leftarrow ux_2 + (1 - u)x_1$.

6. $y \leftarrow uy_2 + (1 - u)y_1$.

7.   "Turn on" the pixel that is closest to the location (x, y).

8.   Increment u by E.

9. End the algorithm.

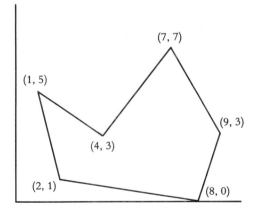

Figure 21.8 We want the computer to shade the inside of the polygon.

below $y = 0$. Our scan would be limited to the range $y = 7$ to $y = 0$.

A crude algorithm for filling this polygon would be

1. $T \leftarrow 7$.

2. $E \leftarrow 0.05$

3. Repeat the following as long as $T \geq 0$.

4.   Draw horizontal line segment(s) at $y = t$ that lie in the polygon.

5. $T \leftarrow T - E$.

6. End the algorithm.

Except for step 4, this algorithm is simple. To draw the horizontal line segments, we must decide which sides of the polygon will be "active," or involved in the determination of the endpoints of the line segment(s). Suppose the sides of the polygon are labeled as shown in Figure 21.9. For $y = 6$, Figure 21.10 shows that only sides 1 and 2 need to be active; whereas for $y = 4$, sides 1, 2, 3, and 4 would be active. Information regarding the sides of

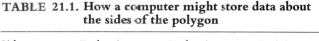

TABLE 21.1. How a computer might store data about the sides of the polygon

Side no.	End point	End point	Activity status
1	(7, 7)	(9, 3)	Off
2	(7, 7)	(4, 3)	Off
3	(1, 5)	(4, 3)	Off
4	(1, 5)	(2, 1)	Off
5	(9, 3)	(8, 0)	Off
6	(2, 1)	(8, 0)	Off

the polygon and their activity status could be stored in the memory of the computer in tabular form, as shown in Table 21.1.

Notice that the table is constructed so that the y coordinates of the vertices associated with a line are arranged in nonincreasing order. Since the filling algorithm starts with $y = 7$, we find two sides (sides 1 and 2) coming out of vertex (7, 7). The activity status of each of these sides is changed to "on." As y is decremented down to 5, the two x coordinates on sides 1 and 2 are computed and arranged in

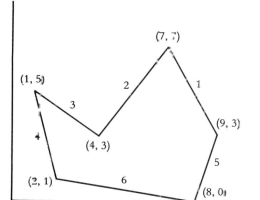

Figure 21.9 The same polygon with its sides labeled.

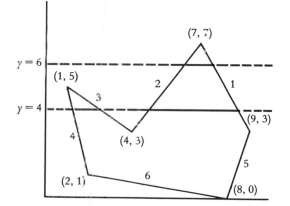

Figure 21.10 The active sides of the polygon for $y = 4$ and $y = 6$.

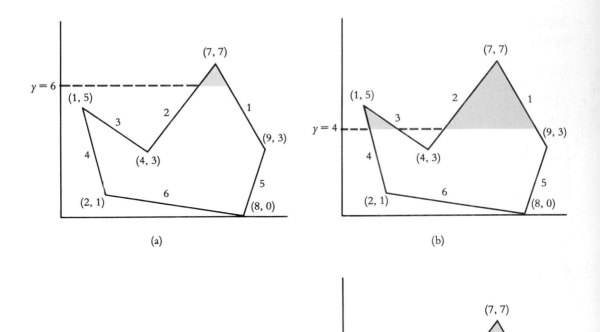

Figure 21.11 (a) When $y = 6$, two sides of the polygon are active. (b) When $y = 4$, four sides are active. (c) When $y = 3$, four sides are active but two of them generate the same value, 4.

increasing order—x_1, x_2—and a line segment is drawn from x_1 to x_2. This is illustrated for $y = 6$ in Figure 21.11a.

When y assumes value 5, sides 3 and 4 also become active, and for values of y between 5 and 3, the four intersection points with the active sides (x_1, x_2, x_3, and x_4), are arranged in nondecreasing order. Horizontal line segments are drawn from x_1 to x_2 and from x_3 to x_4. Figure 21.11b illustrates this for $y = 4$. In general, if the intersection with the active sides

gives $x_1, x_2, \ldots, x_n$ when the sides are arranged in increasing order, then segments from x_1 to x_2, from x_3 to x_4, and so forth are drawn.

In Figure 21.11c, when y assumes value 3, two x values, both equal to 4, are generated because sides 2 and 3 are active. When y becomes less than 3, sides 1, 2, and 3 are changed to "off" and side 5 is changed to "on." This process will continue until y assumes a negative value and all sides have "off" status.

Transformations

We can do a lot even with simple polygonal shapes if we can reconstruct a given picture through movement, magnification, or relocation. For example, we could produce animation (motion of an object on a graphics screen) with a rapid succession of movements of the object. This section considers translation, reflection, scaling, and rotation, four basic forms of motion that can be produced on a graphics screen.

Translation is the movement of a figure from one location to another without altering its shape, size, or orientation. To move a picture, say 2 units to the right and 5 units above its current location, we can look at the coordinates (x, y) of all pixel locations that are "on" and compute the coordinates (X, Y) of the points 2 units to the right and 5 units above (x, y). You can see from Figure 21.12 that this makes $X = x + 2$ and $Y = y + 5$. Now we turn on the pixel at location (X, Y). If we want the original figure to disappear, we can turn off the pixel at (x, y). In general, to re-create a figure h units to the right and k units above its original location, we execute the following algorithm:

1. Input h, k.

2. For all pixels that are "on":

3.   Determine the coordinates (x, y).

4. $X \leftarrow x + h$.

5. $Y \leftarrow y + k$.

6.   Turn on pixel at location (X, Y).

7.   Turn off pixel at location (x, y).

8. End the algorithm.

Note that h or k, or perhaps both, might be negative, in which case the movement will be in the opposite direction.

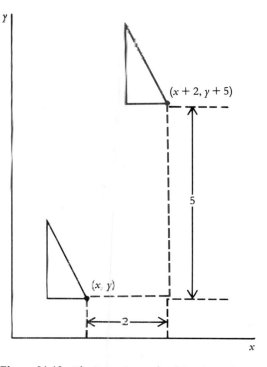

Figure 21.12   The image is translated 2 units to the right and 5 units up.

Another motion the computer can generate is *reflection,* which is the movement of a figure to its mirror image in respect to an axis or the origin.

1. To reflect a figure along the x axis, we compute (X, Y) with $X \leftarrow x$ and $Y \leftarrow -y$ in lines 4 and 5 of the above algorithm (Figure 21.13a).

2. Reflection along the y axis could be achieved with $X \leftarrow -x$ and $Y \leftarrow y$ at lines 4 and 5 (Figure 21.13b).

3. Reflections about the origin are done with $X \leftarrow -x$ and $Y \leftarrow -y$ at lines 4 and 5 (Figure 21.13c).

We can also easily scale — stretch or shrink — a given object in the horizontal or vertical di-

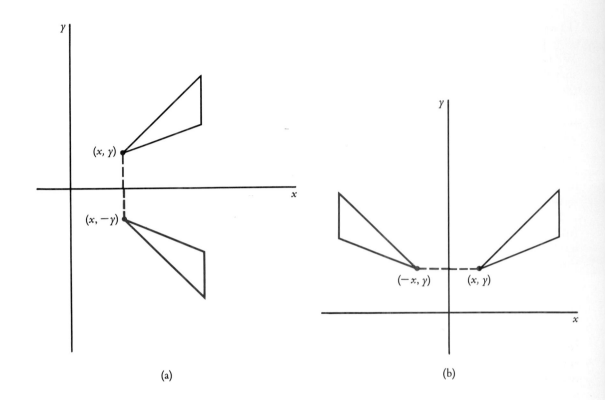

(a)

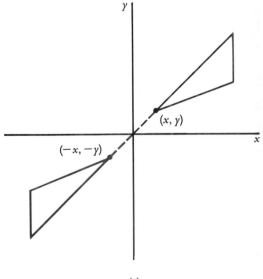

(c)

Figure 21.13   (a) The image is reflected about the x axis. (b) The image is reflected about the y axis. (c) The image is reflected about the origin.

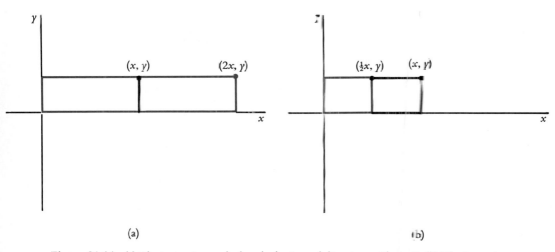

Figure 21.14 (a) The image is stretched in the horizontal direction, with $r = 2$. (b) The image is shrunk in the horizontal direction, with $r = \frac{1}{2}$.

rection. For scaling in the horizontal direction, we take a point on the object as the origin, then for each point (x, y) on the object, we compute the point (rx, y) and turn on the pixel at this point. In Figure 21.14a, the rectangular object is stretched in the horizontal direction; in Figure 21.14b, it is shrunk in the horizontal direction. In general, this type of transformation will have the effect of stretching the figure in the horizontal direction if $r > 1$ and shrinking it if $r < 1$.

We can get similar results in the vertical direction by transforming points (x, y) to points (x, sy) as in Figure 21.15. If we desire scaling in both directions, (x, y) could be transformed to (rx, sy), as in Figure 21.16.

We need a few results from trigonometry to describe how to rotate a figure about a given point. Suppose on a coordinate system we have a point $A = (x, y)$, at distance R from the origin, on a line making an angle α with the x axis. As shown in Figure 21.17, to rotate point A counterclockwise about the origin by an angle β, we need to determine the coordinates

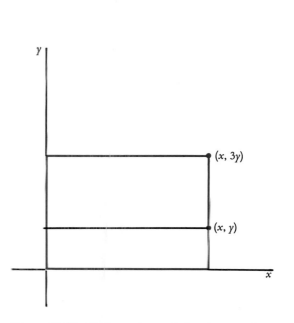

Figure 21.15 The image is stretched in the vertical direction, with $s = 3$.

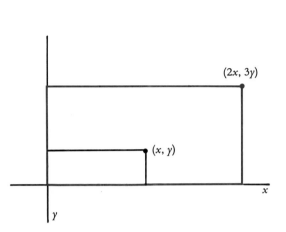

Figure 21.16 The image is scaled in the horizontal and vertical directions, with $r = 2$ and $s = 3$.

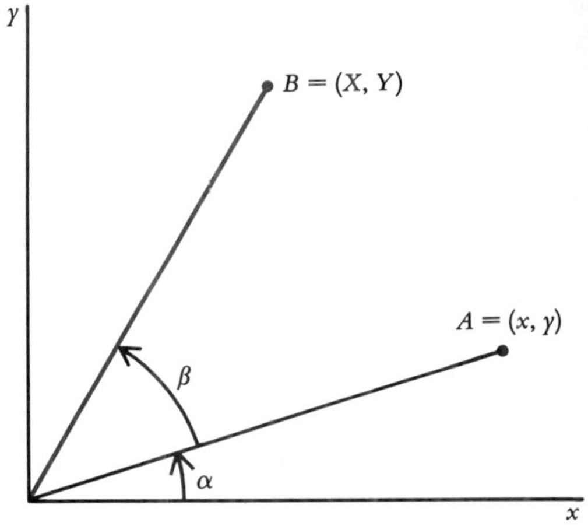

Figure 21.17 The line segment is rotated counterclockwise by an angle β.

of $B = (X, Y)$, which has the same distance R from the origin as A but makes an angle $\alpha + \beta$ with the x axis. The rotation algorithm is developed in Box 21.1.

To rotate a complex figure, we could, of course, rotate each point on the figure, but this process is generally computationally expensive; typically we use more-sophisticated methods.

Windowing

To observe details, we often want to enlarge a certain portion of a complex picture. The process of identifying and magnifying or reducing a rectangular subportion of a picture is called **windowing.** It is useful in this context to distinguish between the actual *object* we are illustrating and the *image* of the object produced on

---

BOX 21.1 How to Rotate a Point Around the Origin

From trigonometry we know that for the points A and B of Figure 21.7

$$\cos \alpha = \frac{x}{R} \quad \text{or} \quad x = R \cos \alpha$$

and

$$\sin \alpha = \frac{y}{R} \quad \text{or} \quad y = R \sin \alpha$$

Similarly

$$X = R \cos (\alpha + \beta)$$
$$Y = R \sin (\alpha + \beta)$$

Using the sum-of-angles identities from trigonometry, we have

$$X = R \cos (\alpha + \beta) = R(\cos \alpha \cos \beta - \sin \alpha \sin \beta)$$
$$Y = R \sin (\alpha + \beta) = R(\cos \alpha \cos \beta + \sin \alpha \cos \beta)$$

Thus, to rotate point (x, y) counterclockwise β degrees about the origin, we can do the following:

1. Compute the distance, R, of (x, y) from the origin, $R = \sqrt{x^2 + y^2}$.

2. Determine α such that $\cos \alpha = x/R$ and $\sin \alpha = y/R$.

3. Compute

$$X = R(\cos \alpha \cos \beta - \sin \alpha \sin \beta)$$
$$Y = R(\cos \alpha \sin \beta + \sin \alpha \cos \beta)$$

4. Turn on the pixel closest to the point (X, Y).

---

the screen. The dimensions of the object are given in inches, feet, meters, millimeters, and so forth. The image of the object, however, must always fit within the graphics screen. We can arbitrarily select a measurement scale that makes the screen 1 unit by 1 unit, with the origin located at the lower left corner of the screen. To produce the image, the dimensions of the object are scaled—magnified or reduced—so that an image whose dimensions are between 0 and 1 is generated.

If a selected window of the object is to be displayed, scaling transformations are applied to the window to fill the screen. In many applications, it is desirable to show the contents of a window in a rectangular subportion of the screen called a **viewport**. It should be noted that the size and location of the viewport are independent of the size and location of the window. Thus, we can have different viewports associated with the same window and, conversely, different windows associated with the same viewport.

To produce the picture of the window at the viewport, we apply the sequence of transformations shown in Figure 21.18b through 21.18d to the original shape (Figure 21.18a):

1. A translation moves the lower left corner of the window to the origin.

2. A scaling transformation makes the size of the window the same as the size of the viewport.

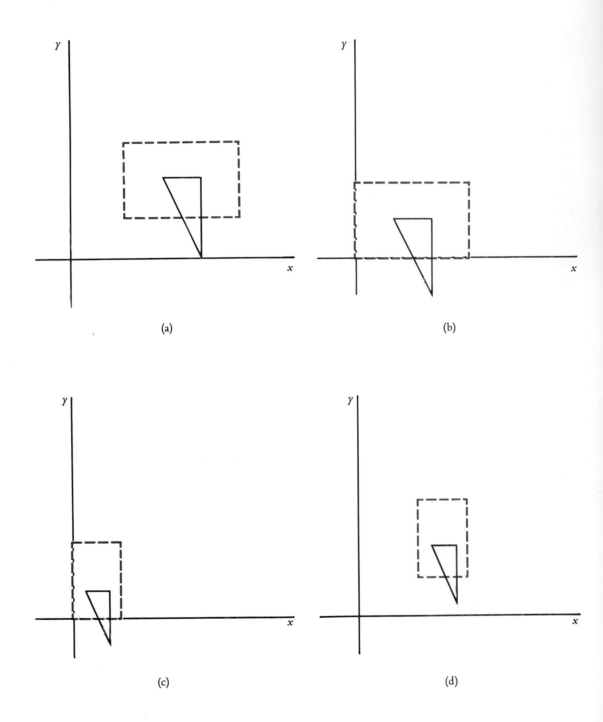

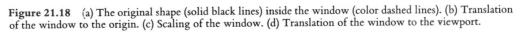

Figure 21.18 (a) The original shape (solid black lines) inside the window (color dashed lines). (b) Translation of the window to the origin. (c) Scaling of the window. (d) Translation of the window to the viewport.

3. A translation moves the window to the viewport location.

Fractals

As we saw in earlier sections, it is easy to describe points, lines, and polygonal shapes so that computer programs can "draw" them on display screens. It is also easy to combine these primitive shapes to form three-dimensional objects of prescribed forms. However, a large class of objects is not so easy to describe. Natural objects such as mountains, clouds, subatomic structures, and planets have forms that are either amorphous or very complex.

The concept of fractional dimension, or **fractal,** was developed in order to describe the shapes of natural objects. In geometry, we consider a line a one-dimensional object and a plane a two-dimensional object. What dimension should a jagged line have? Within the concept of fractals, a jagged line is given a dimension between 1 and 2, with the exact value determined by the line's "jaggedness." An interesting property of fractal objects is that as we magnify a figure, more details appear but the basic shape of the figure remains intact.

The first phase for the construction of a simple fractal object with dimension between 1 and 2 can be described as follows:

1. Start with a line segment AB.

2. Divide the line segment AB into three equal segments, AC, CD, and DB.

3. Construct an equilateral triangle CED on the middle segment and then remove the middle segment CD.

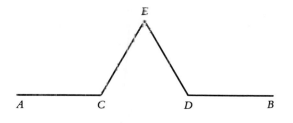

In the second phase, we repeat steps 1, 2, and 3 on each of the four segments, AC, CE, ED, and DB, which yields

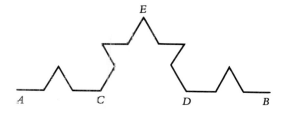

In the third phase, the same three steps are repeated on the 16 segments produced by the previous phase, giving

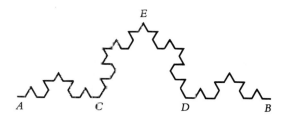

This process can be repeated until we get the desired level of complexity.

In the same way, we can construct a fractal object with dimension between 2 and 3 by starting with a rectangle, subdividing it into four sections, constructing a pyramid above the "middle" section, and repeating this process.

BOX 21.2 Fractal Geometry and Benoit Mandelbrot

Fractal geometry as a serious mathematical endeavor began about thirteen years ago with the pioneering work of Benoit B. Mandelbrot, a Fellow of the Thomas J. Watson Research Center, IBM Corporation. Fractal geometry is a theory of geometric forms so complex they defy analysis and classification by traditional Euclidean means. Yet fractal shapes occur universally in the natural world. Mandelbrot has recognized them not only in coastlines, landscapes, lungs, and turbulent water flow but also in the chaotic fluctuation of prices on the Chicago commodity exchange.

Though Mandelbrot's first comprehensive publication of fractal theory took place in 1975, mathematicians were aware of some of the basic elements during the period from 1875 to 1925. However, because mathematicians at that time thought such knowledge of "fractal dimension" deserved little attention, their discoveries were left as unrelated odds and ends. Also, the creation of fractal illustrations — a laborious and nearly impossible task at the turn of the twentieth century — can now be done quickly and precisely using computer graphics. (Even personal computers can now be used to generate fractal patterns with relative ease.)

The word *fractal* — coined by Mandelbrot — is related to the Latin verb *frangere,* which means "to break." The ancient Romans who used *frangere* may have been thinking about the breaking of a stone, since the adjective derived from this action combines the two most obvious properties of broken stones — irregularity and fragmentation. The adjectival form is *fractus,* which Mandelbrot says led him to fractal.

Mandelbrot also has stated that fractals and fractions are related, especially if one views a "fraction" as a number that lies between integers (whole numbers). Similarly, a fractal set, a precise mathematical term defined by Mandelbrot, can be seen as lying between classical Euclidean shapes.

When dealing with these classical Euclidean shapes (circle, square, sphere, cube) we need only two parameters — position and length — to describe them. Fractals, however, require three parameters. The first is length; the second is "fractal dimension," denoted by D; the third is called "random seed" or "chance." The most interesting and important of the three is parameter two.

In Euclidean geometry a line segment and a square can be described as self-similar shapes; that is, all line segments or squares are simply enlarged or reduced replicas of each other. From these self-similar Euclidean forms, Mandelbrot saw that the ratio $D = \log N/\log (1/r)$ is the same as the form's dimension. (N is any integer; L is the sum of N straight segments of length $r = L/N$.) From this definition Mandelbrot defined a fractal set (the entire family of fractal shapes) as a mathematical set such that D is greater than the topological dimension D_T, or $D > D_T$.

(Modified from "Mandelbrot Gets Franklin Medal," *SIAM News,* vol. 19, no. 3, May 1986, p. 12. Reproduced with permission of The Society for Industrial and Applied Mathematics.)

Concluding Comments

This chapter has focused on some of the basic concepts of computer graphics. The theory as well as the practical applications of computer graphics is a rapidly evolving branch of computer science. Ever-increasing applications of computer graphics have led to the development of better algorithms for generating pictures and better devices for displaying pictures Increasing computation speed has dictated a change in the approach to making pictures. At the same time, decreasing costs have given rise to more diversified techniques for the development of image-making programs. These advances and innovations will no doubt continue to take place.

As a result of intensified research in the discipline of human/computer interfaces, traditional methods of communicating with the computer and interacting with the internal structures of the computer are being replaced with new methods. One method that is increasing in popularity relies completely on graphics. The standard computer terminal is being replaced by a graphics terminal, whose user selects commands and receives feedback by interacting with a graphics menu.

Traditionally the user has typed on a keyboard, and the words and data were transmitted as a series of letters and numerals. Now information can be transmitted in pictures instead. Even computer programs can be written by organizing series of graphic images. In short, computer-generated graphics are playing a much larger role in the information, data processing, and communications fields.

Thanks to research in graphics software, the computer is now able to simulate images that approximate reality, creating more effective presentations of simulated situations that either are not possible to capture in film or real-life scenarios, or are too expensive or dangerous to test. Aerospace simulations, for example, use more realistic visual feedback and thus train pilots and astronauts more effectively. Expensive movie sets are being replaced with computer-generated sets, with the actors composited over them later.

As computer-generated graphics continues to enter the realm of practical situations, it becomes increasingly important for people to gain further understanding and appreciation of its basic mathematical foundations. Through research, the efficiency of computer graphics software will continue to increase while the price of graphics systems decreases. This will make computer graphics a standard tool in almost every area of human activity.

BOX 21.3 Newton's Method and Computer Graphics

Mathematicians continually work to create and refine mathematics, and a discovery made years ago suddenly can have important and unexpected consequences in the solution of a seemingly unrelated problem. Notable examples of the benefits of reworking old mathematical mines are to be found in applied mathematics. Furthermore, mathematicians increasingly are using high-speed computers to investigate certain mathematical phenomena, and these investigations often produce solutions best seen as graphics.

Much of the work of modern scientists and engineers depends on the ability to solve equations. Familiar equations of the form $x^2 + x - 12 = 0$ have two solutions, or roots. Factoring the equation gives us $(x - 3)(x + 4) = 0$; the roots are thus 3 and -4. Equations of degree two or more (for example, $x^2 - 3x - 23 = 0$; $z^4 + z^3 - 5z^2 - 8z + 1 = 0$;

$n^7 + 2n^5 + 12n^2 - 5 = 0$) have as many roots as the highest power to which the initial variable is raised. These complicated equations are not easily factored, so the roots are not readily apparent. And not all of the roots may be real; some of them may be complex numbers.

One of the oldest and best known methods for solving equations is Newton's method. A calculus-based technique, Newton's method first requires us to estimate an approximate root to an equation. Then we insert that estimate into a formula and calculate a new, more accurate estimate of the root. The formula is then applied to this new number and a further refinement is obtained. The process is continued until we are satisfied with the accuracy of the answer.

The formula for Newton's method can be transformed into a computer algorithm that can efficiently find one or more roots for most polynomial equations. The trick, however, is to select an appropriate starting point or first estimate so that the process converges quickly on a specific root. Not all first guesses converge quickly, and some will cause Newton's method to fail under certain conditions. These conditions arise when the chosen point falls on a boundary between regions associated with different roots or when the process gets stuck oscillating between two numbers, neither of which is a root.

Since the solution of many practical problems depends on a scientist's ability to solve equations, knowing when and why methods of solution such as Newton's method will fail or produce strange results can be enormously important. Because of the complexity of effects involved, computer graphics can provide a clearer picture of what is occurring to equations near their boundary regions. For a given equation and starting value, the computer runs the algorithm implementing Newton's method and calculates the root toward which that value converges. If the computer then assigns a position and a color to the resulting point, a graphical picture can be created. Calculating the results for one million initial values and plotting the points in color can build a detailed picture of the mathematical activity (see Color Plates 38–40). Shades of color indicate how fast a point converges on a root.

The large areas of color are known as basins of attraction, and starting points selected from those regions will converge on a root after a reasonable number of repeated calculations, or iterations. For mathematicians the interesting areas are those near or at a boundary. Graphically these border regions appear as complicated swirling patterns. Moving a starting point a slight amount in a boundary area can cause widely divergent results that would adversely affect the accuracy of an equation's solution.

If a portion of the boundary region is magnified, its complexity does not diminish and in fact looks much like the original image of the boundary. Further magnification only shows a repetition of the same pattern. This similarity of an intricate pattern regardless of scale is explained by fractals (see Box 21.2).

Pretty pictures aside, the computer-generated graphics allow mathematicians to see exactly what is happening in the mathematics. Of particular interest to current researchers is the notion of "chaos." Depending on the initial conditions, Newton's method may indicate that there is no basin of attraction for a root; that is, no choice of starting points lead to a root. As the suspected roots become mathematically unstable, chaos occurs.

Because engineers use Newton's method and other equations to solve difficult problems—such as designing supersonic aircraft—it is vital to understand under what conditions these methods will demonstrate chaotic behavior. At high speeds and high temperatures, stresses on materials become acute, and seemingly small design faults can prove disastrous. Thus not only can knowledge of a solution method's quirks determine the success or failure of an engineering project but it can be crucial in safeguarding those humans touched by such projects. Research on chaos is still developing, and it will be some time before the important equations are fully analyzed and their chaotic behavior mapped.

REVIEW VOCABULARY

Cathode ray tube   Commonly used video display device for computer graphics.

Fractal dimension, or fractal   A measurement of the jaggedness of an object.

Pixel   The smallest rendition of a point on a computer graphics device.

Reflection (of an object about an axis)   The mirror image of the object along the axis.

Rotation (of an object)   The circular movement of an object about a given point.

Scaling (of an object)   The stretching or shrinking of an object in a horizontal or vertical direction.

Translation (of an object)   The movement of an object from one location to another without altering its shape, size, or orientation.

Viewport   A rectangular subportion on a graphics screen.

Windowing   The process of magnifying or reducing a rectangular subportion of a picture.

EXERCISES

1. Give a transformation that will move a picture

   a. 4 units to the right

   b. 2 units down

   c. 3 units to the left and $\frac{1}{2}$ unit up

2. Give a sequence of transformations that will change the rectangle with vertices $(1, 1)$, $(2, 1)$, $(2, 3)$, and $(1, 3)$ into the rectangle with vertices $(3, -1)$, $(4, -1)$, $(4, 1)$, and $(3, 1)$.

3. Give a sequence of transformations that will change the rectangle with vertices $(1, 1)$, $(2, 1)$, $(2, 3)$, and $(1, 3)$ into the rectangle with vertices $(3, -1)$, $(4, -1)$, $(4, 3)$, and $(3, 3)$.

4. Give a sequence of transformations that will change the rectangle with vertices $(1, 1)$, $(2, 1)$, $(2, 3)$, and $(1, 3)$ into the rectangle with vertices $(3, -1)$, $(3.5, -1)$, $(3.5, 3)$, and $(3, 3)$.

5. What happens if a negative factor, r, is used in a scaling transformation? What effect is produced if $r < -1$? If $r > -1$?

6. Show how reflections in the lines $y = x$ and $y = -x$ can be accomplished by a scaling operation followed by a rotation. As suggested in Exercise 5, negative scaling factors may be used.

7. When filling the polygon described by Table 21.1, how will it be known when to activate or deactivate the sides of the polygon? Give an algorithm for filling the polygon.

References

Chapter 1

Beltrami, Edward: *Models for Public Systems Analysis,* Academic Press, New York, 1977.

Cozzens, Margaret, and Richard Porter: *Mathematics and Its Applications to Management, Life, and Social Sciences,* D. C. Heath, Lexington, Ky., 1986.

Malkevitch, Joseph, and Walter Meyer: *Graphs, Models, and Finite Mathematics,* Prentice-Hall, Englewood Cliffs, N.J., 1974.

Chapter 2

Cozzens, Margaret, and Richard Porter: *Mathematics and Its Applications to Management, Life, and Social Sciences,* D. C. Heath, Lexington, Ky., 1986

Malkevitch, Joseph, and Walter Meyer: *Graphs, Models, and Finite Mathematics,* Prentice-Hall, Englewood Cliffs, N.J., 1974.

Chapter 3

Graham, Ronald: "Combinatorial Scheduling Theory," in Lynn Steen (ed.), *Mathematics Today,* Springer-Verlag, New York, 1978.

Chapter 4

Cozzens, Margaret, and Richard Porter: *Mathematics and Its Applications to Management, Life, and Social Sciences,* D. C. Heath, Lexington Ky., 1986.

Gass, Saul: *An Illustrated Guide to Linear Programming,* McGraw-Hill, New York, 1970.

Glicksman, Abraham: *An Introduction to Linear Programming and the Theory of Games,* Wiley, New York, 1963.

Malkevitch, Joseph, and Walter Meyer: *Graphs, Models, and Finite Mathematics,* Prentice-Hall, Englewood Cliffs, N.J., 1974.

Chapter 5

Box, George E. P., William G. Hunter, and J Stuart Hunter: *Statistics for Experimenters,* Wiley, New York, 1978, Chapters 4, 7, 8.

Freedman, David, Robert Pisani, and Roger Purves: *Statistics,* Norton, New York, 1978, Chapters 1, 2, 19.

Moore, David S.: *Statistics: Concepts and Controversies,* 2d ed., W. H. Freeman, New York, 1985, Chapters 1, 2.

Chapter 6

Chambers, John M., William S. Cleveland, Beat Kleiner, and Paul A. Tukey: *Graphical Methods for Data Analysis,* Wadsworth, 1983, Chapters 2, 4.

Freedman, David, Robert Pisani, and Roger Purves: *Statistics,* Norton, New York, 1978, Chapters 3, 5, 8–11.

Moore, David S.: *Statistics: Concepts and Controversies,* 2d ed., W. H. Freeman, New York, 1985, Chapters 4–6.

Chapter 7

Mosteller, Frederick, Robert E. K. Rourke, and George B. Thomas: *Probability with Statistical Applications,* Addison-Wesley, 1970, Chapters 2, 5, 8.

Moore, David S.: *Statistics: Concepts and Controversies,* 2d ed., W. H. Freeman, New York, 1985, Chapters 5, 8.

Chapter 8

Freedman, David, Robert Pisani, and Roger Purves: *Statistics,* Norton, New York, 1978, Chapters 20, 21, 23.

Mosteller, Frederick, Robert E. K. Rourke, and George B. Thomas: *Probability with Sta-* *tistical Applications,* Addison-Wesley, 1970, Chapters 9, 10, 12.

Moore, David S.: *Statistics: Concepts and Controversies,* 2d ed., W.H. Freeman, New York, 1985, Chapter 9.

Chapter 9

Arrow, Kenneth J.: *Social Choice and Individual Values,* Cowles Commission Monograph 12, Wiley, New York, 1951; 2d ed., Yale University Press, 1963.

Black, Duncan: *The Theory of Committees and Elections,* Cambridge University Press, 1958; revised version, 1968.

Brams, Steven J.: *Game Theory and Politics,* Free Press, New York, 1975.

Brams, Steven J., and Peter C. Fishburn: *Approval Voting,* Birkhäuser, Boston, 1982.

Brams, Steven J., William F. Lucas, and Philip D. Straffin, Jr. (eds.): *Political and Related Models,* Springer-Verlag, New York, 1983.

Farquharson, Robin: *Theory of Voting,* Yale University Press, New Haven, Conn., 1969.

Malkevitch, Joseph, and Walter Meyer: *Graphs, Models, and Finite Mathematics,* Prentice-Hall, Englewood Cliffs, N.J., 1974.

Peleg, Bezalel: *Game Theoretical Analysis of Voting in Committees,* Cambridge University Press, 1984.

Riker, William H.: *The Theory of Political Coalitions,* Yale University Press, 1962.

Sen, Amartyak K.: *Collective Choice and Social Welfare,* Holden-Day, San Francisco, 1970.

Chapter 10

Banzhaf, John F., III: "One Man, 3.312 . . . Votes: A Mathematical Analysis of the Electoral College," *Villanova Law Review,* 13:304–332, 333–346 (Winter 1968); and 14:86–96 (Fall 1968).

Barrett, Carol, and Hanna Newcombe: "Weighted Voting in International Organizations," *Peace Research Reviews* II, April 1968.

Brams, Steven J., William F. Lucas, and Philip D. Straffin, Jr. (eds.): *Political and Related Models,* Springer-Verlag, New York, 1983, especially Chapters 9, 10, 11.

Dubey, Pradeep, and Lloyd S. Shapley: "Mathematical Properties of the Banzhaf Power Index," *Mathematics of Operations Research,* 4:99–131 (May 1979).

Shapley, Lloyd S.: "A Value for *n*-Person Games," in H. W. Kuhn and A. W. Tucker (eds.), *Contributions to the Theory of Games,* vol. 2, Princeton University Press, 1953, pp. 307–317.

Shapley, Lloyd S.: "Simple Games: An Outline of the Descriptive Theory," *Behavioral Science,* 7:59–66 (January 1962).

Shubik, Martin (ed.): *Game Theory and Related Approaches to Social Behavior,* Wiley, New York, 1964, especially Part 3, "Political Choice, Power, and Voting."

Chapter 11

Brams, Steven J.: *Biblical Games,* MIT Press, 1980.

Brams, Steven J.: *Superpower Games,* Yale University Press, 1985.

Case, James H.: *Economics and the Competitive Process,* New York University Press, 1979.

Dawkins, Richard: *The Selfish Gene,* Oxford University Press, 1976.

Hamburger, Henry: "N-person prisoner's dilemma," *Journal of Mathematical Sociology,* 3:27–48 (1973).

Luce, R. Duncan, and Howard Raiffa: *Games and Decisions,* Wiley, New York, 1957.

McDonald, John: *The Game of Business,* Doubleday, 1975; Anchor, 1977.

Ordeshook, Peter G. (ed.): *Game Theory and Political Science,* New York University Press, 1978.

Rapoport, Anatol, Melvin Guyer, and David Gordon: *The 2x2 Game,* University of Michigan Press, 1976.

Williams, John D.: *The Compleat Strategyst* (sic), McGraw-Hill, New York, 1954; revised edition, 1966.

Chapter 12

Balinski, M. L., and H. P. Young: *Fair Representation: Meeting the Ideal of One Man, One Vote,* Yale University Press, 1982.

Dubins, L. E.: "Group Decision Devices," *American Mathematical Monthly,* 84:350–363 (May 1977).

Dubins, L. E., and E. H. Spanier: "How to Cut a Cake Fairly," *American Mathematical Monthly,* 68:1–17 (January 1961).

Huntington, E. V.: "The Apportionment of Representatives in Congress," *Transactions of the American Mathematical Society,* 30:85–110 (1928).

Kuhn, Harold W.: "On Games of Fair Division," in Martin Shubik (ed.), *Essays in Mathematical Economics,* Princeton University Press, 1968, pp. 29–37.

Steinhaus, H.: *Mathematical Snapshots,* new ed., Oxford University Press, 1960, pp. 65–75.

Chapter 13

Roughly the same level as this book:

Huntley, H. E.: *The Divine Proportion,* Dover Publications, New York, 1970.

O'Daffer, Phares G., and Stanley R. Clemens: *Geometry: An Investigative Approach,* Addison-Wesley, 1975, Chapters 1–5.

Stevens, Peter: *Patterns in Nature,* Little Brown, Boston-Toronto, 1974.

More advanced:

Grünbaum, Branko, and G. C. Shephard, *Tilings and Patterns,* W. H. Freeman, New York, 1987.

Steinhardt, P. J.: "Quasicrystals," *American Scientist,* 74:6 (Nov.-Dec. 1986), pp. 586–598.

Washburn, Dorothy K., and Donald W. Crowe: *Symmetries of Culture: Theory and Practice of Plane Pattern Analysis,* University of Washington Press, Seattle (to appear 1988).

Chapter 14

Galileo Galilei: *Two New Sciences,* English translation by Stillman Drake, University of Wisconsin Press, Madison, 1974.

Haldane, J. B. S.: *On Being the Right Size and Other Essays,* Oxford University Press, New York, 1985.

McMahon, T. A., and J. T. Bonner: *On Size and Life,* Scientific American Books, New York, 1983.

Smith, J. Maynard: *Mathematical Ideas in Biology,* Cambridge University Press, 1968.

Chapter 15

Clark, C. W.: *Mathematical Bioeconomics: The Optimal Management of Renewable Resources,* Wiley, New York, 1976.

Frauenthal, J. C.: *Introduction to Population Modeling,* Birkhauser, Boston, 1980.

Haberman, R.: *Mathematical Models,* Prentice-Hall, Englewood Cliffs, N.J., 1977, chapter on population dynamics–mathematical ecology.

Maki, D. P., and M. Thompson: *Finite Mathematics,* 2d ed., McGraw-Hill, New York, 1983, Chapter 11.

Human population modeling is the subject of demography, and most texts and references on that subject will include some mathematical models together with the application of those models.

Chapter 16

Heath, T. L.: *The Thirteen Books of Euclid's Elements,* vol. 1, Dover Publications, New York, 1956, Introduction and Books I and II.

Layzer, David: *Constructing the Universe,* Scientific American Library, W. H. Freeman, New York, 1984.

van der Waerden, B. L.: *Science Awakening,* Science Editions, Wiley, New York, 1963.

Chapter 17

Cohen, I. Bernard: *The Birth of a New Physics,* revised and updated, Norton, New York, 1985.

Cohen, I. Bernard: *Revolution in Science,* Belknap Press, Cambridge, Mass., 1985.

Greenberg, M. J.: *Euclidean and Non-Euclidean Geometries,* W. H. Freeman, New York, 1980.

Taylor, E. F., and J. A. Wheeler: *Spacetime Physics,* W. H. Freeman, New York, 1966.

Chapter 18

Horowitz, E., and Sahni, S.: *Fundamentals of Computer Algorithms,* Computer Science Press, 1978.

Knuth, D.: "Ancient Babylonian Algorithms," *Comm. ACM,* 15(7):671–677, (July 1972).

Knuth, D.: *The Art of Computer Programming: Sorting and Searching,* vol. 3, Addison-Wesley, New York, 1973.

Sahni, S.: *Software Development in Pascal,* Camelot Publishing, Fridley, Minn., 1985.

Chapter 19

McEliece, Robert J.: "The Probability of Computer Memories," *Scientific American,* 252(1) (January 1985).

MacKenzie. Charles E.: *Coded Character Sets: History and Development,* Addison-Wesley, 1980.

Chapter 20

McEliece, Robert J.: "The Probability of Computer Memories," *Scientific American,* 252(1) (January 1985).

Pohl, Ira, and Alan Shaw: *The Nature of Computation: An Introduction to Computer Science,* Computer Science Press, 1981.

Chapter 21

Demel, John T., and Michael J. Miller: *Introduction to Computer Graphics,* Brooks/Cob, 1984.

Hearn, Donald, and M. Pauline Baker: *Microcomputer Graphics: Techniques and Applications,* Prentice-Hall, Englewood Cliffs, N.J., 1983.

Rodgers, David F., and J. Alan Adams: *The Mathematical Elements of Computer Graphics,* McGraw-Hill, New York, 1976.

Answers to Selected Exercises

Chapter 1

1. Valence of $A = 1$.   Valence of $B = 3$.   Valence of $C = 3$.   Valence of $D = 3$.
   Valence of $E = 0$.

3. Graphs (b) and (c) have Euler circuits. Here is an Euler circuit for graph (b):

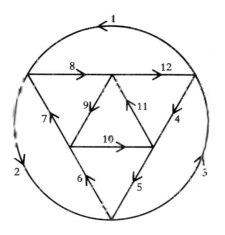

5. If there were an Euler circuit, it would have to use the disconnecting edge twice
   (remember, the circuit must come back to its starting point). But an Euler circuit
   can't use an edge twice. Thus, there is no Euler circuit. Consequently, there is an odd
   valence.

7. Here is an eulerization with 6 added edges. There is no eulerization with fewer than 6.

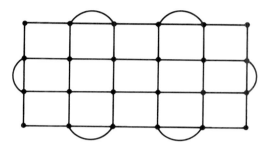

9.

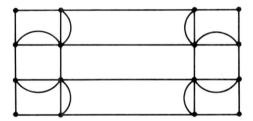

15. The vertical edge whose endpoints have valence 3 can be removed to produce a graph that possesses an Euler circuit.

Chapter 2

1. **a.** Yes.   **b.** No.   **c.** No.

3. **a.** No. Such a tour would have to use edges x_1x_2, x_2x_3, x_6x_4, x_4x_1, x_7x_5, and x_5x_1, which is not possible.   **b.** No.

5. **a.** No Euler circuit. One Hamiltonian circuit is: x_1, x_2, x_4, x_3, x_5, x_6, x_8, x_7, x_1.
   b. Euler circuit and Hamiltonian circuit.   **c.** No Euler circuit. One Hamiltonian circuit is: z_1, z_{12}, z_{11}, z_9, z_3, z_4, z_6, z_7, z_8, z_{10}, z_5, z_2, z_1.   **d.** One Euler circuit is: u_1, u_3, u_7, u_2, u_{13}, u_6, u_3, u_9, u_{10}, u_4, u_5, u_6, u_{15}, u_{14}, u_{10}, u_{11}, u_5, u_2, u_1. There is no Hamiltonian circuit.

7. Traveling salesman problem.

9. $5! = 120$, $6! = 720$, $7! = 5040$, $8! = 40{,}320$. There are approximately 180,000 TSP tours in the complete graph on 9 vertices.

11. **a.** Regardless of which vertex begins the nearest-neighbor algorithm, the same tour of length $18 + 47 + 62 + 71 + 38 = 236$ arises. This same length tour also arises from the sorted-edges algorithm.

15. **a.** The minimum-cost spanning tree uses the edges of weight 1, 2, 3, 4, 5, 8.
 b. The minimum-cost spanning tree is indicated using red lines in the figure:

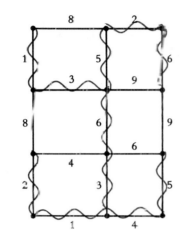

17. In the graph model you would not join these cities by an edge.

21. **a.** If the cheapest edge is unique, T must contain it. If there is a tie for cheapest edge, T need not contain more than one such edge. **b.** False. **c.** True. **d.** False in general.

23. **a.** The earliest completion time is 31. The critical path is T_3, T_6, T_7. **b.** The earliest completion time is 35. There are two critical paths: T_1, T_3, T_5, T_7 and T_1, T_2, T_6, T_8. **c.** The earliest completion time is 29. The critical path is T_2, T_5, T_7.

25. The critical path is T_1, T_5, T_7; thus, if any of these tasks is shortened, the earliest completion time would be reduced. If task 5 is reduced to time length 7, the earliest completion time is not reduced by 3, because T_1, T_4, T_7 has length 28 and thus becomes a new critical path when T_5 has length 7.

Chapter 3

1. Processor 1: T_1 starts 0 finishes 8; T_6 starts 8 finishes 13, idle 13 to 15; T_5 starts 15 finishes 27; T_7 starts 27 finishes 30; T_{11} starts 30 finishes 34, idle 34 to 38; T_{10} starts 38 finishes 45.

 Processor 2: T_2 starts 0 finishes 15; T_9 starts 15 finishes 21, idle 21 to 27; T_8 starts 27 ends 38, idle 38 to 45.

 Processor 3: T_3 starts 0 finishes 6; T_4 starts 6 finishes 15, idle 15 to 45.

2. **a.** Operating rooms, nurses, CAT scan machines, and so forth. **c.** Tasks to turn shuttle plane around between flights, use of runways, pilot schedules, and so forth.

5. For the first list, the completion time using the given list and the list-processing algorithm is 43. If the decreasing-time-list algorithm is used, the completion time is 40. The schedule produced is an optimal schedule but is not unique.

7. **a.** $T_1, T_3, T_2, T_5, T_4, T_6, T_7, T_8, T_{11}, T_{12}, T_9, T_{10}$.

9. **a.** Identical photocopy machines, identical microprocessors, for example.
   b. Runways at an airport, different human providers of a service, and so forth.

17. **a.** Using first fit, the bins contain (listing from the bottom of the bin): bin 1, 80, 20; bin 2, 90, 20; bin 3, 130; bin 4, 60, 30, 30; bin 5, 90, 40.
    b. Using first-fit decreasing, the bins contain (listing from the bottom of the bin): bin 1, 130; bin 2, 90, 40; bin 3, 90, 30; bin 4, 80, 50; bin 5, 60, 30, 20. Either method gives an optimal answer.

19. For 1, 1, 2, 2, . . . , 20, 20 but using first-fit decreasing, 17 bins are filled. Seventeen bins is optimal.

23. Different size bins might occur in a situation where objects were being packed into boxes of different sizes. One possible algorithm would be to rearrange the objects to be packed into a largest-to-smallest order and then place the next item in a bin that would leave the least room left over. However, note that it may no longer make sense merely to minimize the total number of bins used, because all the weights might fit in one large bin but waste a lot of room in it. It might be better to use two smaller bins that had less room left over.

Chapter 4

1.

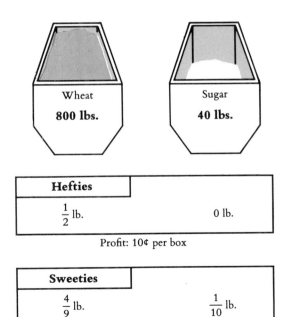

Hefties	
$\frac{1}{2}$ lb.	0 lb.

Profit: 10¢ per box

Sweeties	
$\frac{4}{9}$ lb.	$\frac{1}{10}$ lb.

Profit: 13¢ per box

3. The new profit formula: profit $= 6x + 8y$. The maximum profit mixture doesn't change.

5. (30, 30) is feasible.
   $(-10, 60)$ is not feasible.
   (70, 90) is not feasible.
   (10, 10) is feasible.

7. (0, 750)   0 panels exterior, 750 interior
   (0, 0)   0 panels exterior, 0 interior
   (500, 0)   500 panels exterior, 0 interior
   (500, 200)   500 panels exterior, 200 interior
   (300, 600)   300 panels exterior, 600 interior
   (300, 600) has a profit of $4800, which is the highest profit available.

9. The best profit doubles.

11. If the picture looks like this, the maximum profit is not at a corner point.

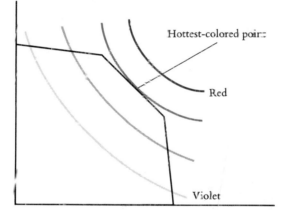

Chapter 5

1. **a.** Registered voters in the Second Congressional District. **b.** All mayonnaise jars.
   c. All 256K chips produced by the supplying company.

3. **a.** Probably higher, because the sample consists entirely of self-selected readers of a health foods magazine, who are not typical of the general population. **b.** Probably lower, because the telephone survey omits viewers without telephones, and these may be disproportionately black.

5. With labels 00–31, choose Drasin, Vlasov, Hoffer, Chan.

7. **a.** Results will vary with the lines in Table 5.1 used. **b.** The histogram shows it is unlikely that no tickets go to women in a fair drawing. (The actual probability is about 5%.)

9. A placebo should be used as the alternative treatment.

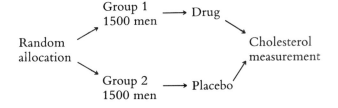

11. Double-blind would probably be used in practice, but it is not essential because the outcome is not subjective.

13. Randomize into two groups of 10 patients, one of which receives the treatment and the other a placebo. Double-blind is essential. Using labels 00-19 in alphabetical order, choose for the treatment: Burt, Washington, Epstein, Lippmann, Chavez, Popkin, Huang, Kidder, Ashley, Williams. The other patients receive the placebo.

15. First choose a simple random sample of 5 for treatment A; then a simple random sample of 5 of the remaining 15 for treatment B; then a simple random sample of 5 of the remaining 10 for treatment C; the remaining 5 get treatment D. It is shorter to relabel after the second step, because the remaining 10 can then get single-digit labels 0–9. The final outcome depends on your choice of labels.

17. A B C   A B C D
     B C A   C D A B
     C A B   D C B A
                   B A D C

Chapter 6

1. **b.** Symmetric.   **c.** No gaps or outliers.

5. Mean = $2323.72; median = $2179.

7. Range = $3626; quartiles = $1955, $2656; deciles = $3049 and $1569. Top 10% = Alaska, D.C., New York, Massachusetts, Oregon; bottom 10% = Alabama, Tennessee, Mississippi, South Carolina, Kentucky.

9. A few multimillion dollar awards pull the mean up but do not affect the median.

15. For $x = 20$, predicted $y = 5.2$ cubic feet; for $x = 40$, predicted $y = 9.2$ cubic feet.

17. **a.** Higher.   **b.** Predicted $y = 461.7$ and residual $= 4.3$ **c.** Predicted math score = 431.8, so the actual score is higher than predicted. Hawaii has many Asian inhabitants whose home language may not be English.

19. **b.** $y = -1.39 + 0.30x$.   **c.** Predicted $y = -0.19$.

Chapter 7

3. A and B are legitimate; C has negative entries and D has sum 1.5, so both are illegitimate.

5. 0.55.

7. 1/3.

9.
Outcome	2	3	4	5	6	7	8	9	10	11	12
Probability	1/36	2/36	3/36	4/36	5/36	6/36	5/36	4/26	3/36	2/36	1/36

13. **a.** 2.5%.   **b.** 64 inches and 74 inches.   **c.** 16%.

15. **a.** 628 or higher.

17. 7.

19.
Heads	0	1	2	3
Probability	0.125	0.375	0.375	0.125

Expected value = 1.5

Chapter 8

1. **a.** Statistic, parameter.   **b.** Statistic, statistic.   **c.** Parameter, parameter.

3. **a.** Mean = 35%, standard deviation = 22.75%.   **b.** Mean = 12 centimeters, standard deviation = 0.002 centimeter.

5. 15% ± 1.8%.

7. 1.18%, 1.26%, 1.29%, 1.26%, 1.18%.

9. For individuals, 0.0008 to 0.0012 inch; for means, 0.0009 to 0.0011 inch.

11. 3.4137 ± 0.0012 grams.

13. **a.** For p, $\hat{p} \pm 1.28 \, [\hat{p}(100 - \hat{p})/n]^{1/2}$; for μ, $\bar{x} \pm 1.28 \, \sigma/\sqrt{n}$.   **b.** 41 ± 1.6%.
    c. 3.4137 ± 0.0007 grams.

15. Center line = 5 grams; control limits = 4.9982 and 5.0017 grams.

19. Women are concentrated in the lower-paying fields (social and behavioral sciences) and are underrepresented in the higher-paying fields (physical sciences and engineering). So the average salary for women would be lower overall even if they earned the same as men in each field alone.

Chapter 9

1. **a.** 3! = 6.   **c.** $n!$.   **d.** 13 for **a**.

3. **a.** E beats H by 7 to 3 and then D beats E by 7 to 3.   **c.** In **a** they could vote for H on the first and second ballot in an attempt to achieve their second choice H: H beats E by 6 to 4 and H beats D by 6 to 4.

5. **a.** V. **b.** B and V tie. **c.** V. **e.** V.

7. **a.** (i) Chinese, (ii) Italian, (iii) Mexican. **b.** Mexican. **c.** (i) Mexican, (ii) Mexican and Chinese tied, (iii) Chinese.

9. **a.** A breaks the tie by selecting a. **b.** B could "compromise" and vote for c, which then beats a by 2 to 1.

Chapter 10

1. **a.** 2. **c.** 12. **e.** 8. **g.** 4. **i.** $2^5 = 32$. **k.** Infinity.

3. **c.** (i) {1, 2, 3}, {1, 2}, {1, 3}, {2, 3}. (ii) {1, 2}, {1, 3}, {2, 3}. (iii) {1}, {2}, {3}, $\varnothing$.
   (iv) None. (v) None.
   e. (iv) 4.
   g. (ii) {1, 2}, {1, 3}, {2, 3}. (iv) 4.
   h. (ii) Any three-person coalition. (iv) None. (v) Any three-person coalition.

5. **c.** $(1/3, 1/3, 1/3)$. **e.** $(1/3, 1/3, 1/3, 0)$ **h.** $(1/5, 1/5, 1/5, 1/5, 1/5)$.
   i. $(1/2, 1/6, 1/6, 1/6)$.

7. **a.** Evenly. **c.** The three-person subgroup has all the power.

9. $[5:3, 1, 1, 1, 1, 1, 1]$.

11. Because {1, 2} and {3, 4} are winning coalitions $w_1 + w_2 \geq q$ and $w_3 + w_4 \geq q$ and thus $w_1 + w_2 + w_3 + w_4 \geq 2q$. However, {1, 3} and {2, 4} are losing coalitions so $w_1 + w_3 < q$ and $w_2 + w_4 < q$, which implies the contradiction $w_1 + w_2 + w_3 + w_4 < 2q$.

13. In the *Canadian Journal of Political Science,* vol. 6, 1973, pp. 140–143, D. R. Miller determined the Shapley power index (which uses permutations rather than combinations) for this voting scheme, as shown in the first column. Eleanor Walther (unpublished) computed the Banzhaf index as given in the second column. She observed that a prairie province has a larger Shapley index but a smaller Banzhaf index than an Atlantic province, and such a nonmonotonicity cannot occur in a weighted voting system.

	Shapley index	Banzhaf index
Ontario or Quebec	0.3155	0.2280
British Columbia	0.1250	0.1244
Alberta, Manitoba, or Saskatchewan	0.0417	0.0570
New Brunswick, Newfoundland, Nova Scotia, or Prince Edward Island	0.0298	0.0622

15. An individual senator's vote is pivotal only when he or she votes in the negative in a tie vote or in the affirmative on a ballot that passes by just one vote. In the former case, which should occur about half the time, the Vice President can vote to break the tie.

Chapter 11

1. **a.** (i), (v), (vi), and (vii).
   b. The optimal strategy for the row player, the column player, and the value of the
   game are: (i) Row 1, column 2, and 5.   (iii) (3/8, 5/8), (5/8, 3/8), and $-(1/8)$.
   (v) Row 3, column 1, and 1.   (vii) Row 1 or 3, column 2 or 4, and 5.
   c. (i) Row 2 and column 1.   (iii) None.   (v) Rows 1 and 2 and column 2.
   (vii) Rows 2 and 4 and columns 1 and 3.

3. (5/8, 3/8), (3/4, 1/4), and the value is 23/40.

5. (1/6, 5/6), (1/4, 3/4), and the value is 3/4.

7. **a.** With a 50% chance of rain, leave your umbrella at home because $0.5(-2) +$
   $0.5(-1) = -1.5 < 0.5(-5) + 0.5(3) = -1$. With a 75% chance of rain, carry your
   umbrella because $0.75(-2) + 0.25(-1) = -1.75 > 0.75(-5) + 0.25(3) = -3$.
   b. Carry your umbrella.   **c.** Carry your umbrella, because it is "optimal" for nature
   to rain.   **d.** Leave your umbrella home.

9. **a.** It is likely that each player will play his or her first strategy because the result
   (5, 5) is a *dominant* equilibrium outcome. The outcome (2, 2) is also in equilibrium.
   c. His choice of boxing and her choice of ballet are dominating strategies. The
   outcomes (4, 0), (0, 0), and (0, 4) are all in equilibrium. They may attempt to alter
   the game and both agree to select the same event by lot, for example, by flipping a
   coin to decide whether to go to boxing or ballet together.

Chapter 12

3. **a.** Stirnweiss by .30854 to .30845.   **c.** Johnson by a margin of .00037.

9. Using the Hamilton method, we obtain: **a.** 2, 2 and 1.   **b.** 2, 1, and 2.   **c.** Yes,
   science.

11. **b.**   (i) Hamilton: 88, 2, 2, 1, 1, 1, 1, 1, 1, 1, 1.
         (ii) Jefferson: 90, 1, 1, 1, 1, 1, 1, 1, 1, 1, 1.
         (iii) Webster: 90, 1, 1, 1, 1, 1, 1, 1, 1, 1, 1.
         (iv) Adams: 80, 2, 2, 2, 2, 2, 2, 2, 2, 2, 2.
    c. Jefferson, Webster-Willcox, and Adams.

13.   (i) Hamilton: 10, 9, 7, 5, 4, 1.
      (ii) Jefferson: 11, 9, 7, 5, 3, 1.
      (iii) Webster: 10, 9, 8, 5, 3, 1.
      (iv) Adams: 10, 9, 7, 5, 3, 2.

Chapter 13

1. **a.** 21 spirals in one direction; 34 in the other.   **b.** Yes, 21 and 34 are adjacent
   numbers in the Fibonacci sequence.

3. **a.** MOM and WOW reflect into themselves by vertical reflection and into each other by horizontal reflection. MUd and bUM reflect into each other by vertical reflection. **b.** The only one is pod. MOM and WOW rotate into each other. SIS rotates into itself.

   c. Some answers could be:

NOW NO
SWIMS
ON MON

CHECK BOOK BOX

T M
H A
A T
T H

OX HIDE

5. **a.** (i) Vertical reflection. (ii) Translation. (iii) Half turn (= rotation by 180°). (iv) Glide reflection.

   b.

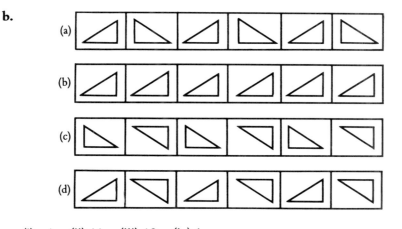

(a)

(b)

(c)

(d)

   c. (i) *m*1. (ii) 11. (iii) 12. (iv) 1*g*.

7. The interior angle at a vertex of a regular *n*-gon has size $(n-2)180/n$ degrees. At a point in the tiling where polygons meet, the total of the interior angles must be 360 degrees. We need to make out a table of interior angles for regular polygons and investigate the possible combinations that add to 360 degrees. Here is a partial table:

n	Interior angle	*n*	Interior angle	*n*	Interior angle
3	60	7	$128\frac{4}{7}$	13	$152\frac{4}{13}$
4	90	8	135	14	$154\frac{2}{7}$
5	108	9	140	15	156
6	120	10	144		
		11	$147\frac{3}{11}$		
		12	150		

We divide the search into the following categories:

1. Three polygons meet at a point.
   a. The largest interior angle is more than 150 degrees.
   b. The largest interior angle is between 120 and 150 degrees, inclusive.

2. Four polygons meet at a point.
   a. The largest interior angle is more than 150 degrees.
   b. The largest interior angle is between 90 and 150 degrees, inclusive.

3. More than four polygons meet at a point.

Because the possibilities for an angle of more than 150 degrees are virtually unlimited (any regular polygon of more than 12 sides will do), we limit our search in case 1a by noting that the two remaining angles must total less than 210 degrees. We therefore can have one polygon with between 5 and 12 sides and the other with either 3 or 4. There are 16 possible such pairings, of which only 5 correspond to a remaining angle that is realized by a regular polygon. The resulting kinds of vertices, which we give in terms of the sides of the accompanying polygons, are

$$3.7.42, \quad 3.8.24, \quad 3.9.18, \quad 3.10.15, \quad 4.5.20$$

In case 1b, the second two remaining angles must total between 210 and 240 degrees, so the larger of them must be at least 105 degrees but no more than 150 degrees. The corresponding polygon must have between 6 and 12 sides. The odd fractions involved with polygons of 7, 9, and 11 sides cannot be augmented to a whole number of degrees by the remaining angles. Each of the possibilities 12, 10, 8, and 6 for the largest angles can be paired with a second largest angle from the same list. Of the 10 pairs, 5 prove feasible:

$$3.12.12, \quad 4.6.12, \quad 5.10.10, \quad 4.8.8, \quad 6.6.6$$

It is easy to show that there are no feasible combinations for case 2a. In case 2b, the largest angle must be between 90 and 150 degrees. The largest two angles cannot total more than 240 degrees, and the second largest angle must be at least 90 degrees. Again, we need not consider polygons with 11, 9, or 7 sides; the pairs of angles we need to consider are

$$4.12, \quad 4.10, \quad 4.8, \quad 6.6, \quad 4.6, \quad 5.5, \quad 4.5, \quad 4.4$$

Of these, four prove feasible:

$$3.3.4.12, \quad 3.3.6.6, \quad 3.4.4.6, \quad 4.4.4.4$$

For case 3, the minimum angle size is 60 degrees, so six polygons meeting at a point must all be triangles; for five, there is a "surplus" of only 60 degrees to parcel to angles larger than 60 degrees. The resulting feasible combinations are

$$3.3.3.3.3.3, \quad 3.3.3.3.6, \quad 3.3.3.4.4$$

This gives a total of 17 *species* of the vertex. We do not consider clockwise versus

counterclockwise ordering of the polygons at the vertex as being distinct, but four of the species each offer two otherwise distinct ways to order the polygons:

3.3.3.4.4	and	3.3.4.3.4
3.3.4.12	and	3.4.3.12
3.4.4.6	and	3.4.6.4
3.3.6.6	and	3.6.3.6

Grünbaum and Shephard (1986, pp. 59–61, 64) depict each of the 21 vertex *types* we have enumerated and prove by elimination that only 11 of these lead to tilings.

8. Tessellations of a sphere correspond to polyhedra, the analogue in three dimensions of polygons. There are five regular polyhedra: the tetrahedron, with 4 triangular faces; the cube, with 6 square faces; the octahedron, with 8 triangular faces; the dodecahedron, with 12 pentagonal faces; and the icosahedron, with 20 triangular faces.

     A proof that there are only these five is relatively simple and can be found in *Excursions into Mathematics,* by Anatole Beck et al., Worth, 1969, pp. 12–13. Photographs of these five, sometimes known as the Platonic solids, can be found in *Polyhedron Models,* by Magnus J. Wenninger, Cambridge University Press, 1970, pp. 14–19. On pages 20–32 Wenninger gives photographs of the 13 semiregular (or Archimedean) polyhedra, which correspond to semiregular tilings of the sphere. The polyhedron corresponding to the soccer ball is officially known as a truncated icosahedron.

Chapter 14

1. **a.** Assuming that Gulliver weighed 180 pounds, the Lilliputians would each weigh $180 \times (\frac{1}{12})^3 = 0.10$ pound $= 1.6$ ounce.
    b. Human infants may be only one foot long at birth, barely twice Lilliputian size. Other mammals—either as infants (e.g., pandas) or as adults (e.g., mice)—are smaller than Lilliputians. So Lilliputian-size people are conceivable; they are not ruled out by area-volume considerations or even by size-shape considerations. However, a Lilliputian scaled down exactly in every respect seems impossible. Certain biological structures, such as brain cells, appear to be of roughly the same size, however large the creature (e.g., elephant brain cells are only twice the size of mouse brain cells). If we abandon perfect geometric similarity and allow that a Lilliputian would have brain cells the same size as those of humans, then the Lilliputian could have only $(\frac{1}{12})^{1/3} = \frac{1}{1728} = 0.06\%$ as many brain cells as a human. We do not know what effect this difference might have on Lilliputian behavior or intelligence.

3. The weight of an object is directly proportional to the amount of gravitational force to which it is subjected. (The mass of the object is the amount of matter it contains, and that quantity is independent of gravity.) A given mountain on earth would weigh only 0.16 times as much if transplanted to the moon, which means that it could be $1/0.16 = 6.3$ times as high before it would crumble. So mountains on the moon could be 6.3 times as high as mountains of the same material on earth. In fact,

however, the highest peaks on the moon are less than 5 miles high, lower than the highest on earth; the reason is that the moon has not experienced the same mountain-building mechanisms (such as tectonic movement) that the earth has. Jupiter, correspondingly, would be able to support mountains only $1/2.6 = 0.38$ times as high as the highest on earth; however, Jupiter in fact consists mostly of gas, and virtually nothing is known about its solid core. Of all the planets, Mars bears the greatest similarity to earth, and for Mars, we see our calculations agree with reality: Mars has a gravity that is 0.38 that of earth, so that mountains could be $1/0.38 = 2.6$ times as high there. Earth's highest mountain is about 6 miles high, so a Martian mountain of the same material could be $6 \times 2.6 = 15.6$ miles high. The highest mountain on Mars, Olympus Mons, is 15 miles high.

5. Let us say that a sparrow is about 4 inches long and that Icarus was 5 feet 4 inches tall, making him $\frac{64}{4} = 16$ times as long. The sparrow's minimum speed is 20 miles per hour, and minimum cruising speed scales as the square root of length; so Icarus must have had a minimum cruising speed of $16^{1/2} \times 20 = 80$ miles per hour. How long would it have taken him to cover the 96 kilometers from Crete to the Greek mainland?

7. The Gossamer Condor, now on display in the Smithsonian Air and Space Museum, weighs only 70 pounds and has a wingspread of 95 feet. The only reason it can fly at all is that it is not a scaled-up bird: its weight is far less than scaling would demand, so that it has a very low ratio of weight to wing area. The Gossamer Condor can fly no faster than 12 miles per hour because its human pilot can generate only one-half horsepower. With that amount of power, it can achieve only one-third the speed of a hang glider—it can fly at 12 miles per hour because it is really a glider, a powered glider, so it is gliding rather than cruising. (See "The man who launched a dinosaur," by Patrick Cooke. *Science 86*, April 1986, pp. 26–35.)

9. MacMahon and Bonner demonstrate theoretically that on log-log paper the line should have a slope of one-ninth, and the theoretical line fits their data points quite neatly. (They use different, older data.)

Chapter 15

1. **a.** 70,246,806. **b.** 65,270,809. **c.** 75,557,190.

3. Brazil and the United States will have equal population sizes in about 42 years. This prediction is based on the data given below.

Brazil	United States	Years after 1985
338,907,905	343,319,280	41
346,567,224	346,374,821	42

5. $100 per day yields $1000 for 10 days.
   $1 for day 1, $2 for day 2, etc., yields $1023.00 for 10 days.

7. $12,000.00 was deposited (assuming deposits made for exactly 10 years). The interest rate was 19.76%.

9. $5,712.40, assuming interest is compounded daily.

11. **a.**

Observation number	Population size
1	10.00
2	17.20
3	28.59
4	44.93
5	64.72
6	82.99
7	94.28
8	98.59
9	99.70
10	99.94

b.

Observation number	Population size
1	110.00
2	101.20
3	100.23
4	100.05
5	100.01
6	100.00
7	100.00
8	100.00
9	100.00
10	100.00

In part **a** the population size increases toward the carrying capacity, and in part **b** the population size decreases toward the carrying capacity.

c.

Observation number	Population size
1	10.00
2	26.20
3	61.00
4	103.82
5	96.68
6	102.46
7	97.92
8	101.58
9	98.69
10	101.02

Observation number	Population size
1	110.00
2	90.20
3	106.11
4	94.44
5	103.89
6	96.61
7	102.50
8	97.88
9	101.61
10	98.66

13. **a.**

Observation number	Population size
1	10.00
2	16.36
3	26.19
4	40.52
5	59.69
6	82.29
7	104.78
8	123.23
9	136.18
10	144.96

b.

Observation number	Population size
1	10.00
2	28.18
3	70.73
4	128.81
5	120.95
6	137.79
7	132.10
8	147.01
9	142.93
10	156.40

c. For $k = 0.7$ the population is increasing. It approaches an upward sloping trend line from below. For $k = 2$ the population oscillates about an upward sloping trend line. In neither case is there an upper bound to the size of the population.

15. $G(x, h) = g(x) + g(x - 1) + g(x - 2) + \cdots + g(x - h + 1)$.

The answers to Exercises 17 and 19 must be determined graphically and therefore only approximate solutions can be obtained.

17. From a population of size 50: 252.
    From a population of size 80: 125.

19. $1.7.

Chapter 16

1. The length of the shadow must be measured from the center of the pyramid's base, which is inaccessible. Also, during the time it would take to measure such a long distance, the length of the shadow will have changed slightly.

3. The length S of the longer shadow satisfies the relation $\frac{12}{16} \times S = 471$. Hence, $S = 471 \times \frac{16}{10} = 754$ feet. For the length s of the smaller shadow, $\frac{10}{16} \times s = 215$. Hence, $s = 215 \times \frac{16}{10} = 344$ feet.

5. Because the length of the woman's shadow is equal to her height, the length of the building's shadow will be equal to its height. The building is therefore 100 feet tall.

7. $20,000 \times 15 = 300,000$ stades. Converting to kilometers, $157.5 \times 300 = 47,250$ kilometers.

9. $\tan 0.25° = MR/ER$. Hence, $MR = 240,000 \times 0.00436 = 1046$ miles.

11. $VS/ES = VS/93,000,000 = 0.73135$. Hence, $VS = 93,000,000 \times 0.73135 =$ about $68,000,000$ miles.

Chapter 17

1. **a.** Diameter of the moon: 2000 miles. Circumference of the earth: 21,991 miles. Circumference of the moon: 6238 miles. **b.** 100 miles   **c.** If P is a point on a circle whose center is O, then the radius OP is perpendicular to the tangent that touches the circle at P.   **d.** 1004.98.   **e.** 4.98.

3. **a.** The dashed line AB represents the equator; the dashed line DE (the tangent) represents the horizon line for the observer.   **b.** Angle DOP.   **c.** Angle BCO.   **d.** They are equal.

11. A single point, a pair of intersecting lines, and two coincident lines.

13. **a.** 32 feet.   **b.** After $(3 + \sqrt{17})/2$ seconds.   **c.** 68 feet.   **d.** The height of the body at $t = 0$.

15. **a.** $p = 3$, $q = 3$.   **b.** Cube: $pF = qV = 24$.   Tetrahedron: $pF = qV = 12$. Octahedron: $pF = qV = 24$.   **c.** Twice the number of edges.   **d.** $V = 20$, $E = 30$.   **e.** $V = 12$, $E = 30$.

Chapter 18

1. 1. *Input N.*
   2. *Sum ← 0.*
   3. *counter ← 1.*
   4. Repeat until *counter > N*:
   5.   [input *X.*
   6.   *sum ← sum + X.*
   7.   *counter ← counter + 1.*]
   8. Print *sum.*
   9. End the algorithm.

3. Following the insertion of the second and third numbers, the list consists of

   12   17   42   98   56   63   34   72   25   83

   The movements associated with the insertion of the fourth through tenth numbers are given below:

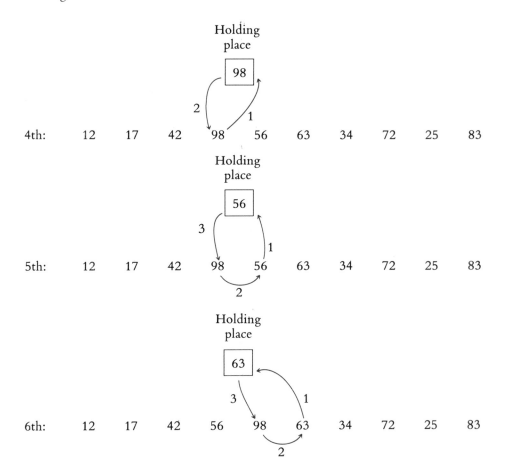

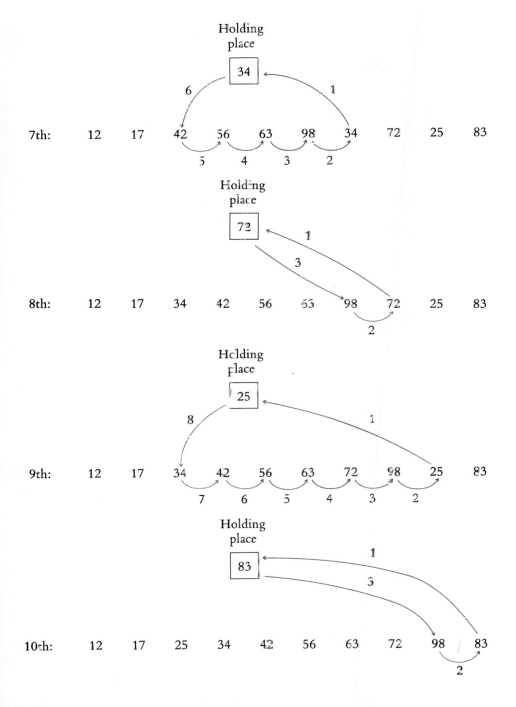

Holding place

34

6   1

7th: 12   17   42   56   63   98   34   72   25   83

5   4   3   2

Holding place

72

1

3

8th: 12   17   34   42   56   63   98   72   25   83

2

Holding place

25

8   1

9th: 12   17   34   42   56   63   72   98   25   83

7   6   5   4   3   2

Holding place

83

1

3

10th: 12   17   25   34   42   56   63   72   98   83

2

5. $\dfrac{N^2 + 4N - 5}{2}$.

7. **a.** Algorithm A should be preferred when $N > 82$; the two algorithms are equivalent when $N = 82$, and algorithm B is superior if $N < 82$.
b. For $N < 70$ algorithm A is faster, and for $N > 70$ algorithm B is faster.
c. The analytical comparisons used in **a** and **b** above become too complicated in this case. We therefore use a different approach: we graph $1000N^3$ and $2^N/5$ on one set of axes and attempt to determine which is the larger by inspecting the graph. Initial calculations for $N = 10$, 20, and 30 yield:

N	$1000N^3$	$2^N/5$
10	10,000	204.8
20	8,000,000	209,715.2
30	27,000,000	214,748,364.8

These results tell us that for small values of N algorithm B is faster. However, for N somewhere between 20 and 30 A begins to significantly outperform B. Trying $N = 26$ and $N = 27$ locates the crossover value between 26 and 27.

N	$1000N^3$	$2^N/5$
26	17,576,000	13,421,772.8
27	19,683,000	26,843,545.6

The following is a graphic illustration of this process.

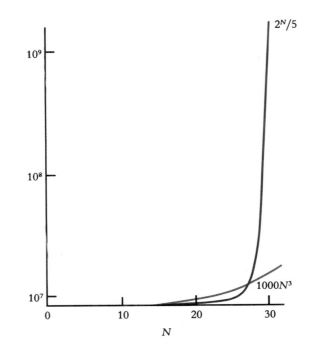

9. BASIC program segment:

```
100   For pass = 1 to N − 1
110       maxpos = 1
120       last = N − pass + 1
130       for I = 2 to last
140           if X(maxpos) < X(I) then maxpos = I
150       next I
160       temp = X(maxpos)
170       X(maxpos) = X(last)
180       X(last) = temp
190   next pass
```

Chapter 19

A	NOT A
T	F
F	T

A	B	A OR B	NOT $(A$ OR $B)$	NOT A	NOT B	(NOT A) AND (NOT B)
T	T	T	F	F	F	F
T	F	T	F	F	T	F
F	T	T	F	T	F	F
F	F	F	T	T	T	T

A	B	A XOR B	NOT A	(NOT A) AND B	NOT B	A AND (NOT B)	((NOT A) AND B) OR (A AND (NOT B))
T	T	F	F	F	F	F	F
T	F	T	F	F	T	T	T
F	T	T	T	T	F	F	T
F	F	F	T	F	T	F	F

7. The error would be detected but not corrected.

9. Due to three errors, the configuration

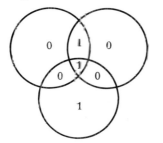

could be transformed to

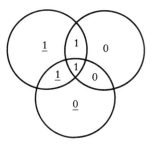

This particular three-error combination will go undetected.

Chapter 20

1. **a.** The representation of   3419 is 0000 1101 0101 1011.
   The representation of   −3419 is 1000 1101 0101 1011.
   The representation of   17,842 is 0100 0101 1011 0010.
   The representation of −17,842 is 1100 0101 1011 0010.
   b. The bit pattern 0000 0011 1010 0110 represents   934.
   The bit pattern 1111 0101 1100 0010 represents −30,146.
   c. $2^{15} − 1 = 32,767$.

3. **a.** The representation of   3419 is 0000 0000 0000 1101 0101 1011.
   The representation of   −3419 is 1111 1111 1111 0010 1010 0101.
   The representation of   17,842 is 0000 0000 0100 0101 1011 0010.
   The representation of −17,842 is 1111 1111 1011 1010 0100 1110.
   b. The bit pattern 0000 0000 0000 0011 1010 0110 represents   934.
   The bit pattern 1111 1111 1111 0101 1100 0010 represents −2622.
   c. $2^{23} − 1 = 8,388,607$.

5. In the sign-magnitude and two's-complement schemes, an integer is divisible by 2 if and only if its rightmost bit is 0. In the one's-complement scheme, an integer is divisible by 2 if and only if its rightmost and leftmost bits are equal.

7. 40,000 bytes of additional memory will be required. The array method requires 1,370,000 bytes; therefore, the percentage increase is

$$\frac{(40,000)(100)}{1,370,000} = 2.9\%$$

Chapter 21

1. **a.** $X \leftarrow x + 4$ **b.** $X \leftarrow x$ **c.** $X \leftarrow x − 3$
    $Y \leftarrow y$ $Y \leftarrow y − 2$ $Y \leftarrow y + \frac{1}{2}$

3. $X \leftarrow x − 1$
   $Y \leftarrow y − 1$

followed by

$$X \longleftarrow x$$
$$Y \longleftarrow 2y$$

followed by

$$X \longleftarrow x + 3$$
$$Y \longleftarrow y - 1$$

5. Consider scaling in the horizontal direction. This would be accomplished through $X \leftarrow rx, Y \leftarrow y$.

- When $r = -1$, the transformation is a reflection about the y-axis.

- When $-1 < r < 0$, the transformation is a reflection followed by a shrinking with a factor of $|r|$.

- When $r < -1$, it is a reflection followed by a stretching with a factor of $|r|$.

7. The algorithm assumes that a table similar to the one given in this chapter has already been constructed for the polygon that is to be filled. E is the amount by which y values are decremented so that the next horizontal line of pixels can be drawn.

   1. $Y \leftarrow$ the y value of the first point in column 2 of the table.

   2. $YMIN \leftarrow$ the y value of the last point in column 3 of the table.

   3. Repeat until $Y < YMIN$:

   4.   Scan column 2; if $Y =$ the y coordinate of a point, then reverse the activity status of the corresponding line.

   5.   Scan column 3; if $Y =$ the y coordinate of a point, then reverse the activity status of the corresponding line.

   6.   Compute the x coordinate of all the points on the active lines when their y coordinates $= Y$.

   7.   Arrange the x values obtained in step 6 in increasing order.

   8.   Draw horizontal line segments between pairs of x values (i.e., between x_1 and x_2, between x_3 and x_4, etc.) at vertical level Y.

   9. $Y \leftarrow Y - E$.

   10. End of algorithm.

Index